LOOKING AT TIME FROM A PHYSICS PERSPECTIVE

LOOKING AT TIME FROM A PHYSICS PERSPECTIVE

Patricio Robles

APPLE ACADEMIC PRESS

First edition published 2024

Apple Academic Press Inc.
1265 Goldenrod Circle, NE,
Palm Bay, FL 32905 USA

760 Laurentian Drive, Unit 19,
Burlington, ON L7N 0A4, CANADA

CRC Press
2385 NW Executive Center Drive,
Suite 320, Boca Raton FL 33431

4 Park Square, Milton Park,
Abingdon, Oxon, OX14 4RN UK

Apple Academic Press exclusively co-publishes with CRC Press, an imprint of Taylor & Francis Group, LLC

Library and Archives Canada Cataloguing in Publication

Title: Looking at time from a physics perspective / Patricio Robles.
Names: Robles, Patricio, author.
Description: First edition. | Includes bibliographical references and index.
Identifiers: Canadiana (print) 20230567118 | Canadiana (ebook) 2023056724X | ISBN 9781774914625 (hardcover) | ISBN 9781774914632 (softcover) | ISBN 9781003414421 (ebook)
Subjects: LCSH: Physics. | LCSH: Time.
Classification: LCC QC28 .R63 2024 | DDC 530.12—dc23

Library of Congress Cataloging-in-Publication Data

CIP data on file with US Library of Congress

ISBN: 978-1-77491-462-5 (hbk)
ISBN: 978-1-77491-463-2 (pbk)
ISBN: 978-1-00341-442-1 (ebk)

This book is dedicated to my wife, Carmen, my daughter, Claudia, and my grandchildren, Leonora and Bruno, and to all those who have encouraged me to carry out this project.

About the Author

Patricio Robles has a long professional and academic career as a physics teacher and researcher, as well as an electrical engineer. Despite being retired, he continues developing research in theoretical physics and teaching courses on electrical projects at Pontifical Catholic University of Valparaíso, Chile. He also works as a consulting engineer in electrical studies related to the analysis of faults in electrical systems and protection schemes. He has published his research on topics about electromagnetic fields and quantum physics in international journals and book chapters. His more recent publications are related to bounded pairs of electrons, massless Dirac fermions, and photon propagation through dispersive media. During the last five years, he has collaborated with international journals of physics and electrical engineering as a referee, reviewing articles submitted for publication in journals such as the *European Journal of Physics* and *Nanotechnology Journal.*

Contents

Abbreviations

CAI	cellular automaton interpretation
CCR	canonical commutation relations
CTC	closed time like curves
EMTR	electromagnetic time reversal
FTL	faster-than-light
GM	gravitomagnetic field
GPS	Global Positioning System
GR	general relativity
GRT	general relativity theory
LQG	loop quantum gravity
LHC	Large Hadron Collider
NRQM	nonrelativistic quantum mechanics
QT	quantum mechanics
RWM	Robertson–Walker metric
SIOF	single-input optical fiber
SOOF	single-output optical fiber
SRT	special relativity theory
ToA	time of arrival
TRM	time-reversal models
TRS	time reversal symmetry

Preface

The expression “how fast time passes” is frequently used in our daily lives. This perception seems to become more real everyday considering the fact, that enormous and sustained development in technology related to communication and information systems is contributing to a considerable reduction in the “time” required for different activities and processes of our society.

There are several publications and books about time. Why would it then be interesting to read another book about this subject? Everyone is familiar with the term *time* as it is incorporated into our lives since birth. Nevertheless, it is worth taking into account that familiar concepts are the most mysterious, and so is the case of time, which is something that can’t be perceived by our five senses, except when one feels that they are becoming older as time “goes by.”

The concept of time is viewed in a different way by a physicist, a biologist, and a historian, for example. In particular, people related to sciences and engineering frequently work with graphics showing the change of some variables with time, for instance, position, velocity, and temperature versus time. But the question, “what is time?”, is seldom formulated and it is rather impossible to obtain an accurate answer from the great number of books and articles that deal with this subject from different perspectives. Someone could say that time is what is measured by clocks. Some could say that time is a kind of parameter, and others could say that the concept of time has sense only under the occurrence of events or changes.

It is worth mentioning, the opinion about the concept of time of the famous theologian, Saint Agustín: “I know if nobody asks me. If someone asks me and I want to explain it, I do not know.” For the French philosopher André Comte-Sponville: “Time is the present. The past is not since it no more exists. The future is not because it is not yet. Only the present exists and it is the only real time”.

Further, almost everybody relates the concept of time with something that flows in one direction and is determined by a sequence of events so that a time arrow can be associated with that sequence. Each area of physics is involved with a specific arrow. For instance, the cosmological arrow is related to the expansion of the universe starting from a singular event known as the Big Bang. The thermodynamic arrow deals with the second law of thermodynamics, which states that a closed system evolves so that a property known as the entropy (defined as a measure of the disorder) irreversibly increases. The radiative arrow of time considers that in the description of radiative processes, only retarded solutions are possible, so that the corresponding waves are not detected before the irreversible extinction of the source.

The present book does not attempt to give an answer to the question about the exact meaning of time and neither covers philosophical aspects or other topics which are out of the field of physics. It is addressed to students who are starting a graduate program in physics or electrical engineering, interested in complementing their studies of relativity theory and quantum physics, applying the knowledge they have acquired about these themes to the analysis of situations where the issue of time measurement is relevant.

There is the case, for example, of clock synchronization, transit time of optical signals through dielectric and absorbing media, lifetime of excited atomic states, among others. These topics, in addition to being of great importance to theoretical physicists, are the basis of many technological developments. For example, global positional systems (GPS) are based on the predictions of relativity theory about time and the effect of gravity over time measurement.

In addition, the book includes the analysis of themes related to time such as causality and the arrows of time, spooky action at distance and Einstein–Podolsky–Rosen paradox, quantum mechanics and teleportation, apparent superluminal velocity, and time reversal. During the last few years and up to now, these themes have been a matter of prolific research with many related publications. The formalism to analyze these themes is, in general, very abstract and requires knowledge of high-level mathematical theories. Using mathematical tools at the level of undergraduate students of science and engineering, these subjects have been included with the aim of bringing them closer to the interested readers and to motivate research on a subject for which until now there is no established theory widely accepted in the scientific community.

The book is divided into six chapters, which analyzes how the concept of time is present in the main fields of physics, such as classical mechanics, electrodynamics, quantum mechanics, and theory of relativity. Illustrative examples are included in each chapter. I hope that this book contributes to scientific analysis on the subject of time and encourages further research on the subject.

—Patricio Robles
Professor of the Electrical Engineering School
Pontifical Catholic University
Valparaíso, Chile

Introduction

ABSTRACT

This chapter is the introduction to the book and presents the main concepts about the look of time from the perspective of physics. We discuss how the close connection between time and the special coordinates is considered in the special relativity theory, unlike what happens in classical physics, in which the time interval between two events has the same value for all observers in uniform relative motion. The principle of causality in this last theory and the time reversal in quantum physics are discussed conceptually. Finally, a brief analysis of the discrete symmetries in the laws of physics is presented, including symmetry under time reversal, and a brief reference to the concept of the arrow of time is made.

In first university courses of electromagnetism, undergraduate students can appreciate through Maxwell equations and their prediction of electromagnetic waves, a close relation between time and spatial coordinates, where a term of the form x-ct appears as an argument of the corresponding solutions in vacuum. In this last expression, x is the spatial coordinate chosen along the x-axis (considering, for simplicity, a one-dimensional case) and c is the light velocity in vacuum with a value of nearly 300,000 km/s. This connection is reinforced in special relativity theory where time appears as a fourth coordinate in Lorentz transformations which relate measurements of time and position made by two observers moving with constant relative velocity (inertial observers).

For describing any event, it is necessary to specify both its location and the corresponding time of occurrence. In Newton mechanics, time has the same value for all observers detecting the event independent of their relative motion, and only the spatial coordinates change according to Galileo's transformation.[1] A drastic change was introduced by the special relativity theory formulated by Albert Einstein in 1905 where time measured by different inertial observers is not the same. This theory has as basic postulates that light velocity is the same for all inertial observers and that there is a finite universal limit c which cannot be exceeded by any object.[2] In a first course of special relativity theory, one usual example presented about differences in time intervals measured by two inertial observers in relative motion, is the case where a subatomic particle with life time τ is produced in some process and goes across a distance $D = vT > v\tau$ before becoming disintegrated. Here T is the time of flight measured by an observer A for whom the particle is in motion with velocity v. There is an apparent contradiction in this case since the measured time of flight T according to observer A is greater than the lifetime of the particle, but this is explained by relativistic time dilation: the time interval T measured by observer A is greater than time interval τ measured by an observer fixed to the particle according to whom this particle at rest.[3] Another case usually discussed, is the case of two twin brothers.[4] One of them (let us call him Robert) initiates a round journey to a distant star in a rocket moving with a velocity comparable with light velocity. The other twin (let us call him John) keeps on earth. When Robert comes back to meet his brother again, John has measured a greater interval of time than that measured by Robert. This paradox could also be explained by relativistic time dilation in a simplified way if in the first instance, the effects of the acceleration of the spacecraft are not considered. It must be considered that during the trip to the planet, the ship must be subjected to an enormous acceleration to reach a speed comparable to that of light and then decelerate when reaching the planet. A similar process must occur to return to earth. Therefore, when considering these accelerations and decelerations, one would expect biological effects that prevent a human observer from being able to make this type of trip and return to earth in the same biological state with which he began his trip. All these indicate that at least until now, it would not be possible for a human to travel to the future through an experiment like that of the two twins.

The postulates of the Special Relativity Theory imply that if two events A and B are causally connected so that the occurrence of event B occurs only if the first event A occurs, the same causality relation exists for all inertial observers. If it were possible, the motion of an object with a velocity greater than c, there would be an inertial frame of reference for which event B (the effect) occurs before event A (the cause) violating the principle of causality. This has been analyzed in several books and publications and up to now, causality has been accepted as a basic principle of nature that cannot be violated (see for example, the book *A Brief History of Time*, by Stephen Hawking). If events A and B are not causally connected, the sequence may change for two inertial observers. For example, one observer can detect these events as occurring simultaneously while for the other, they occur in sequence. These cases can be analyzed graphically in a space–time diagram which is a coordinate system with the horizontal axis representing the position (let us say x in one dimension) and the vertical axis the time. If one specific event A is located at the origin of this coordinate system and two straight lines with slope 1/c are traced symmetrically with respect to the t-axis, this defines the projection of a cone in the plane x-t of space–time, called the light cone, so that another event is (is not) causally connected to event A if it is inside (outside) this cone.[2]

In non-relativistic quantum physics, causality is implicit in Schrödinger's equation and time is considered as a continuous parameter. As learned in the first courses of quantum physics, the dynamical motion of a particle is determined by this last equation giving a solution a wave function of the position and time designated as $\psi(\vec{r},t)$ so that the square of its modulus gives the probability of finding this particle in the neighborhood of the position given by the vector $\vec{r}$. This result is invariant under time reversal, that is, if t is replaced by—t, provided the function ψ is replaced by its conjugate ψ^*. This indicates that time reversal is related to the conjugation of the wave function.[5] In fact, in quantum physics, symmetry operations are considered and one of them is the time-reversal operator which transforms the wave function $\psi(\vec{r},t)$ to $\psi^*(\vec{r},-t)$ implying that the vector velocity $\vec{v}$ gets transformed to $-\vec{v}$.

The combination of non-relativistic quantum physics with special relativity theory and Field concepts gives rise to quantum field theory, which underlies elementary particle physics and constitutes the basis for nuclear physics, atomic physics, condensed matter physics, and astrophysics. This theory presents discrete symmetries with respect to time reversal, charge conjugation, and parity, expressed as operators acting over the solutions of the Dirac equation which is a relativistic equation used instead of Schrödinger's equation.[6] This equations has both solutions with positive and negative energies. These last solutions are associated with antiparticles traveling backwards in time (relativistic quantum theory predicts that every particle has a corresponding antiparticle with the same mass but with opposite electric charge and magnetic moment). In this theory, time reversal consists of reversing the sense of the flow of time by applying an operator designated as T. Charge conjugation is expressed by the action of an operator designated as C transforming a particle into its antiparticle. Parity is expressed by the action of an operator designated as P which changes the sign of the spatial coordinates.

These types of symmetries are known as discrete symmetries, since the application of two successive actions of these operators gives the initial quantity, that is $T^2 = C^2 = P^2 = \mathrm{I}$, where I is the identity operator. The theorem CPT expresses that all interactions are invariant under the combined operation *CPT*, although the result of each one of these operations acting separately may be non-invariant for some processes. According to the results of all the experimental known up to now, the gravitational, electromagnetic and nuclear interactions are symmetric with respect to *T*, C, or *P* acting separately.[7] However the weak interaction governing the radioactive decay processes is not symmetric under the separate action of *C* or *P*, but preserves *CP* and *T*. Processes involving K mesons show *CP* and *T* violations but not *CPT* (meson *K* is a member of a family of subatomic particles).

If the laws of physics are unchanged by the combination of operations *CPT*, one expects that they should be unchanged under operation *T* alone, even though there is a big difference between forward and backward directions of time in ordinary life. As Stephen Hawking says in his book, A Brief History of Time when a cup of water falls off a table and breaks into pieces on the floor, except in a film, in our real life, we cannot see the pieces of a broken cup gathering themselves together off the floor and jumping back onto the table. This is usually explained by the second law of thermodynamics which says that in any closed system, the degree of disorder measured by a quantity called Entropy is always increasing with time.[8]

The increase of entropy with time may be associated with an arrow of time distinguishing the past from the future. This is called the thermodynamic arrow of time. Other arrows considered in the physical research about time are[9]:

- Cosmological arrow: related to the expansion of the universe as time increases.
- Quantum arrow: expresses the fact that before a measurement is made in a quantum system, the wave function corresponds to a superposition of states. When a measurement is made, this superposition disappears and the system takes only one of these states, a process known as the collapse of the wave function.
- Radiative arrow: waves never uniformly converge on a point, but in general, they are seen as uniformly diverging.

This book is addressed to undergraduate students of science and engineering interested in complementing their studies of relativity theory and quantum physics applying the knowledge they have acquired about these themes to the analysis of situations where the issue of time measurement is relevant. It is written in a style trying to make a scientific trip inside a kind of vehicle—let us call *time*—starting from the macroscopic level where classical mechanics and electromagnetic theory are fully valid, to the smallest dimensions where it is essential to apply the quantum theory. In this trip, we will find relativistic effects that cannot be ignored so we will have to use the relativity theory. One of the main objectives of this book is to show the readers how the concept of time manifests itself in the different areas covered by the theoretical framework of physics, from classical mechanics and electrodynamics to quantum mechanics, including in a relevant way, the theory of relativity whose formulation at the beginning of the 20th century radically changed the way the notion of time is considered.

In Chapter 1, we analyze conditions for time-reversal symmetry of the laws of motion in classical mechanics according to the Hamiltonian formalism and show some examples where this symmetry does not exist even for ideal and conservative systems, in the presence of interactions which produce forces dependent on velocity, as is the case of the motion of a charged particle subject to a magnetic field.

In Chapter 2, we analyze the effect of time reversing the equations describing the behavior of the electromagnetic field in different media. One of the main conclusions is that the equations of *macroscopic* electrodynamics are not time-reversal invariant when the system has dissipative elements, but in contrast, in a *microscopic* formalism—with all the light and matter degrees of freedom included in the analysis—the dynamics of the considered system has time-reversal symmetry. In the absence of dissipation, the propagation of an electromagnetic wave is symmetric under time reversal and Maxwell´s equations allow the propagation of a retarded and advanced solution. This last solution can be used in engineering techniques for locating sources of electromagnetic waves, as is analyzed in this chapter.

Chapter 3 discusses time-reversal symmetry for systems whose spatial dimensions require a description based on quantum mechanical theory. We start with the analysis of Schrödinger's equation verifying that for obtaining a time-symmetric description of the dynamics of a particle, the operation of time-reversal must include conjugation represented by the action of an antiunitary operator acting over a wave function. In particular, we analyze quantum systems whose evolution is determined by a set of external parameters varying cyclically, discussing how even though these parameters return to their initial values, the state of the system acquires a geometric phase, which depends on the path followed by the external parameters in the parameter space and which in general is attributable to the broken time-reversal symmetry of the system. Another issue covered in this chapter is related to the demand for Hermiticity of a Hamiltonian in a quantum mechanical system, which can be relaxed if the Hamiltonian is invariant under the combined PT symmetry, where P denotes the parity operation and T denotes time reversal.

Chapter 4 analyzes the way in which Einstein's theory of special relativity radically changed the notion of time and space accepted until the beginning of the 20th century based on Newton's laws. In particular, we revisit the relativistic time dilation effect and discuss examples where the time interval between two events is not the same for two inertial observers in relative motion and how even their sequence can change if these events are not causally connected.

In Chapter 5, we analyze how the presence of massive bodies curves the space time in a way determined by the Einstein field equations and the General Relativity Theory. One of the predictions of this theory is that these effects are such that time goes slower as the lower is the distance of an observer to a massive body. Another important prediction of this theory is the existence of black holes, which corresponds to regions of space with a very high density of mass so that gravity is so strong that even light cannot escape from the interior of the corresponding volume, and the space–time is distorted in such a way that time looks as frozen at points over the so-called "event horizon."

In Chapter 6, we discuss some of the most recent publications related to physical theories about time, such as the cellular automaton interpretation of quantum mechanics formulated by Gerard't Hooft, where information loss is considered a source of time-reversal asymmetry. Then we continue with the thermal time hypothesis of Carlo Rovelli, who asserts that it is possible to formulate classical mechanics in a way in which the time variable is treated on equal footing as other physical variables, and not singled out as the special independent variable. Afterward, we analyze theories which propose the quantization of time, and then continue with the discussion of proposed time operators for non-relativistic and relativistic quantum mechanics. At the end of this chapter, we discuss theories related to time in quantum gravity and then we briefly explore the physics of time travel.

REFERENCES

1. Lerner, L. S. *Physics for Scientists and Engineers*; Jones and Bartlett Publishers: Sudbury, 1996.
2. Jackson, J. D. *Classical Electrodynamics*, 3rd ed., Chapter 11; John Wiley & Sons, Inc.
3. Bosnar, D.; Matić, Z.; Friščic, I.; Žugec, P.; Janči, H. A Simple Setup for Cosmic Muon Lifetime Measurements. *Eur. J. Phys.* **2018,** *39*, 045801.

4. Grøn, Ø. The Twin Paradox in the Theory of Relativity. *Eur. J. Phys.* **2006,** *27*, 885–889.
5. Sakurai, J. J. *Modern Quantum Mechanics*, rev ed.; Addison Wesley Publishing Company, 1994.
6. Greiner, W. *Relativistic Quantum Mechanics—Wave Equations*, 3rd ed.; Springer- Verlag: Berlin, Heidelberg, New York, 2000.
7. Chaichian, M.; Dolgov, A.D.; Novikov, V. A.; Tureanu, A. CPT Violation Does Not Lead to Violation of Lorentz Invariance and Vice Versa. *Phys. Lett.* B **2011,** *699* (3), 177–180.
8. Oppenheim, I. Entropy, Information, and the Arrow of Time. *J. Phys. Chem B* **2010,** 16184–16188.
9. Luminet, J. P. *Matière, Espace, Temps, Contribution au Colloque "Le temps et sa flèche"*, Editions Frontières, 1994.

CHAPTER 1

Time in Classical Mechanics

ABSTRACT

In the absence of the dissipation of energy, one expects that the laws of Classical Mechanics should have time-reversal symmetry. However, this requires careful analysis as made in this chapter where we analyze conditions for time-reversal symmetry of the laws used to predict the movement of objects in classical mechanics according to the Hamiltonian formalism and show some examples. In this chapter, we analyze cases where this symmetry does not exist even for ideal and conservative systems, in the presence of interactions which produce forces dependent on velocity, as is the case of the motion acquired by a charged particle subject to a magnetic field.

1.1 INTRODUCTION

Time and space form a natural frame where we conceive the occurrence of different processes related to our life. For example, daily at 9 a.m., we can be used to be in some specific place of our home or office turning on our personal computer for reviewing the received e-mails, or we must be present in a meeting in some predefined place and starting at a specific hour.

At a microscopic level and in a way imperceptible for us, several physical processes and interactions occur in our environment making possible the life on our planet.

Symmetries play an important role in helping to find the structures of physical theories. Roughly speaking, a symmetry is a transformation (a map) that leaves invariant the structure of a physical theory. In Physics, the existence of symmetries has important consequences. Thus, translation invariance (a consequence of the uniformity of the space) implies the conservation of the total momentum vector $\vec{P}$ of the system. Similarly, isotropy of the

Looking at Time from a Physics Perspective. Patricio Robles (Author)

space implies the conservation of the total angular momentum $\vec{L}$ and time translation invariance implies the conservation of the total energy E.

Thereby, such invariances provide information about the properties of the space. This argument has made philosophers of science see an alternative way of approaching old philosophical problems from a new perspective. The nature of space and time could be better understood by studying the symmetries presented by physical theories.

Philosophical and scientific publications have devoted special efforts to drawing conclusions about the nature of time, in particular, whether or not time has a privileged direction. There are some properties that we intuitively assign to our concept of time, as for example, a flow-like behavior with a privileged direction only toward the future. However, such properties are not necessarily related to the real world. There are two main issues: on the one hand, one may wonder why the predictions obtained from the fundamental dynamical laws should give an asymmetric evolution of our world. On the other hand, one would like to know if the theories of physics imply a fundamental direction of time.

However, it cannot be affirmed that time has a fundamental direction, considering the possibility that the asymmetric behavior of our world could be a result of, for example, a very special set of initial conditions at the beginning of the universe, being possible the existence of time reversed worlds that comply with the same physical laws as ours. In this case, the direction of time is a property that is due to a particular solution (or to a set of solutions) of the fundamental dynamical laws. The second question is concerned with the fact that according to the fundamental dynamical laws, certain evolutions are physically impossible, in particular, those that imply that time goes from the present to the past. When analyzing a physical theory, a relevant aspect is whether it yields solutions for both directions of time, independently of special initial conditions and their corresponding solutions. A fundamental arrow of time should be encoded in the dynamical laws, and should not be just a property of a particular solution or a set of solutions of these laws.

In the analysis of the arrow of time as a property of the fundamental dynamical equations, the concept of symmetry acquires a relevant role, and the property of having a privileged temporal direction may be formally represented in terms of a symmetry transformation. There are two notions that are essential to understand the directed nature of time (or its lack of directionality) in terms of physical symmetries: the first one is that of "reversal of time," as the symmetry transformation that changes the direction of time, and the second notion is that of *"time-reversal invariance,"* as the symmetry that results from the first.

What is the meaning of time-reversal invariance? Commonly, a physical process is said to be time-reversal invariant if its reverse is also allowed by the laws of nature. However, the exact physical meaning of the time-reversal invariance of a process is still a matter of discussion. According to the simplest view, to reverse a process, it suffices to reverse the order as the instantaneous states of the process occur.

A theory is time-reversal invariant if for any infinite sequence of instantaneous states $S(t_o)$, $S(t_1)$, $S(t_2)$ ..... $S(t_N)$ allowed by the theory, the reverse sequence of time-reversed states, $S(t_N)$.... $S(t_2)$, $S(t_1)$, $S(t_o)$ is also allowed.[1]

In the Newtonian formulation of classical mechanics, time reversal is simply the result of reversing the order in which events (positions, in this case) occur along a trajectory $x(t)$. That is, if $x(t)$ is the curve describing the position along the axis x of a particle at every instant t, then the time-reversed motion is given by the curve $x(-t)$, which describes the same positions occurring in the reverse order.

In the Hamiltonian formulation of classical mechanics, time reversal is more than reversing the order of events in a trajectory. Since under the time-reversal transformation, a rightward-moving body becomes a leftward-moving body, the momentum of an instantaneous state must be reversed.

The common dogma that classical mechanics is time-reversal invariant requires careful analysis. There are various ways that it can go awry, even for conservative systems, in the presence of uncommon interactions such as velocity-dependent forces.[2]

In this chapter, we study the time-reversal symmetry or asymmetry of the laws of motion in classical mechanics according to the Hamiltonian formalism and show some examples. By means of these examples, the following conclusions can be obtained:

- Systems whose dynamics is symmetric under time reversal but with energy dissipation are not time-reversal invariant in the sense that the reverse sequence of time-reversed states, $S(t_N)$ .... $S(t_2)$, $S(t_1)$, $S(t_o)$ is not allowed.
- If a system is conservative, this does not necessarily imply that it is time-reversal invariant.

1.2 TIME-REVERSAL SYMMETRY

The traditional way in which time-reversal symmetry is described is related to the form used to make observations of physical phenomena. If from the picture of the dynamics of a mechanical system, it is impossible to decide whether its dynamic behavior occurs in the forward or backward direction in time, this system qualifies as one that has time-reversal symmetry.

When we consider the case of a pendulum oscillating with non-negligible friction, we can identify the difference between a video that shows the dynamics of this pendulum in the forward or backward direction.

As analyzed in Ref. [2], there are several examples showing that time-reversal invariance can fail in classical mechanics.

1.2.1 WHAT DOES TIME-REVERSAL INVARIANCE MEAN?

Suppose we film one harmonic oscillator bobbing back and forth as shown in Figure 1.1, and then play the film in reverse. The result would be a new "reversed" motion of a bob on a spring. This transformation is roughly what will be meant by the *time-reversal transformation.*

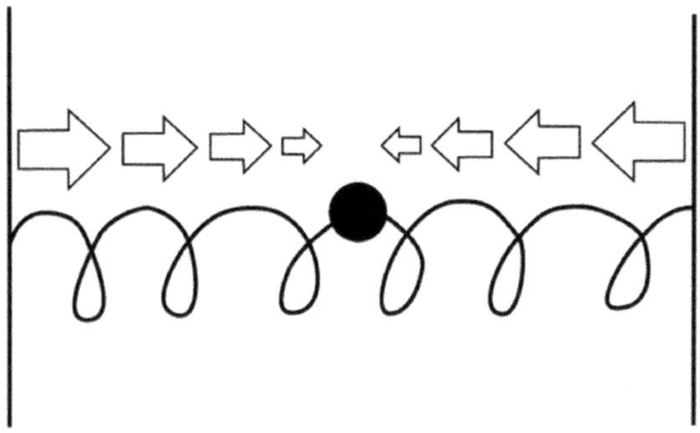

FIGURE 1.1 A particle joined to a harmonically oscillating spring.

How should this transformation be described mathematically? In the Newtonian formulation, it is simply the reversal of the order in which a particle attached to this oscillator (see Figure 1.1) goes through the different positions along a path $x(t)$. That is, if $x(t)$ is the curve describing the position of the bob over time, then the time-reversed trajectory is given by $x(-t)$.

If we do not know the original film, then as observers it would be impossible to affirm with certainty that the movie went backwards. This is explained because the backward movement of the film also corresponds to a possible dynamic of the same system. This means that the reverse motion is described by the same laws as the forward motion. The only difference between moving forward or backward in time is determined by the starting position and speed of the pendulum at the point where the film began to be shown.

If from a movie that shows the dynamics of a mechanical system, it is not possible to decide whether it moves forwards or backwards, we say that it is a physical system with time-reversal symmetry.

Coming back to the case of a swinging pendulum subject to friction, it is possible to identify the difference between a forward and a reverse film of this pendulum. Namely, the original (forward) film will show how the amplitude of the oscillations of the swinging pendulum is decreasing until there is no movement at all. However, the film that is rolled back will show a pendulum oscillating with increasing amplitude with time. The latter film clearly has no physical sense according to our perception of reality since it does not satisfy the natural laws of motion, if we assume that there is no effect of some hidden source of energy supplied to the pendulum. The effect of friction, meaning energy dissipation, breaks the *time-reversal symmetry* of an ideal pendulum which theoretically can oscillate eternally.

The time-reversal symmetry described in the above case is an example of a situation very often studied in books about classical mechanics. Although in nature, it is practically impossible to find mechanical systems with perfect time-reversal symmetry, in theory if a truly isolated pendulum could be built, it would have time-reversal symmetry. The friction and energy transfer result from the coupling of the pendulum with its environment.

1.3 TIME REVERSAL FOR A SIMPLE CASE: THE DYNAMICS OF A PROJECTILE

We start with a simple case corresponding to the well-known example consisting of the throw of a projectile, which could be a small stone or a tennis ball, as shown in Figure 1.2.

Here $\vec{v}_o$ is the initial velocity vector with components $v_{xo} = v_o \cos\alpha$ and $v_y = v_o \sin\alpha$ along the x and y axes as shown and $v - |\vec{v}_o|$. From a kinematical analysis based on Newton's laws, we get that in the absence of any friction, the components of the instantaneous velocity and position of the projectile are given by

$$v_x(t) = v_{xo} \tag{1.1}$$

$$v_y(t) = v_{yo} - gt \tag{1.2}$$

$$x(t) = v_{xo}t \tag{1.3}$$

$$y(t) = v_{yo}t - \frac{1}{2}gt^2 \tag{1.4}$$

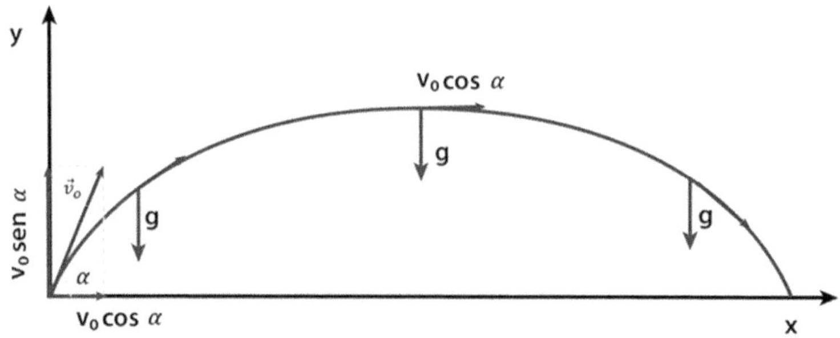

FIGURE 1.2 Trajectory of a projectile thrown with an initial velocity vector that forms an angle α with respect to the x axis.

The time of flight T is calculated from $y(T) = 0$ implying from eq 1.4 that $T = 2v_{yo}/g$. Substituting this value in eq 1.2, we get

$$v_y\left(t = T\right) = v_{yo} - \frac{g2v_{yo}}{g} = -v_{yo} \tag{1.5}$$

From eq 1.3, it can be deduced that the horizontal displacement of the projectile from the initial to the final position is given by $D = 2v_{xo}v_{yo}/g$.

From the above equations, we can appreciate that there is time symmetry because the description does not change if t is replaced by $-t$ and the vector $\vec{v} = (v_x, v_y)$ is replaced by $-\vec{v}$: This is like time is made to run backward so that the initial instant is chosen as the instant when in Figure 1.2, the projectile is at point $(x = D, y = 0)$.

Let us now change the conditions including friction and for simplicity, we assume that there is a small friction in the horizontal direction so that only the x-component of the velocity is affected by viscous friction represented by a coefficient μ according to the following equation:

$$m\frac{dv_x}{dt} + \mu v_x = 0 \tag{1.6}$$

Considering the same initial conditions as above, from eq 1.6, one gets that while the vertical position is still given by eq 1.4, the x-position is now given by

$$x(t) = v_{xo}\tau\left(1 - e^{-t/\tau}\right), \tag{1.7}$$

where $\tau = \dfrac{m}{\mu}$ is the corresponding time constant. If the coefficient μ is small, the following approximation may be used

$$e^{-t/\tau} \approx 1 - \frac{t}{\tau} + \frac{1}{2}\frac{t^2}{\tau^2} \tag{1.8}$$

From the combination of eqs 1.7, 1.8, and 1.4, the resulting kinematics for this case is described by

$$x(t) = v_{xo}t - \frac{v_{xo}\mu}{2m}t^2 \tag{1.9}$$

$$y(t) = v_{yo}t - \frac{1}{2}gt^2 \tag{1.10}$$

From these last two equations, it can be seen that time-reversal symmetry is lost for this case due to the second term in eq 1.9. In fact, if we make $t \to -t$ and $\vec{v}_o \to \vec{v}_o$, we would have

$$x(-t) = (-v_{xo})(-t) - \frac{(-\upsilon_{xo})\mu}{2m}(-t)^2 = v_{xo}t + \frac{v_{xo}\mu}{2m}t^2 \neq x(t) \tag{1.11}$$

This last equation indicates that the time-reversed description of the dynamics of the motion here considered is different from the description with time running forwards, as is familiar to us.

EXAMPLE 1.1

With respect to the dynamics of a projectile discussed in this section, let us consider now a more real case where there is friction only for the motion in the vertical direction so that only the y-component of the velocity is affected by viscous friction represented by a coefficient η according to the following equation

$$m\frac{dv_y}{dt} + \eta v_y = mg \tag{1.12}$$

Obtain the vertical components of the velocity and position and discuss how a loss of time-reversal symmetry occurs due to dissipation.

SOLUTION

Solving the differential eq 1.12, we get that the instantaneous value of the y-component of the velocity vector is given by

$$\upsilon_y(t) = \frac{mg}{\eta} + \left(\upsilon_{yo} - \frac{mg}{\eta}\right)e^{-t/\tau_2} \tag{1.13}$$

where $\tau_2 = m/\eta$ is the time constant for this case.

The vertical position $y(t)$ is obtained by integration of $\upsilon_y(t)$ with initial condition $y(0) = 0$, obtaining

$$(t) = \frac{mg}{\eta}t + \left(\upsilon_{yo} - \frac{mg}{\eta}\right)\tau_2\left(1 - e^{-t/\tau_2}\right) \quad (1.14)$$

If the coefficient η is small, we can make an approximation similar to that of eq 1.8 so that we obtain

$$y(t) \approx \frac{mg}{\eta}t - \left(\upsilon_{yo} - \frac{mg}{\eta}\right)t - \left(\upsilon_{yo} - \frac{mg}{\eta}\right)\eta\frac{t^2}{2m} = \upsilon_{yo}t - \left(\upsilon_{yo} - \frac{mg}{\eta}\right)\eta\frac{t^2}{2m} \quad (1.15)$$

Time reversing the above equation implies making $t \rightarrow -t$ and $\upsilon_{yo} \rightarrow -\upsilon_{yo}$) so that

$$y(-t) = +(-\upsilon_{yo})(-t) - \left(-\upsilon_{yo} - \frac{mg}{\eta}\right)\eta\frac{(-t)^2}{2m} == \upsilon_{yo}t + \left(\upsilon_{yo} + \frac{mg}{\eta}\right)\eta\frac{t^2}{2m} \quad (1.16)$$

The above equation shows that if the friction coefficient $\eta \rightarrow 0$, $y(-t) \rightarrow y(t)$ and the condition of time-reversal symmetry is approached.

1.4 REVERSIBLE LAWS OF MOTION IN CLASSICAL MECHANICS

In Classical Mechanics, for a given set of initial conditions, the dynamical equations of a physical system allow us to predict the motion of the different particles which constitute that system. Having a complete solution to the motion of a system implies that the coordinates of all its constituent particles as functions of time become known.

That means that for a single point particle moving in all three spatial directions, we can obtain its position vector $\vec{r}(t)$ as a function of time. For a system of several particles, the motion is described by a set of functions denoted as $\vec{r}_i(t)$, where the subscript i labels one specific particle in which we are interested. So generally speaking, solving the dynamical equations of a system means that we are able to predict where each particle will be found at any instant of time. In addition, knowing the function $\vec{r}_i(t)$ means that we can take its derivative and obtain the velocity $\vec{\upsilon}_i(t) = d\vec{r}_i(t)/dt$ at any time as well.

The movement of all the components of a system is encoded in a set of differential equations, known as the equations of motion. If quantum and relativistic effects are not significant and can be neglected, these equations can be obtained using Newton's laws of motion which students learn in undergraduate courses of Physics as presented in several standard books, as for example that quoted in Ref. [3].

Newtonian mechanics can be generalized as made by means of the Hamiltonian formulation of classical mechanics that allows a description of a system in terms of generalized coordinates q_k and generalized momenta p_k[4] by means of the Hamiltonian function

$$H(q,p,t)=\sum_k p_k\dot{q}_k-\left[T(q,\dot{q},t)-V(q,t)\right], \tag{1.17}$$

where $T(q, \dot{q}, t)$ represents the kinetic energy, $V(q, t)$ the potential energy, and the index k identifies a specific particle of the considered system. The second term inside brackets on the right side of eq 1.17 is the Lagrangian function $\mathcal{L}(q,\dot{q},t)=T(q,\dot{q},t)-V(q,t)$. The dynamics of a system can be described by means of the Lagrange equations which in the absence of dissipation are:

$$\frac{d}{dt}\left(\frac{\partial\mathcal{L}}{\partial\dot{q}_k}\right)-\frac{\partial\mathcal{L}}{\partial q_k}=0 \tag{1.18}$$

An alternative form for describing the dynamics of a system is given by the Hamilton equations which are given as:

$$\dot{q}_k=\frac{\partial H}{\partial p_k} \tag{1.19}$$

$$\dot{p}_k=-\frac{\partial H}{\partial q_k} \tag{1.20}$$

Under a time-reversal transformation, $t\rightarrow -t$, $p_k\rightarrow -p_k$ and $q_k\rightarrow q_k$, and from the above equations, we obtain

$$\left.\begin{aligned}\frac{dq_k}{d(-t)}&=\frac{\partial H}{\partial(-p_k)}\Rightarrow\dot{q}_k=\frac{\partial H}{\partial p_k}\\ \frac{d(-p_k)}{d(-t)}&=-\frac{\partial H}{\partial q_k}\Rightarrow\dot{p}_k=-\frac{\partial H}{\partial q_k}\end{aligned}\right\}$$

Therefore, the form of the equations is kept under time reversal. One of the effects of time-reversal symmetry is as follows: consider a system is evolved forward in time starting from some initial condition up to an instant t; at this instant, the evolution is stopped, the sign of the velocity of each particle in the system is reversed, and the system is allowed to evolve once again for another time interval of length t; the system will return to its original starting point in phase space so that it will come back to its initial condition.

In this section, we analyze three examples where the absence of dissipation leads to time-reversal symmetry.

EXAMPLE 1.2

Using a simple example, time-reversal symmetry can be verified for an ideal case. To do this, we consider the well-known example of the classical harmonic oscillator consisting of a mass m connected to one extreme of a spring of elastic constant k shown schematically in Figure 1.3. The Hamiltonian for this system shows that there is time-reversal symmetry.

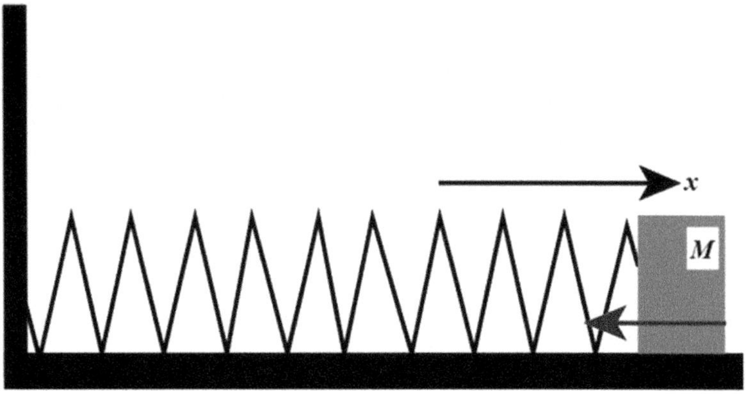

FIGURE 1.3 Classical Harmonic Oscillator. A mass M is fixed to a spring of elastic constant k.

SOLUTION

Considering that the Hamiltonian is the total energy of the system, for this case, we can write directly

$$\mathcal{H} = \frac{p^2}{2m} + kq^2 \tag{1.21}$$

Since this Hamiltonian does not show an explicit dependency on time, there is time-translational symmetry, meaning that energy is conserved. Furthermore, from Hamilton eqs 1.19 and 1.20, we obtain

$$\dot{q} = \frac{\partial H}{\partial p} = \frac{p}{m} \text{ and } \dot{p} = -\frac{\partial H}{\partial q} = -kq$$

Combining these last two equations

$\ddot{q} = \frac{\dot{p}}{m} = -\frac{kq}{m} \Rightarrow \ddot{q} + \frac{k}{m}q = 0$ which is the equation of an ideal classical harmonic oscillator without dissipation, confirming that there is energy conservation and therefore time-reversal symmetry.

EXAMPLE 1.3

Suppose now that the mass *m* of the above example is attached to an extreme of a spring with elastic constant *k* and with the other end fixed to a massless cart moving uniformly with speed υ_o under the action of an external device. See Figure 1.4. Is there time-reversal symmetry in this case? This example is based on Ref. [4].

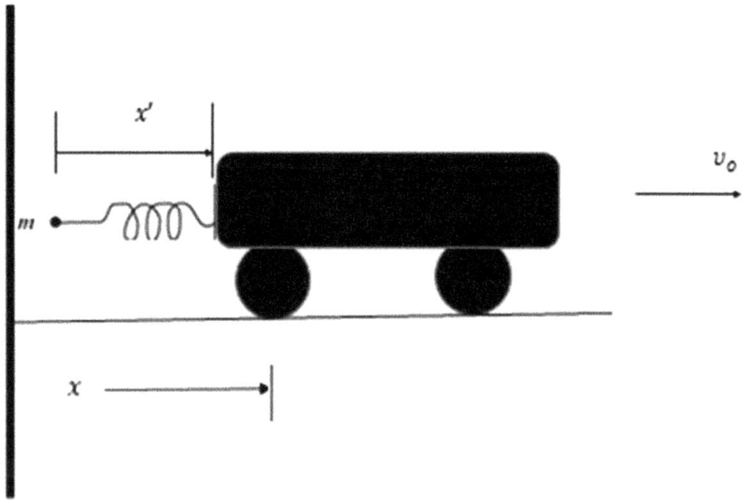

FIGURE 1.4 Harmonic oscillator fixed to a uniformly moving cart.

Source: Adapted from Ref. [4]

SOLUTION

If $\upsilon_o \ll c$, we can use Galileo's transformation between a reference system S fixed to ground and system S' fixed to the cart, so that the coordinates x and x' of the mass *m* referred to the systems S and S', respectively, are related as follows:

$$x' = x - \upsilon_o t \tag{1.22}$$

Equation 1.22 corresponds to the effective compression of the spring, so that the Lagrangian function, which is the difference between the kinetic and potential energy for the system mass-spring, is given by

$$\mathcal{L}(x, \dot{x}, t) = K_{cin} - U_{pot} = \frac{m}{2}\dot{x}^2 - \frac{k}{2}(x - \upsilon_o t)^2 \tag{1.23}$$

The corresponding Hamiltonian is

$$\mathcal{H}(x, p, t) = \frac{p^2}{2m} + \frac{k}{2}(x - \upsilon_o t)^2 \quad (1.24)$$

Since this Hamiltonian has an explicit dependence on time, one could think that it is not a conserved function. However, we can see that it is time-reversal invariant since it does not change under the transformation $t \to -t$ and $\upsilon_o \to -\upsilon_o$.

Furthermore, this Hamiltonian can be written in terms of x' as

$$\mathcal{H}(p', x') = \dot{x}'p' - \mathcal{L}(x, x') \quad (1.25)$$

Considering eq 1.22 and that $\dot{x} = \dot{x}' + \upsilon_o$, the Lagrangian of eq 1.23 can be written in terms of the coordinates x' and p' as

$$\mathcal{L}(x', \dot{x}') = \frac{m}{2}\dot{x}^2 - \frac{k}{2}x'^2 = \frac{m}{2}(\dot{x}' + \upsilon_o)^2 - \frac{k}{2}x'^2 = \frac{m}{2}\dot{x}'^2 + m\upsilon_o\dot{x}' + \frac{m}{2}\upsilon_o^2 - \frac{k}{2}x'^2 \quad (1.26)$$

From eq 1.26, we obtain the conjugate momentum p' as

$$p' = \frac{\partial \mathcal{L}}{\partial \dot{x}'} = m\dot{x}' + m\upsilon_o \to \dot{x}' = \frac{p'}{m} - \upsilon_o \quad (1.27)$$

which combined with eqs. (1.25) and (1.26) gives

$$\mathcal{H}(p', x') = \dot{x}'p' - \mathcal{L}(x, x') = \left(\frac{p'}{m} - \upsilon_o\right)p' - \frac{m}{2}\left(\frac{p'}{m} - \upsilon_o\right)^2$$

$$-m\upsilon_o\left(\frac{p'}{m} - \upsilon_o\right) - \frac{m}{2}\upsilon_o^2 + \frac{k}{2}x'^2$$

$$= \frac{(p' - m\upsilon_o)^2}{2m} - \frac{m}{2}\upsilon_o^2 + \frac{k}{2}x'^2 \quad (1.28)$$

From this last equation, we can see that this new Hamiltonian has no explicit time dependence so that it is a conserved quantity, and therefore, since there is no dissipation, there is a conservation of the total energy. In addition, it can be proved that this system is time-reversal invariant and that the corresponding Hamilton's equation lead to the same dynamic equation:

$$m\ddot{x}' + kx' = 0 \quad (1.29)$$

EXAMPLE 1.4

Prove that the system of example 1.3 is time-reversal invariant and that the dynamics is given by eq 1.29.

SOLUTION

From Hamilton's eqs 1.19 and 1.20 and with the Hamiltonian of eq 1.28

$$\dot{x}' = \frac{\partial H}{\partial p'} = \frac{p' - m\upsilon_o}{m}$$

$$\dot{p}' = -\frac{\partial H}{\partial x'} = -kx'$$

Combining these last two equations we obtain:

$\ddot{x}' = \frac{\dot{p}'}{m} = \frac{-kx'}{m} \Rightarrow m\ddot{x}' + kx' = 0$ which is the same equation as 1.29 corresponding to a classical harmonic oscillator without dissipation, so that there is time-reversal symmetry.

1.5 DISSIPATION AND TIME-REVERSAL ASYMMETRY

For a system with dissipation losses due to friction, there is no time-reversal symmetry as can be seen in the following example based on Ref. [4] (problem 16 of Chapter 2) dealing with a one-dimensional system where frictional effects are incorporated directly in the Lagrange function as follows:

$$\mathcal{L}(q,\dot{q},t) = e^{\gamma t}\left(\frac{m\dot{q}^2}{2} - \frac{kq^2}{2}\right) \tag{1.30}$$

The corresponding Hamiltonian $H(q,p,t)$ is obtained from eq 1.17 with the canonical momentum $p = \frac{\partial \mathcal{L}}{\partial \dot{q}}$ given for this case as

$$p = m\dot{q}e^{\gamma t} \Rightarrow \dot{q} = \frac{p}{m}e^{-\gamma t} \tag{1.31}$$

Therefore, in terms of the canonical quantities q, p, t, the Hamiltonian for this case is

$$\mathcal{H}(q,p,t) = p\dot{q} - \mathcal{L}(q,\dot{q},t) = \frac{p^2}{m}e^{-\gamma t}$$

$$-e^{\gamma t}\frac{m}{2}\frac{p^2}{m^2}e^{-2\gamma t} + e^{\gamma t}\frac{kq^2}{2} = \frac{p^2}{2m}e^{-\gamma t} + e^{\gamma t}\frac{kq^2}{2} \tag{1.32}$$

From eq 1.32, we get

$$\dot{p} = -\frac{\partial H}{\partial q} = -kqe^{\gamma t} \tag{1.33}$$

Combining Eqs. (1.31) and (1.33):

$$\ddot{q} = \frac{\dot{p}}{m}e^{-\gamma t} - \gamma\frac{p}{m}e^{-\gamma t} = -\frac{kq}{m} - \gamma\dot{q} \tag{1.34}$$

eq 1.34 can be recast as

$$\ddot{q} + \gamma\dot{q} + \frac{k}{m}q = 0 \tag{1.35}$$

This equation describes the motion acquired by a damped harmonic oscillator starting, for example, from an initial position, and with zero speed, and whose dynamics has no time-reversal symmetry due to the term $\gamma\dot{q}$ which represents the damping of the oscillations.

EXAMPLE 1.5

Starting from eq 1.32 and with initial conditions $(0) = q_o, \dot{q}(0) = \dot{q}_o = 0$ show that for the undamped case, there is no time-reversal symmetry and that the generalized position is given by

$$q(t) = e^{-\frac{\gamma}{2}t}\left(q_o cos\sqrt{\frac{k^2}{m^2} - \frac{\gamma^2}{4}}t + \frac{\dot{q}_0 + \frac{\gamma}{2}q_o}{\sqrt{\frac{k^2}{m^2} - \frac{\gamma^2}{4}}} sin\sqrt{\frac{k^2}{m^2} - \frac{\gamma^2}{4}}t\right) \tag{1.36}$$

SOLUTION

From the expression of the Hamiltonian given by eq 1.32, it can be seen that there is no time-reversal symmetry since

$$\mathcal{T}\left[\mathcal{H}(q,p,t)\right] = \mathcal{H}(q,-p,-t) = \frac{(-p)^2}{2m}e^{-\gamma(-t)} + e^{\gamma(-t)}\frac{kq^2}{2} = \frac{p^2}{2m}e^{+\gamma t} + e^{-\gamma t}\frac{kq^2}{2} \neq H$$

(q, p, t), where the symbol $\mathcal{T}$ means the time-reversal operator. The time-reversal asymmetry for this system is confirmed if one determines the temporal evolution as follows:

Assuming a solution of the form e^{st} with s a complex parameter to be determined, from eq 1.35 we get

$$s^2 + \gamma s + \frac{k}{m} = 0 \tag{1.37}$$

Solving for s, we obtain $s = -\frac{\gamma}{2} \pm i\Omega$ where $\Omega = \sqrt{\frac{k}{m} - \frac{\gamma^2}{4}}$.

Therefore, we get that the general solution of eq 1.35 is

$$q(t) = e^{-\frac{\gamma}{2}t}\left\{Ae^{i\Omega t} + Be^{-i\Omega t}\right\} \tag{1.38}$$

where A and B are constants that can be obtained from the initial conditions, which are $q(0) = q_o, \dot{q}(0) = \dot{q}_o = 0$. From eq 1.38, we obtain the following system of equations for determining the constants A and B:

$$\left.\begin{aligned} A + B &= q_o \\ A - B &= -i\frac{\gamma q_o}{2\Omega} \end{aligned}\right\} \tag{1.39}$$

From the system of Eqs. (1.39), we obtain

$$A = \frac{q_o}{2} - i\left(\frac{\gamma q_o}{4\Omega}\right) \tag{1.40}$$

$$B = \frac{q_o}{2} + i\left(\frac{\gamma q_o}{4\Omega}\right) \tag{1.41}$$

Substituting the expressions for A and B in eq 1.38 and after some algebra, one obtains eq 1.36. Clearly, this expression gives a time evolution that has no time-reversal symmetry since

$$q(-t) = e^{-\frac{\gamma}{2}(-t)}\left(q_o cos\sqrt{\frac{k^2}{m^2} - \frac{\gamma^2}{4}}(-t) + \frac{\frac{\gamma}{2}q_o}{\sqrt{\frac{k^2}{m^2} - \frac{\gamma^2}{4}}} sin\sqrt{\frac{k^2}{m^2} - \frac{\gamma^2}{4}}(-t)\right)$$

$$= e^{+\frac{\gamma}{2}t}\left(q_o cos\sqrt{\frac{k^2}{m^2} - \frac{\gamma^2}{4}}t - \frac{\frac{\gamma}{2}q_o}{\sqrt{\frac{k^2}{m^2} - \frac{\gamma^2}{4}}} sin\sqrt{\frac{k^2}{m^2} - \frac{\gamma^2}{4}}t\right) \neq q(t)$$

EXAMPLE 1.6

Analyze the dynamics of a particle of mass m which is left to fall from a height h and is subject to a frictional damping force which varies proportionally with the magnitude of the velocity. Discuss about the effect of dissipation over the loss of time-reversal symmetry. This problem is based on problem 23, Chapter 1 of Ref. [4].

SOLUTION

The Lagrangian for the particle without friction is

$$\mathcal{L}_o = \frac{1}{2} m\dot{z}^2 - mgz \tag{1.42}$$

The frictional effect in the Lagrangian formalism is included by adding the dissipation function $\frac{k\dot{z}^2}{2}$, so that the Lagrangian is

$$\mathcal{L} = \mathcal{L}_o + F = \frac{1}{2} m\dot{z}^2 - mgz + \frac{k\dot{z}^2}{2} \tag{1.43}$$

Therefore, the equation of motion is obtained from the Lagrange equation which in this case is

$$\frac{d}{dt}\left(\frac{\partial \mathcal{L}}{\partial \dot{z}}\right) - \frac{\partial \mathcal{L}}{\partial z} + \frac{\partial F}{\partial \dot{z}} = 0 \Rightarrow m\ddot{z} = mg - k\dot{z} \tag{1.44}$$

With $\dot{z} \equiv \frac{d\upsilon}{dt}$ eq 1.44 may be recast as

$$m\frac{d\upsilon}{dt} + k\upsilon = mg \tag{1.45}$$

Equation 1.45 is a linear differential equation of the first order and if the initial condition is $\upsilon(0) = 0$, the solution is

$$\upsilon(t) = \frac{mg}{k}\left(1 - e^{-\frac{k}{m}t}\right) = \upsilon_\infty\left(1 - e^{-\frac{t}{\tau}}\right), \tag{1.46}$$

where $\upsilon_\infty = \frac{mg}{k}$ is the final velocity acquired by the particle and $\tau = m/k$ is the time constant. This case is quite similar to the behavior of an R–L circuit (a resistor R connected in series with an inductor L, as shown in Figure 1.5) where if a constant voltage V is applied at the terminals, the evolution of the current is given by the equation

$$L\frac{di}{dt} + i(t)R = V \tag{1.47}$$

and the current $i(t)$ is given by

$$i(t) = \frac{V}{R}\left(1 - e^{-\frac{t}{\tau}}\right), \tag{1.48}$$

where the time constant for this case is $\tau = L/R$.

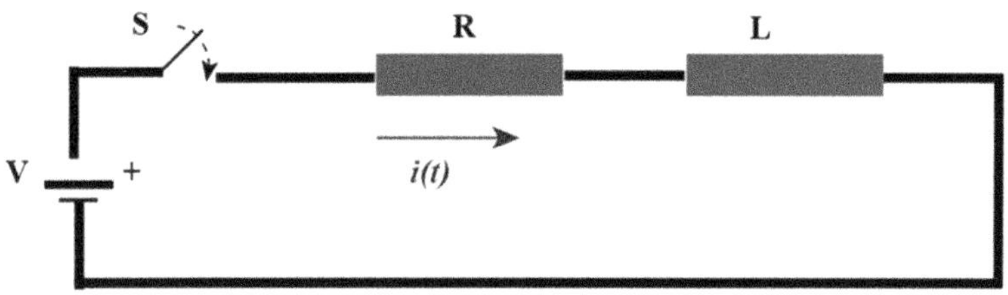

FIGURE 1.5 An RL circuit whose dynamical behavior is analog to that of the mass falling due to gravity and subject to a friction force proportional to velocity. This circuit is energized at $t = 0$ when switch S closes.

Clearly, the dynamics of both systems, the falling particle subject to friction and the RL circuit, are not time-reversal invariant due to energy dissipation.

EXAMPLE 1.7

Repeat example 1–6 but with frictional damping proportional to the square of the velocity, that is $F_{friction} = -A\upsilon^2 sign(\upsilon)$ where A is a constant and *sign* $(\upsilon) = +1$ if $\upsilon > 0$ and -1 if $\upsilon < 0$.

SOLUTION

Directly from Newton's laws, we have that the dynamics of a falling particle subject to this kind of friction force is given by

$$\dot{\upsilon} = \frac{d\upsilon}{dt} = g - \frac{A}{m}\upsilon^2 \tag{1.49}$$

At first sight and from a mathematical point of view, one would be tempted to say that there is time-reversal symmetry since applying the time-reversal operator to both sides of eq 1.49, we would have

$$\mathcal{T}[\dot{\upsilon}] = \frac{d(-\upsilon)}{d(-t)} = \frac{d\upsilon}{dt}$$

$$\mathcal{T}\left[+g-\frac{A}{m}\upsilon^2\right]=g-\frac{A}{m}(-\upsilon)^2=g-\frac{A}{m}\upsilon^2$$

Therefore, we would think that both sides of eq 1.49 are invariant under time reversal. To verify this statement, let us first examine the evolution of the velocity and the vertical position. First, we note that the particle reaches a steady-state velocity υ_∞ which in this case is given by $\upsilon_\infty=\pm\sqrt{mg/A}$. In terms of this velocity, the equation of motion can be written as

$$\frac{d\upsilon}{dt}=\frac{g}{\upsilon_\infty^2}\left(\upsilon_\infty^2-\upsilon^2\right)\Rightarrow\frac{d\upsilon}{\upsilon_\infty^2-\upsilon^2}=\frac{g}{\upsilon_\infty^2}dt \quad (1.50)$$

The left side of this last equation can be expressed as

$$\frac{d\upsilon}{\upsilon_\infty^2-\upsilon^2}=\frac{d\upsilon}{2\upsilon_\infty}\left[\frac{1}{\upsilon_\infty-\upsilon}+\frac{1}{\upsilon_\infty+\upsilon}\right] \quad (1.51)$$

From Eqs. (1.50) and (1.51) and after some calculus of integrals we obtain

$$\upsilon(t)=\upsilon_\infty tanh\left(\frac{gt}{\upsilon_\infty}\right) \quad (1.52)$$

Solving eq 1.52 for the vertical position $y(t)$ with initial condition $y(0)=h$ and zero initial velocity, we obtain

$$y(t)=h-\frac{\upsilon_\infty{}^2}{g}ln\left[Cosh\left(\frac{g}{\upsilon_\infty}t\right)\right]$$

The mass m reaches the ground at an instant t_g given by

$$\frac{\upsilon_\infty{}^2}{g}ln\left[Cosh\left(\frac{g}{\upsilon_\infty}t_g\right)\right]=h$$

Under time reversal $\mathcal{T}\left[\upsilon(t)\right]=-\upsilon(t)=-\upsilon_\infty tanh\left(\frac{gt}{\upsilon_\infty}\right)$ and solving again for the vertical position $y(t)$ with initial condition (0) = 0, we obtain

$$y(-t)=\frac{\upsilon_\infty{}^2}{g}ln\left[Cosh\left(\frac{g}{\upsilon_\infty}t\right)\right]$$

From this last equation, it can be seen that in the absence of dissipation, at time $-t_g$, the particle returns to its initial height h over the ground. However, although the dynamical equations show time-reversal symmetry, during the free fall of the particle, there occurs dissipation of energy due to the frictional force. This implies that on reaching the ground, the particle has a kinetic energy K equal to its initial potential energy MgH minus the energy Q dissipated by friction. In order to return to its initial state, that is, to be situated again at a height H with zero speed, it must be launched by an external agent with a kinetic energy $K'=K+Q > K$. Therefore, this example shows that without an external agent, the analyzed process is irreversible, even though the dynamic equations show symmetry under time reversal.

1.6 TIME REVERSAL OF A CONSERVATIVE SYSTEM

In general, it is said that a system is conservative if the work $\int_{\vec{r}(t_1)}^{\vec{r}(t_2)} \vec{F} \cdot d\vec{r}$ is path independent, that is the integral that represents the work done by the force $\vec{F}$ is independent of the chosen path to join the points corresponding to the initial position $\vec{r}(t_1)$ and the final position $\vec{r}(t_2)$. This is valid when the force $\vec{F}$ depends only on the position. However, it is not necessarily well-defined in the presence of velocity-dependent forces. This is because there can be many velocity-dependent force vectors $\vec{F}(\vec{r}, \vec{\upsilon})$ at the same position $\vec{r}$, corresponding to the different values of $\vec{\upsilon}$. A path integral of this force along a given curve between the initial al final positions in the space, is not well-defined in such cases, since a position $\vec{r}$ is not enough to know the force $\vec{F}(\vec{r}, \vec{\upsilon})$ at this place.

As a specific case of a system conservative according to the above definition but non-time-reversal symmetric, we consider a particle subject to a force of the form $\vec{F} = K\vec{r} \times \vec{\upsilon}$, where K is a constant with units such this expression has units of force. If this is the only force acting over this particle, the corresponding work made in a motion along some path Σ connecting two points P_1 and P_2 is

$$W = \int_{P_1}^{P_2} \vec{F} \cdot d\vec{s} = \int_{t_1}^{t_{2-t_1+\Delta t}} \vec{r} \times \vec{\upsilon} \cdot \vec{\upsilon}\, dt \tag{1.53}$$

where Δt is the traveling time for going from P_1 to P_2. Since the vector product $\vec{r} \times \vec{\upsilon}$ gives a vector perpendicular to $\vec{\upsilon}$ it can be concluded that for any path Σ this work is null.

In particular, for the case of any closed path, this work is null and one could say that this system is conservative. Nevertheless, there is no time-reversal symmetry. In fact, under time reversal, $\vec{r} \rightarrow \vec{r}, \vec{\upsilon} \rightarrow -\vec{\upsilon}$, so that $\vec{F} \rightarrow -\vec{F}$ in this case. As an illustration, we consider a particle placed at a given distance r from the origin of the system of coordinates. If initially the particle is at rest, under the action of the force $K\vec{r} \times \vec{\upsilon}$, the particle will acquire a velocity with components υ_θ and υ_ϕ according to a dynamics described by the following equations:

$$\frac{d^2\theta}{dt^2} = \frac{Kr}{m}\upsilon_\phi \tag{1.54}$$

$$\frac{d^2\phi}{dt^2} = -\frac{Kr}{m}\upsilon_\theta, \tag{1.55}$$

where m denotes the mass of the particle. In this case, r is the constant since there is no radial acceleration.

From these equations, we get that the dynamics of the rotation in the plane xy is given by

$$\frac{d^3\phi}{dt^3} + \left(\frac{Kr}{m}\right)^2 \frac{d\phi}{dt} = 0 \tag{1.56}$$

The solution of this third-order differential equation can be written as

$$\phi(t) = A - \frac{m}{Kr} B\cos\left(\frac{Kr}{m}t\right) + \frac{m\omega_o}{Kr} C\sin\left(\frac{Kr}{m}t\right), \tag{1.57}$$

with A, B, and C constants dependent on the initial condition.

If we assume that at $t = 0$, $\phi(0) = \phi_o \dot{\phi}(0) = \omega_o$, and $\ddot{\phi}(0) = 0$, we obtain the following solution for the constants

$$\mathrm{A} = \phi_o \tag{1.58}$$

$$B = 0, C = \frac{m}{Kr}\omega_o \tag{1.59}$$

Therefore, we get

$$\phi(t) = \phi_o + \frac{\omega_o}{r/m} \sin\frac{Kr}{m}t \tag{1.60}$$

Applying the time-reversal operator we get

$$\mathcal{T}\left[\phi(t)\right] = \phi(-t) = \phi_o + \frac{\omega_o}{Kr/m} \sin\frac{Kr}{m}(-t) = \phi_o - \frac{\omega_o}{r/m} \sin\frac{Kr}{m}t \neq \phi(t) \tag{1.61}$$

Therefore, this confirms that even though there is no dissipation of energy, there is no time-reversal symmetry for this case. In the following example, we show the break of time-reversal symmetry for a charged particle subject to a magnetic field.

EXAMPLE 1.8

In this example, we analyze the motion of a charged particle subject to the action of an electromagnetic field. Let us consider a particle of mass m and electric charge q moving in a plane under the combined influence of a central potential $V(r)$ and a constant uniform magnetic field $\vec{B}$, perpendicular to the plane and generated by a static vector potential $\vec{A}$, $= \vec{B} \times \vec{r}/2$. Is there time-reversal symmetry and time-translational invariance in this case?

SOLUTION

The motion of the particle can be described by the following equation:

$$m\frac{d^2\vec{r}}{dt^2} = q\left\{-\nabla \mathrm{V}(r) + \frac{d\vec{r}}{dt}\times\vec{B}\right\} \tag{1.62}$$

If the magnetic field is externally controlled, it does not vary under time reversal, and from the direct inspection of eq 1.62, we conclude that there is no time-reversal symmetry due to the second term inside brackets. In fact, under time reversal, this is the only term that changes to $-\frac{d\vec{r}}{dt}$, and therefore, this symmetry is lost. This statement is confirmed if we analyze the Hamiltonian for this system, as follows:

The Lagrangian is given by[5]

$$\mathcal{L} = \frac{1}{2}m\left(\dot{\vec{r}}\right)^2 + q\dot{\vec{r}}\cdot\vec{A} - q\mathrm{V}(r) \tag{1.63}$$

From the identity $\vec{a}\cdot\vec{b}\times\vec{c} = \vec{b}\cdot\vec{c}\times\vec{a} = \vec{c}\cdot\vec{a}\times\vec{b}$, we can write the second term at the right of eq 1.63 as $q\dot{\vec{r}}\cdot\vec{A} = \frac{q}{2}\vec{r}\times\dot{\vec{r}}\cdot\vec{B}$. Assuming that $\vec{B}$ is oriented along the z-axis, we obtain that $\vec{r}\times\dot{\vec{r}}\cdot\vec{B} = x\dot{y}B - y\dot{x}B$. Therefore, the components of the canonical momentum $\vec{P} = \vec{p} + q\vec{A}$ are given by

$$P_x = m\dot{x} - \frac{1}{2}qyB \tag{1.64}$$

$$P_y = m\dot{y} + \frac{1}{2}qxB \tag{1.65}$$

$$P_z = m\dot{z} \tag{1.66}$$

The Hamiltonian is obtained from $\mathcal{H} = P_x\dot{x} + P_y\dot{y} + P_z\dot{z} - \mathcal{L}$. Using equations (1.63)–(1.66) and after making some algebra we obtain

$$\mathcal{H} = \frac{1}{2m}\left\{\left(P_x + \frac{1}{2}qyB\right)^2 + \left(P_y - \frac{1}{2}qxB\right)^2 + P_z^2\right\} + V(r) \tag{1.67}$$

From equations (1.64) to (1.66), it can be seen that the canonical momenta are not invariant under time reversal. Therefore, this Hamiltonian has no time-reversal symmetry, but since it does not depend explicitly on the time $\frac{\partial \mathcal{H}}{\partial t} = 0$, the energy of this system is conserved and there is time-translation invariance.

1.7 TIME-REVERSAL SYMMETRY OF THE DYNAMICS OF A RELATIVISTIC PARTICLE

We begin with a free relativistic particle of mass m whose dynamics is described by the following Hamiltonian[4]:

$$\mathcal{H} = \frac{mc^2}{\sqrt{1-\beta^2}} \tag{1.68}$$

with $\beta = \upsilon/c$ where υ is its speed and c is the light velocity in vacuum.

Clearly, the Hamiltonian of eq 1.68 is time-reversal symmetric since it does not depend on time and it does not change under the transformation $\upsilon \rightarrow -\upsilon$. In terms of the momentum p and considering for simplicity one-dimensional motion along the x direction, this Hamiltonian can be written as

$$\mathcal{H} = \sqrt{c^2 p^2 + m^2 c^4} \tag{1.69}$$

From Hamilton equations (1.19) and (1.20) we get

$$\dot{x} = \upsilon = \frac{\partial H}{\partial p} = \frac{pc^2}{\sqrt{c^2 p^2 + m^2 c^4}} \tag{1.70}$$

and $\dot{p} = -\frac{\partial H}{\partial x} = 0$ for this case.

From eq 1.70 and some little algebra, we obtain the expression for the momentum of a relativistic particle in free motion

$$p = \frac{m\upsilon}{\sqrt{1-\beta^2}} \tag{1.71}$$

Therefore, for this simple case, we conclude that the particle moves both with constant velocity and momentum and that there is time-reversal symmetry.

EXAMPLE 1.9

How does the above result change if the particle is subjected to a constant decelerating force F parallel to the direction of its motion?

SOLUTION

In this case, the Hamiltonian changes to

$$\mathcal{H} = \sqrt{c^2 p^2 + m^2 c^4} - Fx \tag{1.72}$$

This Hamiltonian is also time-reversal invariant since it does not change if $p \to -p$ and the second term in this case is even under time reversal.

From Hamilton's equation (1.20) $\dot{p} = -\frac{\partial H}{\partial x} = F$ which corresponds to the second law of Newton. Using the expression for the momentum of eq 1.71, we have

$$\dot{p} = \frac{dp}{dt} = \frac{d}{dt}\left(\frac{mc\beta}{\sqrt{1-\beta^2}}\right) = F \tag{1.73}$$

If initially the particle was at rest, the integration of this last equation gives

$$\beta(t) = \frac{\frac{F}{mc}t}{\sqrt{1+(Ft/mc)^2}} \tag{1.74}$$

From this last equation, it can be seen that the velocity is odd under time reversal and the same occurs for the momentum. Defining the characteristic time $\tau \equiv mc/F$, this equation can be expressed in terms of the normalized time $\tilde{t} = t/\tau$ as

$$\beta(\tilde{t}) = \frac{\tilde{t}}{\sqrt{1+\tilde{t}^2}} \tag{1.75}$$

By integration of eq 1.74 and assuming as initial position $x(0) = 0$, we obtain that the instantaneous position is given by

$$x(\tilde{t}) = c\tau\left\{\sqrt{1+\tilde{t}^2} - 1\right\} \qquad (1.76)$$

Figure 1.6 shows plots of $\beta(\tilde{t})$ and $x(\tilde{t})$ observing that the velocity and position are odd and even, respectively, under time reversal.

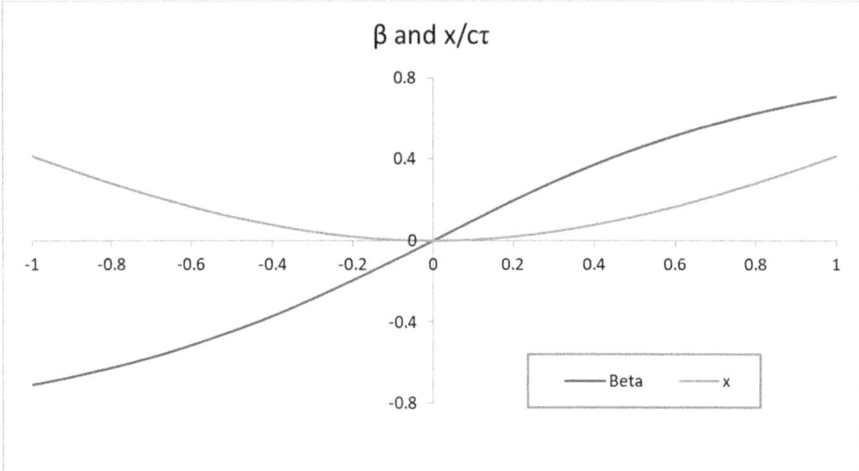

FIGURE 1.6 Velocity and position of a relativistic particle subject to a decelerating force F. The velocity is expressed in terms of c and the position is in units of τc.

1.8 TIME REVERSAL OF A MECHANICAL WAVE

As a further example, we consider the propagation of a mechanical wave that could be for example a transversal wave in an elastic medium like a spring, an acoustic, or a seismic wave. As known,[3] in general, the propagation of a wave satisfies an equation of the form

$$\nabla^2\vec{\xi} - \eta\frac{\partial\vec{\xi}}{\partial t} - \frac{1}{\upsilon^2}\frac{\partial^2\vec{\xi}}{\partial t^2} = 0 \qquad (1.77)$$

where $\vec{\xi}$ is the dynamic variable associated with the energy transmitted by the wave (it could be, for example, a displacement or a velocity, in the case of the wave transmitted in a river when a small stone is thrown over it), η is a parameter representing dissipation effects and υ is the speed of propagation of the wave without dissipation. This speed depends on the properties of the medium, such as elasticity and density. For example, for waves propagating in a spring $\upsilon = \sqrt{T/\mu}$, where T is the applied tension and μ is the linear mass density.

For simplicity, we will consider a plane and monochromatic wave propagating along the z-direction. Trying with a solution of eq 1.77 of the form $\vec{\xi}(z,t) = \vec{\xi}_o exp\left[i(kz - \omega t)\right]$, we obtain the following equation for the magnitude of the propagation vector $\vec{k}$:

$$k^2 - i\eta\omega - \frac{1}{\upsilon^2}\omega^2 = 0 \tag{1.78}$$

From the above equation, we can see that due to the dissipation effects in the medium, the magnitude of the propagation vector is imaginary and if we write it as $k = \alpha - i\beta$, we obtain the following system of equations for α and β:

$$\beta^2 - \alpha^2 + \frac{1}{\upsilon^2}\omega^2 = 0 \tag{1.79}$$

$$2\alpha\beta + \eta\omega = 0 \tag{1.80}$$

The solution of this system of equations gives four combinations of values for the pair of values α and β, which are:

$$\alpha_1 = \pm\frac{1}{\sqrt{2}}\frac{\omega}{\upsilon}\left\{1 + \sqrt{1 + \frac{\eta^2\upsilon^4}{\omega^2}}\right\}^{1/2}, \ \beta_1 = \mp\frac{\eta\omega}{2\alpha_1} \tag{1.81}$$

$$\alpha_2 = \pm\frac{1}{\sqrt{2}}\frac{\omega}{\upsilon}\left\{1 - \sqrt{1 + \frac{\eta^2\upsilon^4}{\omega^2}}\right\}^{1/2}, \ \beta_2 = \mp\frac{\eta\omega}{2\alpha_2} \tag{1.82}$$

It can be seen that the set of solutions α_1, β_1 give real values while the set α_2, β_2 give imaginary values. Let us first examine the solutions α_1, β_1 which for parameters of the medium such that $\eta\upsilon^2 \ll \omega$ can be expressed as

$$\alpha_1 = \pm\frac{1}{\sqrt{2}}\frac{\omega}{\upsilon}\left\{2 + \frac{\eta^2\upsilon^4}{2\omega^2}\right\}^{1/2} = \pm\alpha \tag{1.83}$$

$$\beta_1 = \mp\frac{\eta\upsilon}{\sqrt{2}\left\{2 + \frac{\eta^2\upsilon^4}{2\omega^2}\right\}^{1/2}} = \mp\beta \tag{1.84}$$

With this approach, the solutions for $\vec{\xi}(z, t)$ are

$$\vec{\xi}(z,t)_+ = \vec{\xi}_o e^{i(+\alpha z - \omega t)} e^{-\beta z} \tag{1.85}$$

$$\vec{\xi}(z,t)_- = \vec{\xi}_o e^{i(-\alpha z - \omega t)} e^{+\beta z} \tag{1.86}$$

The solution given by eq 1.85 represents a wave with decreasing amplitude as going away from the source. In turn, the solution corresponding to eq 1.86 represents a wave propagating radially toward the source with increasing amplitude as becoming closer. Nevertheless, this has no physical sense considering that in the propagation of the wave from the source to a point at a distance z, there are energy losses so that if this wave is reflected, for instance by a mirror, it is not possible that in coming back to the source it will have the same amplitude it had during its emission. Therefore, this solution is discarded.

Let us now examine the waves corresponding to solutions α_2, β_2 which being imaginary gives

$$k = \pm i|\alpha_2| - i(\mp i|\beta_2|) = \pm i|\alpha_2| \mp |\beta_2| \tag{1.87}$$

With similar arguments to that previously given for the waves corresponding to the set α_1, β_1, it can be seen that taking only the upper signs for α_2 and β_2 in eq 1.82 leads to a wave with decreasing amplitude as going away from the source.

Nevertheless, it is worth to comment that time reversal of acoustic waves is applied to locate the sources which produce these waves. Time-reversal models (TRM) of acoustic wave propagation have been widely used in seismology and exploration geophysics. For example, time-reversal source imaging was proposed by McMechan[6] to backpropagate seismograms from receivers to find the location of sources.

In the case of electromagnetic waves, time reversal has been proposed as a method for locating lightning strikes and faults in power networks.[7,8]

In general, TRM for studying wave propagation consist of two steps: (1) time reversal of recorded data and (2) back propagating the time-reversed data as sources from receiver locations through an appropriate model of the physical system. The basic assumption of these models is that the wave equation is time-invariant in the case of propagation through a media in which there is no attenuation. In the case of propagation through the Earth, where attenuation is not negligible, there is no time-invariance. A solution to this problem in the case of seismic waves has been proposed by T. Zhu,[9] by means of a TRM approach that compensates for attenuation and dispersion effects during the wave propagation in attenuating media. This approach is based on a viscoacoustic wave equation which explicitly separates the effects of attenuation and dispersion.

1.9 TIME TRANSLATIONAL SYMMETRY

In classical and quantum physics, there are two basic kinds of time symmetry, that is, two ways to transform time coordinates in equations such that certain properties do not change.

Time-translational symmetry means that the laws of physics do not change as time runs forward or backward. So if somebody made an experiment 10 years ago, and if under the same conditions is made again today, or in 40 years more, the corresponding results shall not change.

Time-reversal symmetry means that the physical phenomena do not change if the clocks run forwards or backwards, that is, if the variable t is changed by $-t$.

The time-translation symmetry means that it does not matter where we are with respect to time. The transformation corresponding to time translation means to change t by $t' = t + s$, where s is just a real number. Here transformed time goes in the same direction, we just shift the system (or time beginning) along the time axis, like performing the experiment on Thursday instead of Monday. This transformation is parameterized by the real number s, and s = 0 which means no transformation. If this transformation keeps the Lagrangian unchanged, which means what happens with our system does not depend directly on clocks, then this symmetry holds and as a consequence, we have energy conservation working.

EXAMPLE 1.9

Consider a system described by the following Lagrangian which is time-dependent

$$\mathcal{L} = \frac{m}{2}\left[\dot{q}^2 sin^2 \omega t + q\dot{q}\omega sin2\omega t + q^2\omega^2\right] \tag{1.88}$$

Obtain the corresponding Hamiltonian. What can be said about energy conservation and time-reversal symmetry?

SOLUTION

Observing this Lagrangian, one would expect that the corresponding Hamiltonian is time-dependent, so there would not be energy conservation, and we will suspect that in consequence, there is no time-reversal symmetry.

Nevertheless, as suggested in problem 27, chapter 8, of Ref. [3], the explicit dependence on time can be eliminated if we define a new coordinate = $q\ sin\omega t$, leading to the following expression for the lagrangian:

$$\mathcal{L} = \frac{m}{2}\left[\left(\dot{Q} - \omega q cos\omega t\right)^2 + 2q\omega\left(\dot{Q} - \omega q cos\omega t\right)cos\omega t + \omega^2 q^2\right]$$

$$= \frac{m}{2}\left[\dot{Q}^2 + \omega^2 Q^2\right] \tag{1.89}$$

where we have made some algebra and considered that

$$Q = q\,sin\omega t \rightarrow \dot{Q} = \dot{q} sin\omega t + \omega q cos\omega t \rightarrow \dot{q} sin\omega t = \dot{Q} - \omega q cos\omega t$$

From eq 1.17, the Hamiltonian is obtained from

$$\mathcal{H}(P,Q) = P\dot{Q} - \mathcal{L}(\dot{Q},Q) \tag{1.90}$$

where P is the conjugate momentum given by

$$P = \frac{\partial \mathcal{L}}{\partial \dot{Q}} = m\dot{Q} \rightarrow \dot{Q} = \frac{P}{m} \tag{1.91}$$

By means of the two above equations, we obtain

$$\mathcal{H}(P,Q) = \frac{P^2}{2m} - \frac{m}{2}\omega^2 Q^2 \tag{1.92}$$

This Hamiltonian is not explicitly time-dependent and considering that $\frac{dH}{dt} = -\frac{\partial \mathcal{L}}{\partial t}$ is zero in this case, we conclude that energy is conserved and that there is time-reversal invariance. This last statement can be confirmed from the dynamical evolution of the coordinate momentum q which is obtained from Lagrange's equations, which for this one-dimensional case is:

$$\frac{d}{dt}\left(\frac{\partial \mathcal{L}}{\partial \dot{Q}}\right) - \frac{\partial \mathcal{L}}{\partial Q} = 0 \rightarrow \frac{m}{2}\frac{d}{dt}\dot{Q}^2 - \frac{m}{2}\omega^2 2Q = 0 \rightarrow \ddot{Q} = \omega^2 Q \tag{1.93}$$

Assuming as initial condition $Q(0) = 0$, the solution of this last equation is

$$Q(t) = sinh\omega t \rightarrow q(t) = \frac{sinh[\omega t]}{sin\omega t} \tag{1.94}$$

It can be readily verified that this solution is time-reversal invariant since

$$TR\left[q(t)\right] = \frac{sinh\left[\omega(-t)\right]}{sin\omega(-t)} = \frac{sinh[\omega t]}{sin\omega t} \tag{1.95}$$

1.10 CONCLUSIONS

The issue of symmetry is relevant to physical theories, since they are typically taken as guides to look into the structure of physical theories. Roughly speaking, a symmetry is a transformation (a mapping) that does not change the structure of the theory.

In the Newtonian formulation of classical mechanics, time reversal is simply the reversal of the order of events in a given trajectory $\vec{r}_i(t)$.

In the formulation of classical mechanics based on Hamilton's equations, for time reversal, it is not enough to reverse the sequence with which the analyzed events occur. Since time reversal transforms a rightward-moving body into a leftward-moving body, in addition, the velocity vector and corresponding momentum of an instantaneous state must be reversed. If one looks a film showing the dynamics of a mechanical system, and it is not possible to decide whether it is shown in the forward or reverse direction, the system is said to have symmetric behavior in the face of time reversal.

Time-reversal symmetry in processes described by classical mechanics requires careful analysis. This symmetry cannot exist even for ideal and conservative systems, in the presence of interactions where there are velocity-dependent forces.

KEYWORDS

- **reversibility**
- **invariance**
- **symmetry**
- **conservative system**
- **dissipation**
- **generalized coordinates**

REFERENCES

1. David, A. *Time and Chance*; Harvard University Press, 2000.
2. Roberts, B. W. When We Do (and Do Not) Have a Classical Arrow of Time. *Phil. Sci.* **2013,** *80* (5), 1112–1124.
3. Tippler, P. A. Mosca. G. *Physics for Scientists and Engineers*, 6th ed.; W.H. Freeman, 2007.

4. Goldstein, H.; Poole, C.; Safko, J. *Classical Mechanics*, 3rd ed.; Addison Wesley, 2000.
5. Jackson, J. D. *Classical Electrodynamics*, 3rd ed.; John Wiley & Sons, Inc.: Berkeley, 1999.
6. McMechan, G. A.Determination of Source Parameters by Wavefield Extrapolation. *Geophys. J. R. Astr. Soc.* **1982,** *71*, 613–628.
7. Lugrin, G.; Mora Parra, N.; Rachidi, F.; Rubinstein, M.; Diendorfer, G. On the Location of Lightning Discharges Using Time Reversal of Electromagnetic Fields. *IEEE Trans. Electromagn. Compatib.* **2014,** *56* (1).
8. Razzaghi, R.; Lugrin, G.; Manesh, H.; Romero, C.; Paolone, M.; Rachidi, F. An Efficient Method Based on the Electromagnetic Time Reversal to Locate Faults in Power Networks. IEEE Trans. Power Deliv. **2013,** *28* (3).
9. Zhu, T. Time-Reverse Modelling of Acoustic Wave Propagation in Attenuating Media. *Geophys. J. Int.* **2014.**

CHAPTER 2

About Time in Electromagnetic Theory

ABSTRACT

In this chapter, we analyze the effect of time reversing the equations describing the behavior of the electromagnetic field in different media. One of the main conclusions is that the equations of *macroscopic* electrodynamics are not time-reversal invariant when the system has dissipative elements, but in contrast, in a *microscopic* formalism—with all the light and matter degrees of freedom included in the analysis—the dynamics of the considered system are time-reversal symmetric. In the absence of dissipation, the propagation of an electromagnetic wave is symmetric under time reversal and Maxwell's equations allow the propagation of a retarded and advanced solution. This last solution can be used in engineering techniques for locating sources of electromagnetic waves.

2.1 INTRODUCTION

One of the most relevant attempts to deny the existence of any objective temporal in the physical universe was made in 1930 by G.N. Lewis.[1] He proposed the idea that the concept of "time's arrow" is almost entirely due to the phenomena of consciousness and memory. Lewis analyzed the role of time in optics and electromagnetism. In his view, the laws of optical are entirely symmetrical with respect to the emission and absorption of light. If time is imagined as reversed, emitting and absorbing objects exchange roles but the optical laws are unchanged. Radiation from an emitting particle, such as an atom making a transition from an excited to ground state, seems, however, to be in direct conflict with this idea of temporal symmetry. Lewis admitted that the emission of a continuous spherical shell of energy is essentially irreversible. All parts of the shell move outward until they meet

Looking at Time from a Physics Perspective. Patricio Robles (Author)

absorbing particles, but there is a significant difference in the instants when this occurs. Lewis argued that the concept of symmetrical time leads to the conclusion that the base radiation process must be one in which a single emitting particle sends its energy to only one absorbing particle.

In the case of the electromagnetic theory, Maxwell's equations in vacuum, like the equations of classical mechanics in the absence of dissipation or of a force perpendicular to its velocity (see Section 1.6 of Chapter 1), are unchanged when the time direction is reversed. How is this compatible with the theory of radiation, which in itself implies a temporal asymmetry in the emission–absorption process? The derivation depends on the fact that of the two symmetrical solutions which arise in the mathematical analysis, only it is accepted that the retarded potential has physical sense. Lewis argued that if the advanced potential is selected and the retarded potential is discarded, an electromagnetic theory would result also in agreement with the experimental facts, but in this case, we should interpret the absorbing particle as the active agent, sucking in energy from all parts of space in a spherical shell contracting with the velocity of light.

The electromagnetic arrow was the subject of a lively discussion by Einstein and Ritz in 1909.[2,3] Einstein argued that the retarded and advanced descriptions of radiation processes occurring in any finite region are equivalent since the equations of wave propagation are symmetric with respect to time, but that the auxiliary conditions giving the precise circumstances of emission and absorption are very different. In the retarded description, it is sufficient if all the macroscopic sources are known whereas in the advanced description, all the absorption processes must be known, but unlike the former, they must be fully specified in microscopic detail. In practice, information about the absorption processes is not available and so it is necessary to choose the retarded description.

Einstein's point about the difference between macroscopic sources and microscopic absorbers is compatible with the argument given by G.N. Lewis[1] and extended by Karl Popper,[4] so that the temporal inverse of the expansion of a spherical wave from a point source would, in general, require the co-operation of a large number of sources scattered throughout space so as to produce a coherent wavefront that converges to annihilation at a particular point and at a particular instant. This coherence is so highly improbable that it could not as a rule occur spontaneously.

The idea of the conditional independence of influences emanating from different parts and spatial directions was postulated by Penrose and Percival[5] in an attempt to account for the common direction in time of various irreversible processes. This assumption of conditional independence was introduced as a fundamental asymmetric condition supplementing the temporally symmetric laws of dynamics so as to account for the irreversibility of most physical systems. According to this hypothesis, influences coming "from infinity" in different directions are statistically independent. From their principle, Penrose and Percival derived the time direction of some irreversible physical processes, for example, the irreversibility of the waves that an oscillating dipole radiates toward the environment, and that it is associated with retarded potentials.

The dependence of the electromagnetic arrow of time on the thermodynamic arrow is a feature of the absorber theory of radiation formulated by Wheeler and Feynman.[6] They introduced the hypothesis that every photon has an absorber as well as an emitter. In their theory, an accelerated charge particle emits radiation equally into the past and future, which is retarded and advanced waves are generated symmetrically. If the radiation is confined to an opaque enclosure, so that all of it is absorbed, the waves striking the walls will cause the charged particles therein to radiate likewise into both the past and the future. Wheeler and Feynman showed that if the enclosure is fully opaque, the advanced waves emitted by the walls will just cancel those from the source particle and only the retarded waves will be left. The origin of this time asymmetry lies in the mechanism of absorption where the energy of the waves absorbed by the walls is converted into heat.

The fundamental equations are time-symmetrical and indicate no preference between the familiar retarded fields of experience, which diverge at a finite speed from their associated source charges, and the corresponding advanced fields, which converge on source charges at the same speed.[7] Retarded fields are selected in practice because they appear to correspond to reality, whereas advanced fields are discarded on the grounds that they are contrary to experience. However, this selection has been widely discussed by several physicists. It is worth mentioning a paper by Dirac of 1938[8] where he shows that in order to express the empirically well-established formula for radiation damping in terms of a covariant electromagnetic field, it is necessary to employ *both* the *retarded and advanced* solutions. In Ref. [7], J.E. Hogarth explores the hypotheses that Dirac's result is not fortuitous, arguing that in spite of common metaphysical presuppositions about the nature of time and causality, the advanced solutions are physically significant and must be taken seriously, that in fact, retarded and advanced fields exist in perfect symmetry, and that radiation does not arise through intrinsic self-action of elementary charge on itself. A consequence of these assumptions is, of course, that the retarded nature of macroscopic electrodynamic observations must be explained by some other interrelated physical process. An electrodynamic arrow-of-time must be established from considerations other than the solutions of the field equations. In Ref. [9], John Earman discusses a possible relation between that arrow and the cosmological and thermodynamic arrows.

We start this chapter with the analysis of time reversing Maxwell's equations in vacuum, showing that in empty space (without charges and current external sources), these equations are symmetric under time reversal if in addition to replacing t by—t the magnetic field vectors change their sign. In this case, Maxwell's equations predict the propagation of retarded and advanced electromagnetic waves. This last type of wave is considered in technological applications to locate sources of electromagnetic or sound disturbances.

Next, the effect of time inversion on the propagation of electromagnetic waves in material media with dissipation is analyzed. We show that a description based on a model which only includes the electromagnetic degrees of freedom leads to the conclusion that the equation of macroscopic electrodynamics is not time-reversal invariant. However, we show that if all the light and matter degrees of freedom are considered in the analysis, the dynamic of the system has time-reversal symmetry.

2.2 TIME-REVERSAL INVARIANCE OF MAXWELL EQUATIONS IN VACUUM

At least at the classical level, the basic laws of physics are invariant under change of the sense of time. This means that under the time-reversal transformation $t \rightarrow t' = -t$, the related physical quantities transform so that the form of the equation is kept. For the case of Classical Electrodynamics, the forms of Maxwell's equations are invariant under rotations, space inversion, and temporal inversion.[10]

Let us analyze the time-reversal symmetry of Maxwell's equations starting from Gauss Law, which relates the charge distribution with the resulting electric displacement vector:

$$\nabla \cdot \vec{D} = \rho \tag{2.1}$$

Since the electric charge is invariant under Lorentz transformations and is a scalar under rotations, it can be assumed that is even under time reversal and the same occurs for the charge density ρ. Therefore, from eq. 2.1, it is reasonable to assume that the displacement vector $\vec{D}$ is also time-reversal invariant. Considering that vector $\vec{D}$ is related to the electric field vector $\vec{E}$ by the relation $\vec{D} = \varepsilon\vec{E}$ where ε is the dielectric permittivity of the medium which macroscopically is also time-invariant, it can be concluded that $\vec{E}$ is even under time reversal.[10]

We consider now the Faraday Law:

$$\nabla \times \vec{E} = -\frac{\partial \vec{B}}{\partial t} \tag{2.2}$$

The typical argument is to say that under time reversal with $t \rightarrow t' = -t$ and assuming that $\vec{E}$ is time-reversal invariant, for keeping the form of this law, the magnetic field density vector $\vec{B}$ must be odd under the time-reversal operation, that is $\mathcal{T}[\vec{B}] = -\vec{B}$. Nevertheless, it is worth to go in deep in the analysis and consider that the fact that in the presence of a magnetic field varying in time, according to eq. 2.2 $\nabla \times \vec{E} \neq 0$, and therefore, the electric field $\vec{E}$ is not conservative since $\oint \vec{E} \cdot d\vec{l}$ is not zero along any path whose surface is subject to a time varying magnetic flux. One argumentation for solving, this apparent inconsistency is based on the fact that according to one of the Maxwell's equations $\nabla \cdot \vec{B} = 0$ implying that the magnetic field density $\vec{B}$ can be expressed as $\vec{B} = \nabla \times \vec{A}$ where $\vec{A}$ is the magnetic potential vector. Combining this with eq. 2.2 we can write

$$\nabla \times \vec{E} = -\frac{\partial \left(\nabla \times \vec{A}\right)}{\partial t} \tag{2.3}$$

If for our purposes, it is assumed that the sequence of time and space derivation can be inverted, we can write

$$\nabla \times \vec{E} = -\frac{\partial \left(\nabla \times \vec{A}\right)}{\partial t} = -\nabla \times \frac{\partial \vec{A}}{\partial t} \Rightarrow \nabla \times \left(\vec{E} + \frac{\partial \vec{A}}{\partial t} \right) = 0 \tag{2.4}$$

From eq. 2.4, the term $\vec{E} + \frac{\partial \vec{A}}{\partial t}$ can be expressed in terms of the gradient of a scalar field that it is precisely the electric potential field Φ so that

$$\vec{E} = -\nabla \Phi - \frac{\partial \vec{A}}{\partial t} \tag{2.5}$$

Since the electric potential field Φ depends on the charge distribution inside a given volume V and accepts as a postulate that the charge is invariant under time reversal, the first term at the right side of eq. 2.5 is time-reversal invariant. Furthermore, the magnetic potential vector $\vec{A}$ depends on the distribution of currents inside the same volume V and accepts as a second postulate that currents are odd under time reversal, denoting $\mathcal{T}$ as the time-reversal operator we have

$$\mathcal{T}\left[\vec{E}(t)\right]=-\mathcal{T}\left[\nabla\ddot{\mathrm{O}}(t)\right]-\mathcal{T}\left[\frac{\partial\vec{A}}{\partial t}\right]=-\nabla\Phi(-t)-\frac{\partial\vec{A}(-t)}{\partial(-t)}=-\nabla\Phi(+t)-\frac{\partial\left(-\vec{A}(t)\right)}{\partial(-t)}$$

$$=-\nabla\Phi(t)-\frac{\partial\left(\vec{A}(t)\right)}{\partial(t)} \tag{2.6}$$

Therefore, we conclude that $\mathcal{T}[\vec{E}(t)]=\vec{E}(t)$ confirming that the electric field is even under time reversal. In addition, we can see that the fact that the rotor of any vector field is different from zero does not necessary imply that there is the dissipation of energy, as discussed in Section 1.6 of Chapter 1.

Now we consider Ampere's law relating the magnetic field $\vec{H}$ with the conduction current density $\vec{J}$ and with the so-called displacement current associated with the time variation of the vector $\vec{D}$:

$$\nabla\times\vec{H}=\vec{J}+\frac{\partial\vec{D}}{\partial t} \tag{2.7}$$

For keeping the form of this law, invariant under time-reversal operation $t\rightarrow t'=-t$ and since $\vec{D}$ and $\vec{H}$ are even and odd, respectively, we conclude that the current density must be odd under time-reversal operation, that is $\mathcal{T}\left[\vec{J}\right]=-\vec{J}$.

With respect to the Maxwell's equation corresponding to the null divergence of $\vec{B}$, that is $\nabla\cdot\vec{B}=0$, clearly this equation is time-reversal symmetric since

$$\mathcal{T}\left[\nabla\cdot\vec{B}\right]=\nabla\cdot\left(-\vec{B}\right)=-\nabla\cdot\vec{B}=0 \tag{2.8}$$

The same time-reversal symmetry exists for the law of charge conservation:

$$\nabla\cdot\vec{J}+\frac{\partial\vec{D}}{\partial t}=0 \tag{2.9}$$

In fact, applying the time-reversal operation: $t\rightarrow t'=-t$, $\mathcal{T}[\vec{D}]=\vec{D}$ and $\mathcal{T}[\vec{J}]=-\vec{J}$ so that this last equation is invariant under time reversal.

2.3 TIME REVERSAL OF ELECTROMAGNETIC WAVES IN VACUUM

As known, from Maxwell's equations in vacuum one obtains the wave equations for the electric and magnetic field vectors, as follows:

Applying the rotational operator to both sides of equation (2.2)

$$\nabla \times \nabla \times \vec{E} = -\nabla \times \frac{\partial \vec{B}}{\partial t} \tag{2.10}$$

Using identities of vector calculus, the left side of eq. 2.10 is written as $\nabla(\nabla \cdot \vec{E}) - \nabla^2 \vec{E}$. For vacuum $\nabla \cdot \vec{D} = \varepsilon_o \nabla \cdot \vec{E} = 0$ and assuming that the order of spacial and temporal derivatives can be reversed, the right side can be written as $-\frac{\partial}{\partial t}\nabla \times \vec{B}$. Considering that for vacuum $\vec{B} = \mu_o \vec{H}$, $\vec{D} = \varepsilon_o \vec{E}$, $\vec{J} = 0$ and combining eqs. (2.10) and (2.9), one gets the following wave equation for the electric field vector

$$\nabla^2 \vec{E} - \frac{1}{c^2}\frac{\partial^2 \vec{E}}{\partial t^2} = 0 \tag{2.11}$$

Similarly, one obtains the following wave equation for the magnetic field vector

$$\nabla^2 \vec{H} - \frac{1}{c^2}\frac{\partial^2 \vec{H}}{\partial t^2} = 0 \tag{2.12}$$

Clearly, these equations are symmetrical under time reversal and this implies that there are two possible solutions of the form $f(r-ct)$ and $g(r+ct)$, where $c = 1/\sqrt{\varepsilon_o \mu_o}$ and r is the radial distance from any point with respect to the origin chosen as the point where the source is located. The first one represents a wave moving away from the source (retarded solution) while the second represents a wave traveling toward the source (advanced solution). In particular, if the source of these waves is an electric dipole oscillating with angular frequency ω, for points very far away from the source the solutions can be expressed by means of factors of the form $\frac{1}{r} Re[e^{i\omega\left(t \pm \frac{r}{c}\right)}]$.

2.4 TIME REVERSAL OF MATERIAL RELATIONS

Let us consider, the time-reversal of an electromagnetic field in a stationary dielectric material. The dielectric is assumed to be isotropic, non-uniform, and with magnetic permeability equal to that of a non-magnetic material (μ_o). Assuming spatially local response, the electric displacement vector $\vec{D}(\vec{r},t) = \varepsilon_o \vec{E}(\vec{r},t) + \vec{P}(\vec{r},t)$ in the time domain is related to the electric field acting on the material by

$$\vec{D}(\vec{r},t)=\varepsilon_o\int_0^{\infty}\varepsilon(\vec{r},\tau)\vec{E}(\vec{r},t-\tau)d\tau=\varepsilon_o\vec{E}(\vec{r},t)+\varepsilon_o$$

$$\int_0^{\infty}\chi(\vec{r},\tau)\vec{E}(\vec{r},t-\tau)d\tau, \tag{2.13}$$

where $\vec{P}(\vec{r},t)=\varepsilon_o\int_0^{\infty}\chi(\vec{r},\tau)\vec{E}(\vec{r},t-\tau)d\tau$ is the resultant polarizability of the medium due to the applied electric field, $\varepsilon(\vec{r},\tau)=\delta(\tau)+\chi(\vec{r},\tau)$ with $\chi(\vec{r},\tau)$ being the dielectric susceptibility. The eq. 2.13 expresses a nonlocal behavior in time of the polarization since at time t, it depends on the electric field at other instants. However, there is a locality in space if we assume as valid the approximation that the spatial variation of the applied fields has a scale large compared with the dimensions involved in the processes of atomic or molecular polarization.

Use of the Fourier transform yields the well-known material relations for an isotropic dielectric without spatial dispersion:

$$\vec{D}(\omega)=\varepsilon_o\varepsilon(\omega)\vec{E}(\vec{r},\omega), \tag{2.14}$$

where $\varepsilon(\vec{r},\omega)=\int_0^{\infty}\varepsilon(\vec{r},\tau)e^{-i\omega\tau}d\tau.$

By applying the Fourier transform to both sides of the Maxwell's equations in vacuum, we obtain their frequency-domain version:

$$\nabla\times\vec{E}(\omega,\vec{r})=i\omega\mu_o\vec{H}(\omega,\vec{r}) \tag{2.15}$$

$$\nabla\times\vec{H}(\omega,\vec{r})=-i\omega\varepsilon_o\varepsilon(\vec{r},\omega)\vec{H}(\omega,\vec{r}) \tag{2.16}$$

$$\nabla\cdot\left[\varepsilon(\omega,\vec{r})\vec{E}(\omega,\vec{r})\right]=0 \tag{2.17}$$

$$\nabla\cdot\left[\vec{H}(\omega,\vec{r})\right]=0, \tag{2.18}$$

where it has been assumed that the free currents and charge densities are zero.

Combining Eqs. 2.15 and 2.16 and after doing some algebra, we obtain the inhomogeneous wave equation

$$\nabla^2\vec{H}(\vec{r},\omega)+\frac{\omega^2}{c^2}\varepsilon(\vec{r},\omega)\vec{H}(\vec{r},\omega)=i\omega\varepsilon_o\varepsilon(\vec{r},\omega)\nabla\varepsilon(\vec{r},\omega)\times\vec{E}(\vec{r},\omega) \tag{2.19}$$

If the dielectric permittivity does not vary with the position so that $\nabla\varepsilon(\vec{r},\omega)=0$, this last equation reduces to the homogeneous Helmholtz equation.

Next, let us investigate how time reversal applies to the magnetic field $\vec{H}$ in the frequency domain. The frequency spectrum of the field in the original process is given by the Fourier transform

$$\vec{H}(\omega,\vec{r}) = \frac{1}{2\pi}\int_{-\infty}^{\infty} \vec{H}(t,\vec{r})e^{-i\omega t}\,dt \tag{2.20}$$

Under time reversal, the field in the time domain transforms as

$$\mathcal{T}\left\{\vec{H}(t,\vec{r})\right\} = -\vec{H}(-t,\vec{r}) \tag{2.21}$$

The corresponding Fourier transform in the time reversed process is

$$\mathcal{T}\left\{\vec{H}(\omega,\vec{r})\right\} = \frac{1}{2\pi}\int_{-\infty}^{\infty} -\vec{H}(-t,\vec{r})e^{+i\omega t}\,dt = -\frac{1}{2\pi}\int_{-\infty}^{\infty} \vec{H}(t,\vec{r})e^{+i\omega t}\,dt = -\vec{H}(-\omega,\vec{r})$$

$$= \vec{H}^*(\omega,\vec{r})^* \tag{2.22}$$

Similarly, for the electric field and considering that it is even under time-reversal, we get

$$\mathcal{T}\left\{\vec{E}(\omega,\vec{r})\right\} = \vec{E}(-\omega,\vec{r}) = \vec{E}^*(\omega,\vec{r}) \tag{2.23}$$

EXAMPLE 2.1

Prove that the time-reversal symmetry of field equations implies the following rule for the time reversal of the complex permittivity

$$\mathcal{T}\left\{\varepsilon(\omega)\right\} = \varepsilon^*(\omega) \tag{2.24}$$

SOLUTION

In the frequency domain, the electric displacement $\vec{D}$ is related with the electric field $\vec{E}$ according to

$$\vec{D}(\omega,\vec{r}) = \varepsilon_o\varepsilon(\omega)\vec{E} \tag{2.25}$$

Applying the time reversal operator to both sides of this equation, we have

$$\mathcal{T}\left\{\vec{D}(\omega)\right\} = \mathcal{T}\left\{\frac{1}{2\pi}\int_{-\infty}^{+\infty}\vec{D}(\vec{r},t)e^{-i\omega t}\,dt\right\} = \frac{1}{2\pi}\mathcal{T}\left\{\int_{-\infty}^{+\infty}\vec{D}(\vec{r},t)e^{-i\omega t}\,dt\right\}$$

$$= \frac{1}{2\pi}\int_{-\infty}^{+\infty} \vec{D}(\vec{r},t)e^{+i\omega t}\,dt = \vec{D}(-\omega) = \vec{D}^*(\omega) \tag{2.26}$$

In a similar way, we obtain

$$\mathcal{T}\left\{\vec{E}(\omega)\right\} = \vec{E}(-\omega) = \vec{E}^*(\omega) \tag{2.27}$$

Combining eqs. (2.26) and (2.27)

$$\underbrace{\mathcal{T}\left\{\vec{D}(\omega)\right\}}_{=\vec{D}^*(\omega)} = \mathcal{T}\left\{\varepsilon_o\varepsilon(\omega)\vec{E}(\omega)\right\} = \varepsilon_o\,\mathcal{T}\left\{\varepsilon(\omega)\vec{E}(\omega)\right\} = \varepsilon_o\,\mathcal{T}\left\{\varepsilon(\omega)\right\}\underbrace{\mathcal{T}\left\{\vec{E}(\omega)\right\}}_{=\vec{E}^*(\omega)} \tag{2.28}$$

Equation 2.25 implies that $\vec{D}^*(\omega) = \varepsilon_o\varepsilon^*(\omega)\vec{E}^*(\omega)$, and therefore, eq. 2.28 leads to the result that

$$\mathcal{T}\left\{\varepsilon(\omega)\right\} = \varepsilon^*(\omega) \tag{2.29}$$

The result corresponding to eq. 2.29 implies that under time reversal lossy media become active and vice versa. This can be interpreted considering that time reversal involves the global reversal of the process with correct microscopic initial conditions. If the direct process was lossy, in the reversed process, the phonons of the dielectric lattice will oscillate in such a way that their energy will be transformed back into the energy of electromagnetic waves (we must remember that a phonon is a collective excitation in condensed matter, particularly in solids). Naturally, this exact process reversal is impossible in practice, due to the enormous number of microscopic conditions to be satisfied. We are able to reverse only the macroscopic conditions, sending the wave in the opposite direction without reversing the lattice vibrations. Then the dielectric permittivity will remain lossy and the reversed process will be different from the original one.

We will illustrate this condition with the following examples.

EXAMPLE 2.2

The response of a polarizable particle to an electric field oscillating harmonically with time with angular frequency ω can be described by a Drude classical model (Refs. [10] and [11]) which considers a mechanical oscillator with elastic constant k and damping γ. The driving force is due to the electric force acting over each electron (mass m) so that the corresponding dynamics equation is

$$m\frac{d^2\vec{r}}{dt^2} + \gamma\frac{d\vec{r}}{dt} + k\vec{r} = e\vec{E} \tag{2.30}$$

Using this model to show that the dielectric function can be expressed as

$$\varepsilon(\omega) = 1 + \frac{\omega_{pe}^2}{\omega_{oe}^2 - \omega^2 - i\Gamma_d\omega}, \tag{2.31}$$

where $\omega_{oe}^2 \equiv k/m, \omega_p$ is the plasma frequency dependent on the density of free electrons and Γ_d is a parameter representing the dielectric dissipation. Discuss the effect of time reversal and relate your result with eq. 2.29.

SOLUTION

Considering a solution of the form $\vec{r} = \vec{r}_0 e^{-i\omega t}$ and inserting in eq. 2.30 we obtain

$$\vec{r}_0 = \frac{e}{k - m\omega^2 - i\gamma\omega}\vec{E}_0. \tag{2.32}$$

With $\omega_{oe}^2 \equiv k/m$ and $\Gamma_d \equiv \gamma/m$ the above equation may be recasted as

$$\vec{r}_0 = \frac{e/m}{\omega_{oe}^2 - \omega^2 - i\Gamma_d\omega}\vec{E}_0. \tag{2.33}$$

For an atom with Z electrons, the corresponding dipolar moment is $\vec{p} = Ze\vec{r} = \alpha\vec{E}$ where α is the atomic polarizability, and therefore, the above equations lead to

$$\alpha(\omega) = \frac{Ze^2/m}{\omega_{oe}^2 - \omega^2 - i\Gamma_d\omega} \tag{2.34}$$

If n is the concentration of atoms, each one with Z electrons, the polarization vector is

$$\vec{P} = n\alpha\vec{E} = \frac{nZe^2/m}{\omega_{oe}^2 - \omega^2 - i\Gamma_d\omega}\vec{E} \tag{2.35}$$

The displacement vector $\vec{D}$ is related with the electric field $\vec{E}$ as

$$\vec{D} = \epsilon_o\vec{E} + \vec{P} = \epsilon_o\vec{E} + n\alpha\vec{E} = \epsilon(\omega)\epsilon_o\vec{E} \tag{2.36}$$

Therefore, the dielectric function $\varepsilon(\omega)$ is given by

$$\varepsilon(\omega) = 1 + \frac{n\alpha}{\epsilon_o} = 1 + \frac{nZe^2/m\epsilon_o}{\omega_{oe}^2 - \omega^2 - i\Gamma_d\omega} = 1 + \frac{\omega_{pe}^2}{\omega_{oe}^2 - \omega^2 - i\Gamma_d\omega}, \tag{2.37}$$

where ω_{pe} is the plasma frequency given by $\omega_{pe}^2 = nZe^2/m\varepsilon_0$.

From eq. 2.30, it can be seen that under time reversal, the term $\gamma\frac{d\vec{r}}{dt}$ changes its sign, and therefore, the dynamics described for each electron is not time-reversal invariant since the oscillations increase with time instead of being damped.

Therefore, $\mathcal{T}\{\varepsilon(\omega)\} = \varepsilon^*(\omega) = 1 + \frac{\omega_{pe}^2}{\omega_{oe}^2 - \omega^2 + i\Gamma_d \omega}$ meaning that the phonons of the dielectric lattice will oscillate in such a way that their energy will be transformed back into the energy of the electromagnetic waves, a process which as known is impossible in practice. We conclude that in this case due to dissipation there is a time-reversal asymmetry both at a microscopic as for a macroscopic scale. In fact, at a microscopic level if we use a damped harmonic oscillator for representing the oscillations of the electrons of a medium subject to an oscillating electric field, as expected there is not time-reversal symmetry. This asymmetry also occurs at a macroscopic scale since the dielectric medium, which is a macroscopic quantity, has an imaginary part that implies absorption of energy from the electromagnetic field as analyzed in the following example.

EXAMPLE 2.3

Consider an isotropic metallic sphere of radius R subject to an electromagnetic wave with a wavelength $\lambda \gg R$ so that the associated electric field can be considered as uniform (this consideration is usually called the long wavelength approach). The sphere is made of a material with dielectric permittivity dependent on the angular frequency ω according to a dielectric function described by a Drude model of the form given by eq. 2.31 but with $\omega_{oe} = 0$.

$$\varepsilon(\omega) = 1 - \frac{\omega_p^2}{\omega^2 + i\Gamma_d \omega} \tag{2.38}$$

where ω_p is the plasma frequency dependent on the density of free electrons and Γ_d is a parameter representing the dielectric dissipation.

Obtain an expression for the power absorbed by the sphere from the electromagnetic field and discuss the effect of time reversal.

SOLUTION

As shown in Ref. [10], the electric field acting over the sphere induces an electric dipolar moment which is related with the electric field as

$$\vec{p}(\omega,t) = 4\pi\varepsilon_o R^3 \frac{\varepsilon(\omega)-1}{\varepsilon(\omega)+2} \vec{E}_o e^{-i\omega t} = \alpha(\omega)\vec{E}_o e^{-i\omega t}, \tag{2.39}$$

where $\alpha(\omega)$ is the polarizability of the material of which the sphere is made.

In addition, the electrons of the material of the sphere acquire an oscillatory motion along the same direction of the applied electric field, and therefore, the power absorbed from the electromagnetic field is

$$P_{abs}(t) = \vec{E}_o e^{-i\omega t} \cdot \frac{d\vec{p}}{dt} \tag{2.40}$$

The time average value of this power is

$$\overline{P}_{abs} = \frac{\omega}{2} E_o^2 \, Im\{\alpha(\omega)\} = 12\pi\varepsilon_o R^3 \frac{\Gamma_d \omega \omega_p^2}{\left(3\omega^2 - \omega_p^2\right)^2 + 9\Gamma_d^2 \omega^2} \tag{2.41}$$

From eqs. (2.30) and (2.31), the effect of time reversal, the dielectric function is equivalent to change the sign of the parameter Γ_d implying that the dielectric sphere instead of absorbing energy, it emits it, what certainly is not an observed effect.

EXAMPLE 2.4

Obtain the wave equation for a medium with dielectric function $\varepsilon(\omega) = \varepsilon'(\omega) + i\varepsilon''(\omega)$ and the corresponding retarded and advanced solutions.

SOLUTION

In absence of charge and current distributions, Faraday and Ampere equations in the frequency domain are

$$\nabla \times \vec{E}(\omega,\vec{r}) = i\omega\mu_o\mu(\omega)\vec{H}(\omega,\vec{r}) \tag{2.42}$$

$$\nabla \times \vec{H}(\omega,\vec{r}) = -i\omega\varepsilon_o\varepsilon(\omega)\vec{E}(\omega,\vec{r}) \tag{2.43}$$

Applying the rotor operator to each side of eq. 2.42 and assuming that the dielectric and magnetic functions depend only on the frequency we obtain:

$$\underbrace{\nabla \times \nabla \times \vec{E}(\omega,\vec{r})}_{=\nabla\left(\nabla\cdot\vec{E}\right) - \nabla^2\vec{E}} = i\omega\mu_o \nabla \times \mu(\omega)\vec{H}(\omega,\vec{r}) = i\omega\mu_o\mu(\omega)\underbrace{\nabla \times \vec{H}(\omega,\vec{r})}_{=-i\omega\varepsilon_o\varepsilon(\omega)\vec{E}(\omega,\vec{r})} \tag{2.44}$$

In this case, $\nabla \cdot \vec{E} = 0$ and if we assume that the propagation is through a non-magnetic medium, so that $\mu(\omega) = 1$ we obtain the following wave equation in the frequency domain

$$\nabla^2 \vec{E} + \frac{\omega^2}{c^2}\varepsilon(\omega)\vec{E} = 0, \tag{2.45}$$

where w $w = 1/\sqrt{\varepsilon_o \mu_o}$ is the speed of light in vacuum.

For a medium with dielectric function $\varepsilon(\omega) = \varepsilon'(\omega) + i\varepsilon''(\omega)$, the above equation can be recast as

$$\nabla^2 \vec{E} + \frac{\omega^2}{c^2}\varepsilon'(\omega)\vec{E} + i\frac{\omega^2}{c^2}\varepsilon''(\omega)\vec{E} = 0 \tag{2.46}$$

The solution of this equation has spatial and time dependence given by $e^{i(\vec{k}(\omega)\cdot\vec{r} - \omega t)}$. If for simplicity, we assume propagation along the z direction, we have

$$e^{i(\vec{k}(\omega)\cdot\vec{r} - \omega t)} = e^{i(k(\omega)z - \omega t)}$$

where $\vec{k}(\omega) = k(\omega)\hat{z}$ is the wave vector with

$$k(\omega) = \pm\frac{\omega}{c}\sqrt{\varepsilon'(\omega) + i\varepsilon''(\omega)} = \pm k'(\omega) \pm ik''(\omega) = \pm\frac{\omega}{c}|\varepsilon(\omega)|(cos\alpha + isin\alpha), \tag{2.47}$$

where $|\varepsilon(\omega)| = \sqrt{\varepsilon'(\omega)^2 + \varepsilon''(\omega)^2}$ and $\alpha = tan^{-1}\left[\frac{\varepsilon''(\omega)}{\varepsilon'(\omega)}\right]$.

Therefore, eq. 2.45 has a retarded and an advanced solution given, respectively, by

$$e^{-k''(\omega)z}e^{i\{k'(\omega)z - \omega t\}} \tag{2.48}$$

$$e^{+k''(\omega)z}e^{-i\{k'(\omega)z + \omega t\}} \tag{2.49}$$

Equations 2.48 is the retarded solution and represents a wave with decreasing amplitude (damped) as it propagates through the medium along the $+z$ direction, whereas eq. 2.49 is the advanced solution representing a wave with increasing amplitude as propagates through the medium along the $-z$ direction.

Applying the time-reversal operator to both terms of equation (2.45), we obtain

$$\nabla^2 \vec{E}^* + \frac{\omega^2}{c^2}\varepsilon^*(\omega)\vec{E}^* = \nabla^2 \vec{E}^* + \frac{\omega^2}{c^2}\varepsilon'(\omega)\vec{E}^* - i\frac{\omega^2}{c^2}\varepsilon''(\omega)\vec{E}^* = 0 \tag{2.50}$$

For this case, $\varepsilon^*(\omega) = \varepsilon'(\omega) - i\varepsilon''(\omega)$ and the above equation give retarded and advanced solutions equal to that of eq. 2.45 and given by eqs. 2.48 and 2.49.

The advanced solution is mathematically possible but its probability of occurrence makes it physically impossible, due to the enormous number of microscopic conditions to be satisfied.

EXAMPLE 2.5

Discuss the effect of time reversal of eq. 2.19 and conditions for time-reversal symmetry.

SOLUTION

Applying the time reversal operator to each side of eq. 2.19 and considering eqs. 2.22, 2.23, and 2.29, we obtain

$$\nabla^2\vec{H}^*(\vec{r},\omega)+\frac{\omega^2}{c^2}\varepsilon^*(\vec{r},\omega)\vec{H}^*(\vec{r},\omega)=-i\omega\varepsilon_o\varepsilon^*(\vec{r},\omega)\nabla\varepsilon^*(\vec{r},\omega)\times\vec{E}^*(\vec{r},\omega)$$

This equation has the same form of eq. 2.19 but with the conjugates of all involved quantities. If the dielectric function $\varepsilon(\vec{r},\omega)$ is real, there is time-reversal symmetry. However, this is not the case if this function is imaginary and of the form $(\vec{r},\omega)=\varepsilon'(\vec{r},\omega)+i\varepsilon''(\vec{r},\omega)$. In fact, in this case since $\varepsilon^*(\vec{r},\omega)\neq\varepsilon(\vec{r},\omega)$, this equation is different from eq. 2.19, and therefore, there is not time-reversal symmetry.

2.5 EFFECT OF DISPERSION IN TIME-REVERSAL SYMMETRY

In this section, we analyze how the optical dispersion characteristics of the medium through which an electromagnetic wave propagates affect the time-reversal symmetry. We show that depending on the dielectric function $\varepsilon(\omega)$ of the medium, there are frequency ranges where even in the absence of dissipation time-reversal symmetry is lost.

The interaction between light and medium is modeled as a set of harmonic oscillators excited by the electromagnetic wave and this is represented by the following dielectric function which has M resonance frequencies ω_{oj}[11]

$$\varepsilon(\omega)=1+\sum_{j=1}^{M}\frac{f_j\omega_p^2}{\omega_{oj}^2-\omega^2},\tag{2.51}$$

where ω_p is the plasma frequency given by and $f_j = \rho_j/\rho$ with ρ_j, the density of oscillators corresponding to the mode j and ρ, the total density of these oscillators.

As an example, we consider a medium whose dielectric function can be represented by two oscillators with strengths $f_1 = 0.6$ and $f_2 = 0.4$ and with resonance frequencies $\omega_{01} = 0.3\omega_p$ and $\omega_{02} = 0.5\omega_p$. Considering that for time-reversal symmetry, $k = \frac{\omega}{c}\sqrt{\epsilon(\omega)}$ must be real, the frequency must be in the following ranges:

$$0 < \omega < 0.3\omega_p,\ 0.424\omega_p < \omega < 0.5\omega_p \text{ and } \omega > 1.08\omega_p$$

Out of these ranges of the frequencies, the wave number k takes imaginary values and due to the damping of the amplitude of the electromagnetic waves, time-reversal symmetry is lost as has been previously discussed.

2.6 TIME REVERSAL OF PROCESSES WITH TRANSFER OF ENERGY

The laws that rule the propagation of light in free space are invariant under the "time reversal" operation. This means that when the arrow of time is flipped—so that the dynamics of the electromagnetic wave are reversed in time similar to a movie played backwards—the time-reversed wave remains compatible with the laws of electromagnetism, that is, it still satisfies the Maxwell's equations.

An extraordinary consequence of the time-reversal invariance is that the propagation of light in typical materials is inherently bi-directional. For example, if a wave can go through some channel with no back-reflections, then the time-reversed wave can go through the same channel but propagating in the opposite direction. This is a consequence of the Lorentz reciprocity law,[12,13] which is more general than the time-reversal invariance. Indeed, the Lorentz reciprocity theorem can be applied to *macroscopic* systems which are not invariant under a time-reversal transformation since they have dissipative elements. This will be analyzed in Section 2.11 of this chapter.

In Ref. [14], it is shown that dissipative dielectrics have a hidden time-reversal symmetry. In fact, equations of *macroscopic* electrodynamics are not time-reversal invariant when the system has dissipative elements. This is so because the description provided by macroscopic electrodynamics is incomplete, as it only models the time evolution of the electromagnetic degrees of freedom. In contrast, in a microscopic formalism—with all the light and matter degrees of freedom included in the analysis—the system dynamics is time-reversal symmetric. Thus, in some sense, macroscopic dissipative systems (e.g., lossy dielectrics) have a *hidden time-reversal symmetry*.

In what follows, we present two examples of processes where there is a transfer of energy between an electromagnetic field and a system of moving particles, and for which at the macroscopic level, electromagnetic absorption does not necessarily imply time-reversal asymmetry for the whole process.

EXAMPLE 2.6

We consider two identical spheres of radius a subject to an external oscillating electric field with angular frequency ω. The direction of the electric field vector $\vec{E}_o e^{-i\omega t}$ associated with the laser is at an angle θ with respect to a line along the centers of the spheres as shown in Figure 2.1.

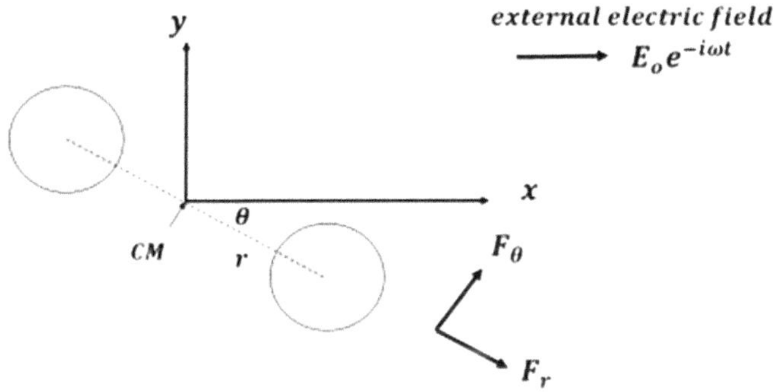

FIGURE 2.1 Two polarizable spheres excited by a fluctuating electric field along the x direction. Directions of reference are indicated for the radial and angular components of the interaction force.

We assume that the particles are separated by a distance r such that the dipolar approach is acceptable so that the electric field can be considered spatially uniform.

The spheres are made of a polarizable material with dielectric function

$$\varepsilon(\omega) = 1 - \frac{\omega_{pe}^2}{\omega^2 + i\Gamma_d} \tag{2.52}$$

where ω_{pe} is the plasma frequency and $\Gamma_d = \gamma/m$ is the absorption parameter.

Due to the incident field, the spheres become polarized and dipole moments are induced on them.[15] As consequence electric forces appear, with time-averaged radial and angular components which in MKS units are given by.[16]

$$F_r(\vec{r}) = \frac{4\pi\varepsilon_o a^2 E_o^{\,2}}{96\sigma^4}\left[\frac{sin^2\theta}{|n_R - u|^2} - \frac{2cos^2\theta}{|n_A - u|^2}\right] \tag{2.53}$$

$$F_\theta(\vec{r}) = \frac{4\pi\varepsilon_o a^2 E_o^{\,2}}{96\sigma^4} Re\left[\frac{sin2\theta}{(n_A - u)(n_R - u)^*}\right] \tag{2.54}$$

Here, $\sigma = r/2a$ is a dimensionless radial coordinate, such that $2\sigma a$ represents the particle–particle separation, and n_A, n_R are attractive and repulsive depolarization factors, respectively, that in the dipole approach are given by the following expressions[16]:

$$n_A = \frac{1}{3}\left(1 - \frac{1}{4\sigma^3}\right) \tag{2.55a}$$

$$n_R = \frac{1}{3}\left(1 + \frac{1}{8\sigma^3}\right) \tag{2.55b}$$

The electromagnetic properties of the material with which the spheres are made appear in the spectral variable $u = 1/[1 - \varepsilon(\omega)$ through the dielectric function $\varepsilon(\omega)$.

Notice that the radial force (2.53) is attractive up to a critical angle whose value in the limit of large separation is 54.74°. This can be seen from eq. 2.53 since for this value of the angle, the term inside the square bracket vanishes if n_A, $n_R = 1/3$.

For $\theta = 0$, there is only an attractive radial force given by

$$F_r(\sigma) = -\frac{4\pi\varepsilon_o (E_o a)^2}{48\sigma^4}\frac{1}{|n_A - u|^2} \tag{2.56}$$

If the spheres have an initial separation corresponding to $\sigma(0) = \sigma_o$ are left free to move they will begin to become closer and neglecting any other form of dissipation of energy except that corresponding to electromagnetic dissipation, the dynamics of the motion will be given by

$$\frac{d^2\sigma}{dT^2} = -\frac{1}{\sigma^4\left\{(n_A - u')^2 + (u'')^2\right\}} \tag{2.57}$$

where the time is denoted by a dimensionless variable $T = t/\tau$ with $\tau \equiv \sqrt{\frac{96ma}{4\pi\varepsilon_o}}/E_o a$ representing a characteristic time that depends on the mass

m of each sphere, their radii and the magnitude E_o of the applied electric field.

Since the dielectric function given by eq. 2.52 has an imaginary part which represents the optical absorption of the material of the spheres, there is dissipation of electromagnetic energy so that it is incorrect to obtain the induced electric force as the gradient of a potential.[17] Nevertheless, this does not necessarily imply time-reversal asymmetry. In fact, eq. 2.57 can be written as

$$\frac{d^2\sigma}{dT^2} = -\frac{1}{\sigma^4 \left(n_A - u'\right)^2} + \frac{\left(u''\right)^2}{\sigma^4 \left\{\left(n_A - u'\right)^2 + \left(u''\right)^2\right\}\left(n_A - u'\right)^2} \tag{2.58}$$

The second term of the right side of eq. 2.58 represents the effect of the absorption of electromagnetic energy from the laser field which is equivalent to a force opposed to the motion of the spheres for becoming in contact. Under time reversal according to eq. 2.29, $\mathcal{T}\{\varepsilon(\omega)\} = \varepsilon^*(\omega)$ implying that $\mathcal{T}\{u(\omega)\} = u^*(\omega)$. Nevertheless, this term does not change since it depends on $(u'')^2$, and therefore, the equation of motion of the pair of particles described by eq. 2.58 is time-reversal invariant. Indeed, according to this equation, due to the electric force of interaction, if initially the spheres are separated and are released, they will approach each other radially. However, due to the absorption of electromagnetic energy, the reverse process is not possible, that is, if the spheres are initially in contact, it is not expected that due to the action of the same electric field, they move away without any external source of energy.

EXAMPLE 2.7

In this example, we discuss the behavior under time reversal of the process corresponding to the transfer of angular momentum from light to classical nanoparticles. As shown in Ref. [18], an unbounded circularly polarized electromagnetic plane wave of angular frequency ω transfers angular momentum to an absorbing isotropic sphere of radius $\boldsymbol{a}$ so that a torque is developed in this object originating its rotation. This torque is given by

$$\Gamma_z = I_{inc}\pi a^2 \frac{Q_{abs}}{\omega} \tag{2.59}$$

where z is the direction of propagation of the laser beam, I_{inc} is the intensity of the incident wave, and Q_{abs} is the Mie absorption efficiency. For an idealized uniform lossless dielectric sphere $Q_{abs} = 0$, no torque would be transferred by a circularly polarized electromagnetic wave. The objective of this example is to obtain this result by means of a semiclassical model in terms of photons and to analyze whether or not there is symmetry in this process under time reversal.

SOLUTION

Considering that each photon in a right-/left-circularly polarized beam of light carries $\hbar$ units of angular momentum pointing along/against the direction of propagation, the generation of a torque on a particle through absorption may be interpreted as follows: by angular momentum conservation the extinction of a photon with energy $\hbar\omega$ transfers to the absorbing particle $\hbar$ units of angular momentum in the direction of propagation. The torque generated is just $\hbar$ times the number of photons absorbed by a unit of time. For an isotropic spherical particle, the torque is only generated by absorption in a form compatible with energy, linear, and angular momentum conservation.

If F_z is the rate of photons incident over the sphere and traveling in the z direction (number of photons per unit of time, each one with energy $\hbar\omega$), A is the photon absorption rate and G is the total rate of scattered photons in all directions, and assuming elastic scattering, conservation of energy requires,

$$F_z\left(\hbar\omega\right) = A\left(\hbar\omega\right) + G\left(\hbar\omega\right) \tag{2.60}$$

As known, for an unbounded electromagnetic plane wave, the electric field vector is in a plane perpendicular to the propagation direction, so in our case, this field has only components oscillating in the x and y directions which produce oscillating electric dipole vectors also in these directions. By symmetry and taking into account that the radiated intensity of an oscillating dipole is null in a direction parallel to its direction and maximum for a perpendicular direction, one may expect that in the x and y direction, there are no scattered photons, and therefore, no angular momentum or linear momentum is transferred in these directions.

Considering that each incident photon carries momentum $\hbar\omega/c$ in the $+z$ direction and denoting G_z as the net rate of photons scattered in $+z$ direction, conservation of linear momentum requires:

$$F_z\left(\frac{\hbar\omega}{c}\right) = \frac{dP_z}{dt} + G_z\left(\frac{\hbar\omega}{c}\right) \tag{2.61}$$

where P_z is the linear momentum acquired by the particle in the z direction.

In addition, each incident photon carries an angular momentum $\hbar$ and conservation of angular momentum requires:

$$\hbar F_z = \frac{dL_z}{dt} + \hbar G_z \tag{2.62}$$

where L_z is the z component of the angular momentum acquired by the particle.

From eqs. 2.60 to 2.62, the torque acting over the particle may be recast as:

$$\Gamma_z = \hbar\{A + (G - G_z)\} \tag{2.63}$$

For an isotropic spherical particle of radius a comparable with the wavelength λ of the incident radiation, the analysis of the angular distribution of the scattered radiation shows that the scattering is highly peaked in the forward direction.[10] Therefore, $G_z = G$ and eq. 2.63 gives

$$\Gamma_z = \frac{dL_z}{dt} = A\hbar = \frac{Absorbed\ Power}{\omega} \tag{2.64}$$

On the other hand, for a small spherical particle such that $a/\lambda << 1$, it can be shown that the scattering may be analyzed in the Rayleigh regime what implies that the effect of scattering may be neglected in comparison with the absorption, so eq. 2.64 leads also to the result given by equation (2.59).

Therefore, in an idealized lossless dielectric and isotropic sphere, no torque would be developed.

Now, let us go to the issue of time reversal of this process. Considering that the angular momentum is odd under time reversal and that the angular frequency is fixed by the source of the incident electromagnetic field, according to eq. 2.64 for time-reversal symmetry, the absorbed electromagnetic energy should be returned from the sphere to the source, or equivalently, the absorbed photons should be re-emitted with a rate equal to that of the incident photons. Clearly, this is an unlikely process since the photons absorbed by the sphere produce other microscopic excitations in the corresponding medium such as transitions between atomic states or vibration modes, which make impossible that the system sphere-electromagnetic field comes back to its initial state before the electromagnetic field was switched on. Therefore, this is not a system with time-reversal symmetry.

2.7 DOPPLER EFFECT AND TIME REVERSAL

According to this effect, the frequency of the waves measured by two inertial frames in relative motion is not the same, so that when the source moves away from the detector, the detected frequency is lower than the emitted one. The inverse effect occurs when the detector is moving toward the emitter. In this section, we consider this effect in the context of time reversal of electromagnetic waves studying the case of a source fixed to a reference frame S emitting a train of rectangular electromagnetic pulses with amplitude A and frequency $\nu = 1/T$ as shown in Figure 2.2. We will consider that the source is moving with respect to the detector with relative velocity $\vec{\upsilon}_{rel}$ and magnitude $|\vec{\upsilon}_{rel}| \ll c$ so that relativistic effects are not considered. The relativistic Doppler effect will be studied in chapter 4.

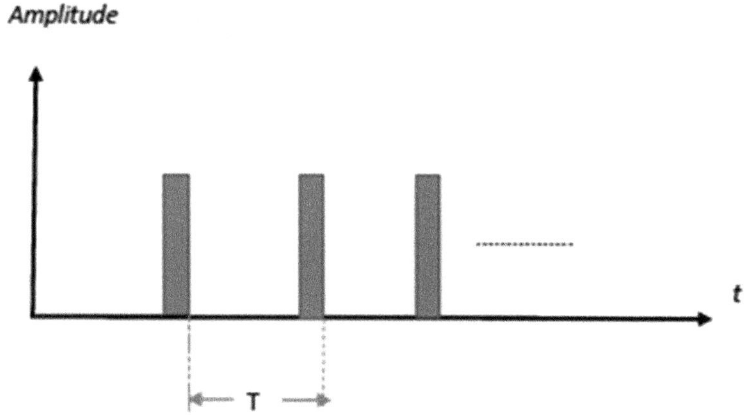

FIGURE 2.2 Train of pulses emitted by a moving source toward a detector.

As previously discussed in this chapter, under time reversal an electromagnetic wave propagating toward the source is a valid solution of Maxwell's equations in empty space or in a medium without dissipation of energy. This type of solution is called ***advanced solution.***

We start our analysis, by revisiting the classical Doppler effect (Example 2.9) and then we analyze how the time reversal of this effect can be obtained if the medium has both negative permittivity and magnetic permeability, as occurs in left-handed materials (Example 2.10).

EXAMPLE 2.9

Consider, a light source moving away from the detector with constant speed υ along the x axis. Determine the relation between the detected and emitted frequency.

SOLUTION

We consider the following sequence of events between two consecutive pulses:

At instant t_1, the source emits the first pulse when the detector is at a distance x_1 from the source.

At instant t'_1, the detector receives this first pulse when it is placed at a distance $x'_1 > x_1$ from the source.

At instant $t_2 = t_1 + T$, the source emits the second pulse when the detector is at a distance x_2 from the source.

This second pulse is received by the detector at the instant t'_2 when it is at a distance $x'_2 > x_2$ from the source.

If the medium has an index of refraction n_r, we have the following relations between the instants of emission and detection of such pulses:

$$t'_1 = t_1 + \frac{x'_1}{c} n_r \tag{2.65}$$

$$x'_1 = x_1 + \upsilon\left(t'_1 - t_1\right) \tag{2.66}$$

$$t'_2 = t_2 + \frac{x'_2}{c} n_r \tag{2.67}$$

$$x'_2 = x_2 + \upsilon\left(t'_2 - t_2\right) \tag{2.68}$$

From these relations and after a bit of algebra, we obtain that for $\upsilon \ll c$, the detected period of the pulses is

$$T' = t'_2 - t'_1 = \frac{T}{1 - \frac{\upsilon}{c} n_r} \approx T\left(1 + \frac{\upsilon}{c} n_r\right) \tag{2.69}$$

Therefore, the detected period of the pulses is greater than that of the period measured at the source. In terms of frequency, since $\nu = 1/T$ we get that

$$\nu' = \frac{1}{T'} = \frac{\nu}{1 + \frac{\upsilon}{c} n_r} \approx \nu\left(1 - \frac{\upsilon}{c} n_r\right) < \nu \tag{2.70}$$

that is, the detected frequency decreases as the detector moves away from the source.

Under time reversal of this process, the velocity υ changes sign so that the source moves toward the detector and additionally, since an advanced wave travels from the detector toward the source, the net effect is that the frequency of this wave measured at the reception point is greater than the value measured by the detector. This effect is called inverse Doppler effect.

EXAMPLE 2.10

Analyze the range of frequencies for which the inverse Doppler effect can occur in a medium having both negative dielectric permittivity and negative permeability, as predicted by V. Veselago in 1968[19] and experimentally observed according to several publications, as for example in Refs. [20, 21]. For that purpose, consider that the dielectric permittivity and magnetic permeability are dependent on the frequency in the form:

$$\in(\omega) = \in_o\left(1 - \frac{\omega_p^{\ 2}}{\omega^2}\right),\ \mu(\omega) = \mu_o\left(1 - \frac{\omega_m^{\ 2}}{\omega^2}\right),$$

where $\omega = 2\pi v$ and ω_p, ω_m are characteristic frequencies associated to the optical properties of the medium, such that $\omega_m < \omega_p$.

SOLUTION

The analysis for this case is as follows:

First, let us remind that the index of refraction of a medium with dielectric permittivity $\in$ and magnetic permeability μ is

$$n_r = \sqrt{\left(\frac{\in}{\in_o}\right)\left(\frac{\mu}{\mu_o}\right)} \tag{2.71}$$

If both $\in$ and μ are negative, so that $\in = -|\in|$ and $\mu = -|\mu|$, eq. 2.80 can be written as

$$n_r = \sqrt{\frac{(-|\in|)}{\in_o}\frac{(-|\mu|)}{\mu_o}} = i\sqrt{\frac{|\in|}{\in_o}}\,i\sqrt{\frac{|\mu|}{\mu_o}} = i^2|n| = -|n| \tag{2.72}$$

It seems that the above equation has been obtained from a mathematical trick, but as shown by Veselago et al.,[21] the appearance of a negative index of refraction requires rewriting of eq. 2.72 in a more general way as

$$n_r = \pm\sqrt{\left(\frac{\in}{\in_o}\right)\left(\frac{\mu}{\mu_o}\right)} \tag{2.73}$$

where the positive sign is used for the usual case, whereas the negative sign is used when $\in$ and μ are both negative.

According to eq. 2.70 and for $n_r < 0$, the frequency of the wave received at the detector is

$$\nu' = \nu\left(1 + |n_r|\frac{\upsilon}{c}\right) \tag{2.74}$$

Equation 2.83 implies that with $n_r < 0$, the frequency measured by the detector is greater than the value measured at the emission point, as is the expected behavior for the time-reversed Doppler effect.[22] For this specific case, this occurs for frequencies such as $\omega < \omega_m$.

2.8 CAUSALITY IN ELECTRODYNAMICS

Causality is one of the fundamental principles in physics and establishes that an effect cannot precede its corresponding cause. A crucial aspect related to this principle is, as we shall see in Chapter 4, that the speed of propagation of any signal cannot exceed a finite limit which corresponds to light velocity in empty space $c = \frac{1}{\sqrt{\varepsilon_o \mu_o}} = 3 \cdot 10^8$ m/s. From the equations of the special relativity theory, it can be deduced that if this limit could be exceeded it would be possible to have situations where one observer could detect an event before the occurrence of its cause. For example, this observer could detect a light signal before its emission by a source.

In this section, based on Ref. [23] and considering the references there quoted, we analyze the transit time of electromagnetic waves through dielectric barriers. The meaning of this time is analyzed and the issue of superluminal propagation (propagation with a velocity higher than light velocity) is discussed in relation to the principle of causality.

It is well known that in a non-dispersive medium with index of refraction $n > 1$, light propagates slower than in vacuum.[24] Therefore, for a monochromatic wave perpendicularly incident on a dielectric and non-dispersive slab of thickness D, the transit time is expressed as nD/c with c/n being the phase velocity that is, in this case, equal to the group velocity.

If one deals with an electromagnetic pulse of finite duration traveling in a dispersive medium so that the refraction index is a function of the frequency;

according to Ref. [10], the velocity of propagation is traditionally expressed by means of the group velocity given by

$$\upsilon_g = \frac{c}{n(\omega)+\omega(dn/d\omega)} \tag{2.75}$$

This velocity is associated with the velocity of transport of electromagnetic energy. For normal dispersion $\frac{dn}{d\omega} > 0$ and $n > 1$, so that this velocity is less than c. Therefore, the transit time of the pulse throughout the medium along a distance D would be $\tau_{transsit} = D/\upsilon_g$. But in a media with anomalous dispersion, $dn/d\omega$ can become large and negative resulting in values of $\upsilon_g > c$ or even negative. In this last case, the transit time $\tau_{transsit}$ also would be negative so that it has no physical meaning.

The question concerning the transit time of an electromagnetic pulse through a dielectric slab, even if dispersion may be neglected, is not trivial and has originated several publications as quoted in Ref. [23]. It is generally assumed that information encoded on the pulse propagates with a velocity corresponding to the group velocity. However, some confusion has arisen with respect to the meaning of group velocity in situations where it exceeds the value c corresponding to the speed of light in a vacuum or where it becomes negative. The conventional viewpoint has been that in such situations group velocity has no longer any appreciable physical significance, or may even lack precise meaning.

One of the usual definitions for this transit time is given in terms of the derivative of the phase shift of the transmitted wave with respect to the frequency. Another point of view considers the dwell time defined in the case of electromagnetic waves as the relation between the time-average stored energy within the barrier and the input power.

In this section, we analyze the model based on the group delay in transmission, which mainly states that this quantity may be obtained from the energy derivative of the transmission phase shift. In a simplistic way, this can be understood considering the phase angle of the transmitted wave with respect to that of the incident wave, which may be expressed as

$$\phi(\omega) = k_s d_{eff} \tag{2.76}$$

where k_s is the wavenumber in the dielectric medium and d_{eff} is an effective distance of traveling of the light inside the slab considering consecutive reflections in both faces of the slab. Taking the derivative with respect to ω of each member of the above equation, one obtains

$$\frac{d\phi}{d\omega} = \frac{dk_s}{d\omega} d_{eff} = \frac{d_{eff}}{d\omega / dk_s} \tag{2.77}$$

The last equality in eq. 2.77 has dimensions of time and can be interpreted as time transit.

For the special case of a monochromatic wave passing through a non-dispersive dielectric slab of thickness D with constant refraction index n, from the phase $\phi(\omega)$ of the transmission coefficient and its first derivative with respect to ω, one obtains that the transit time is given by

$$\tau_{transit} = \frac{D}{c/n} f(n, D) \tag{2.78}$$

where

$$f(n, D) = \frac{4n(1+n^2)}{1+6n^2+n^4-(n^2-1)^2 cos2\beta} \tag{2.79}$$

and $\beta \equiv n\omega D/c$.

The analysis of eq. 2.78 indicates that the transit time is the product of two terms. The term $D/(c/n)$ is the time for the passage of an electromagnetic wave through a distance D in a medium with refraction index n and with a propagation velocity equal to the phase velocity c/n. The second term is a function that determines the deviation of the transit time with respect to the above-expected value. In an equivalent way, we can interpret that due to consecutive reflections in both faces of the slab, the effective displacement of the light inside the slab is $d_{eff} = Df(n.D)$.

From the above expression, one obtains that in particular for $= \lambda_s/4$, where $\lambda_s = \lambda_o/n$ is the wavelength inside the slab, the transit time has a minimum value given by:

$$\tau_{transit} = \frac{nD}{c}\frac{2n}{n^2+1} = \frac{nD}{c}\frac{2n}{(n-1)^2+2n} < \frac{nD}{c} \tag{2.80}$$

From this result, it can be deduced that for this thickness of the slab, the transit time is lower than the expected value nD/c, and this would correspond to a case of superluminal propagation effect. For example, if the refractive index is $n = 1.5$, the transit time is calculated by eqs. (2.78) and (2.79), is approximately 92% of the time $D/(c/n)$ that one would expect for a non-dispersive medium.

Does the result given by eq. 2.80 imply that superluminal velocity is possible?

To answer this question, first of all, it must be considered that as shown in Ref. [25], this delay time is actually the lifetime of the stored energy leaking through both ends of the slab and since that represents a bidirectional flow of energy it cannot be associated with a forward traversal time. In addition, it must be considered that for $D = \lambda_s/4$ according to eqs. (2.78) and (2.79), the transit time takes its minimum values that is given by eq. 2.80 which is lower than the light velocity implying superluminal propagation. Nevertheless, considering the average value of $f(n,D)$ we obtain

$$\langle f(n,D)\rangle_{D=\lambda_s/4} = \frac{1}{D}\int_0^D f(n,z)\,dz = \frac{1}{D}\int_0^D dz\left[\frac{4n\left(1+n^2\right)}{1+6n^2+n^4-\left(n^2-1\right)^2 \cos 2k_s z}\right]$$

$$= \left\{\frac{1}{k_s D} ArcTan\left[\frac{1+n^2}{2n} Tan\left(k_s D\right)\right]\right\}_{D=\lambda_s/4} = 1 \qquad (2.81)$$

Therefore, eqs. (2.78) and (2.81) lead to a transit time equal to nD/c which is the value expected for a long wave train $\frac{\Delta\omega}{\omega} \ll 1$ through a window glass, for example, and there is no conflict with the special relativity theory which implies that for causality no signal can travel faster than light.

As a numerical example, Figure 2.3 shows the transit time of a monochromatic wave through a dielectric slab with an index of refraction n assuming that dispersion and absorption effects may be neglected. Results are obtained using eqs. 2.78 and 2.79 for three values of n, showing that the use of this model leads to propagation with velocity greater than the phase velocity for some combinations of values of n, D, and ω represented by the parameter $\beta = n\omega D/c$. Nevertheless, the use of eq. 2.78 and the average value $\langle f(n, \omega, D)\rangle$ instead of $f(n, \omega, D)$ gives values of the transit time greater than nD/c, as shown in Figure 2.4. In particular, for $= \lambda_s/4$, one obtains that the transit time is nD/c what it is in agreement with what is expected for a monochromatic wave propagating through a transparent and non-dispersive slab.

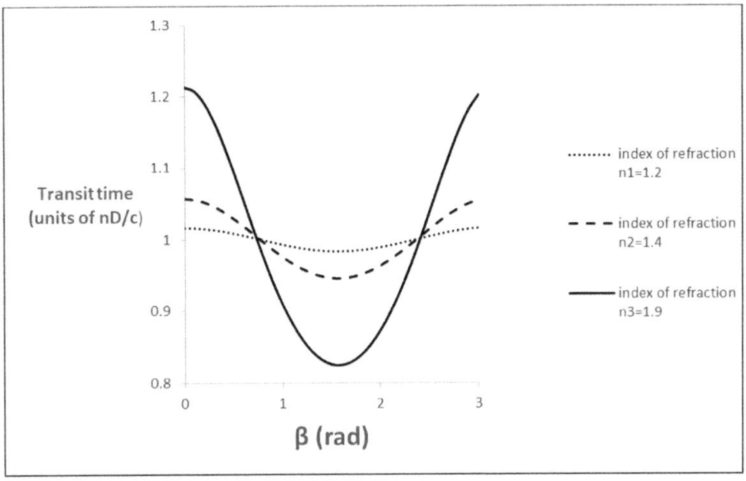

FIGURE 2.3 Transit time of a monochromatic wave through a dielectric slab made of a non-dispersive material versus the variable $\beta = n\omega D/c$ for three values of the index of refraction: $n_1 = 1.2$ (dotted curve), $n_2 = 1.4$ (dashed curve) and $n_3 = 1.9$ (solid curve).

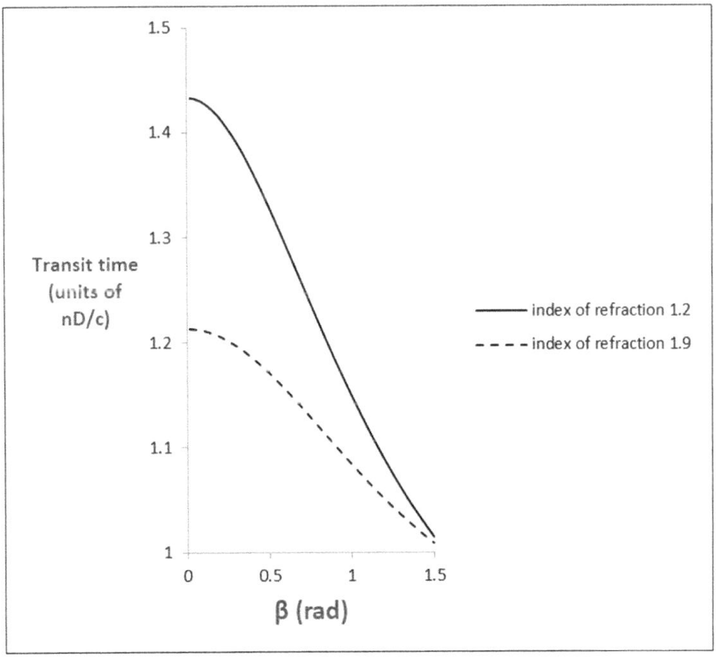

FIGURE 2.4 Average transit time of a monochromatic wave through a dielectric slab made of a non-dispersive material versus the variable $\beta = n\omega D/c$ for index of refraction 1.2 (solid line) and 1.9 (dashed line).

2.9 AN ARROW OF TIME FOR ELECTROMAGNETIC RADIATION

As expressed by J.G. Cramer, an electromagnetic direction of time means that the retarded radiation dominates over the advanced radiation in the universe.[26] Well-known textbooks of classical electrodynamics state that the advanced potentials are an equally valid solution to Maxwell's equations as well as for the retarded potentials.[10,27]

The electromagnetic wave equation in source-free space can be expressed as:

$$\nabla^2 \vec{F} = \frac{1}{c^2}\frac{\partial^2}{\partial t^2}\vec{F} \tag{2.82}$$

where $\vec{F}$ represents the electric or magnetic field vectors, $\vec{E}$ or $\vec{B}$, respectively. Since this differential equation is second order both in time as in space, there are two solutions which for the case of plane electromagnetic waves produced by charges oscillating with angular frequency ω can be expressed as:

$$\vec{F}_1(\vec{r},t) = \vec{F}_o exp\left[i\left(\vec{k}\cdot\vec{r} - \omega t\right)\right] \tag{2.83}$$

$$\vec{F}_2(\vec{r},t) = \vec{F}_o exp\left[i\left(\vec{k}\cdot\vec{r} + \omega t\right)\right] \tag{2.84}$$

where $\vec{k}$ is a vector parallel to the direction of propagation of the wave with magnitude $|\vec{k}| = \omega/c$.

Applying the Hamiltonian operator $\mathcal{H} \equiv i\hbar\,\partial/\partial t$ to the waves described by these last two equations, we obtain that the energy of the photon associated to $\vec{F}_1(\vec{r},t)$ is $+\hbar\omega$ while for $\vec{F}_2(\vec{r},t)$ this energy is $-\hbar\omega$. By analogy with Dirac's theory where an antiparticle has negative energy and it is considered that travels backwards in time, the solution $\vec{F}_2(\vec{r},t)$ is associated to an advanced potential, while the solution $\vec{F}_1(\vec{r},t)$ is associated with a retarded potential. This can be understood more easily if we consider waves propagating in one dimension, for instance along the x axis. In this case, eqs. 2.83 and 2.84 can be rewritten as:

$$\vec{F}_1(x,t) = \vec{F}_o exp\left[-i\omega\left(t - \frac{x}{c}\right)\right] \tag{2.85}$$

$$\vec{F}_2(x,t) = \vec{F}_o exp\left[-i\omega\left(t + \frac{x}{c}\right)\right] \tag{2.86}$$

These last two equations can be interpreted as retarded and advanced potentials, respectively, let us consider the following example: an electric charge placed at $x = 0$ oscillating with angular frequency ω produces an electromagnetic signal. According to eq. 2.85, the signal detected at point x at instant t is not that produced by the oscillating charge just at this instant, but it corresponds to the signal produced by the source at a previous time $t - x/c$. This case occurs, for example, when we look at the light coming from a distant star, which may have already collapsed into a black hole. In that case, what we see is the light that it emitted before this collapse and that, given the enormous distance, is still traveling toward the earth.

On the contrary for the case of the wave described by eq. 2.86, the signal detected at point x at instant t corresponds to a future signal that will be produced at a time $t + x/c$. This contradicts the principle of causality discussed in the previous section and which states that an effect (detection of a signal in this case) cannot occur before its cause (emission of this signal).

In terms of photons, for the case of a retarded potential they travel forward in time, whereas for an advanced potential they travel backwards in time.

Therefore, the advanced potentials are generally discarded as unphysical on causal grounds.[27] Nevertheless as we will discuss in the present section, the advanced potentials can be combined with the retarded potentials for obtaining a theory of absorption as made by J.A. Wheeler and R. Feynman in a famous paper of 1945.[28]

The conventional interpretation of solutions given by eqs. 2.83 and 2.84 is that the retarded solution corresponds to the process of emission of electromagnetic radiation, for example, from an accelerated charge, and the advanced solution describes the absorption of electromagnetic radiation, that is

$$\vec{F}_{rad}(\vec{r},t) = \vec{F}_1(\vec{r},t) \tag{2.87}$$

$$\vec{F}_{abs}(\vec{r},t) = \vec{F}_2(\vec{r},t) \tag{2.88}$$

This can be understood considering the arguments given in the next subsection.

2.9.1 ENERGY CARRIED BY ADVANCED AND RETARDED ELECTROMAGNETIC WAVES

In this subsection, we analyze the energy carried by the electromagnetic waves produced by an oscillating electric dipole oriented in the z direction and expressed as:

$$\vec{p}(t) = \hat{z}p_o \cos\omega t \tag{2.89}$$

This dipole is a radiating source and according to Ref. [27] at large distances from this source (radiation zone), the retarded solutions for the electric and magnetic field vectors are spherical waves given by

$$\vec{E}(\vec{r},t) = -\mu p_o \frac{\omega^2}{4\pi}\left(\frac{sin\theta}{r}\right)cos\left(\omega t - \frac{\omega}{c_m}r\right)\hat{\theta} \tag{2.90}$$

$$\vec{B}(\vec{r},t) = -\mu p_o \frac{\omega^2}{4\pi c_m}\left(\frac{sin\theta}{r}\right)cos\left(\omega t - \frac{\omega}{c_m}r\right)\hat{\phi} \tag{2.91}$$

where μ is the magnetic permeability of the medium considered as isotropic and with dielectric permittivity ε, so that the refraction index is $n_m = \sqrt{\frac{\varepsilon}{\varepsilon_o}\frac{\mu}{\mu_o}}$ and $c_m = c/n_m$ is the corresponding light velocity.

The time-average intensity of the emitted radiation is given by:

$$\langle\vec{S}\rangle = \langle\vec{E}\times\vec{B}/\mu\rangle = \left(\frac{\mu p_o^2\omega^4}{32\pi^2 c_m}\right)\frac{sin^2\theta}{r^2}\hat{r} \tag{2.92}$$

Under time reversal $t \rightarrow t' = -t$ and

$$\mathcal{T}\left[\vec{E}(t)\right] = \vec{E}(-t) = -\mu p_o \frac{\omega^2}{4\pi c_m}\left(\frac{sin\theta}{r}\right)cos\left(-\omega t - \frac{\omega}{c_m}r\right)\hat{\theta}$$

$$= -\mu p_o \frac{\omega^2}{4\pi c_m}\left(\frac{sin\theta}{r}\right)cos\left(\omega t + \frac{\omega}{c_m}r\right)\hat{\theta} \tag{2.93}$$

$$\mathcal{T}\left[\vec{B}(t)\right] = -\vec{B}(-t) = +\mu p_o \frac{\omega^2}{4\pi c_m}\left(\frac{sin\theta}{r}\right)cos\left(-\omega t - \frac{\omega}{c_m}r\right)\hat{\phi}$$

$$= +\mu p_o \frac{\omega^2}{4\pi c_m}\left(\frac{sin\theta}{r}\right)cos\left(\omega t + \frac{\omega}{c_m}r\right)\hat{\phi} \tag{2.94}$$

The electric and magnetic waves corresponding to the above two equations represent advanced waves and the time-average intensity is in this case

$$\langle\mathcal{T}\left[\vec{S}\right]\rangle = \langle\vec{E}\times(-\vec{B})/\mu\rangle = -\left(\frac{\mu p_o^2\omega^4}{32\pi^2 c_m}\right)\frac{sin^2\theta}{r^2}\hat{r} \tag{2.95}$$

This last equation represents a flow of energy returning to the source and in the absence of absorption, there is time-reversal symmetry, as expected.

2.9.2 WHEELER–FEYNMAN ABSORBER THEORY

The Wheeler–Feynman absorber theory was originally conceived as a time symmetric alternative to conventional electromagnetism which, unlike the latter, imposes no *ad hoc* time direction on the electromagnetic processes. It is essentially a set of boundary-condition rules arising from the requirement of time symmetry which according to ref. [29] are expressed verbatim as follows:

"*(1) The process of emission produces an electromagnetic wave consisting of a half-amplitude retarded wave and a half-amplitude advanced wave which lie along the same four-vector axis but with opposite time directions; (2) The process of absorption is identical to that of emission and occurs in such a way that the wave produced by the absorber is 180° out of phase with the wave incident on it from the emitter; and (3) An advanced wave may be reinterpreted by an observer as a retarded wave by reversing the signs of the energy and momentum (and therefore the time direction) of the wave, and likewise a retarded wave may be reinterpreted as an advanced wave.*"

The above statements (1) and (2) can be expressed as:

$$\vec{F}_{rad}(\vec{r},t)=\frac{1}{2}\vec{F}_{ret}(\vec{r},t)+\frac{1}{2}\vec{F}_{adv}(\vec{r},t) \tag{2.96}$$

$$\vec{F}_{abs}(\vec{r},t)=-\vec{F}_{rad}(\vec{r},t)=-\frac{1}{2}\vec{F}_{ret}(\vec{r},t)-\frac{1}{2}\vec{F}_{adv}(\vec{r},t) \tag{2.97}$$

where $\vec{F}_{ret}(\vec{r},t)$ and $\vec{F}_{adv}(\vec{r},t)$ represent the retarded and advanced solutions, respectively.

According to statement (2) in the process of absorption a retarded wave is produced which is exactly in phase opposition with respect to the incident radiation, so that the incident wave "stops" at the observer. This is illustrated schematically using a Minkowski diagram, as shown in Figure 2.5.[29] The emitter ***E*** produces a retarded half-amplitude wave R_e and an advanced half-amplitude wave A_e. The absorber ***A*** produces a half-amplitude retarded wave R_a, which cancels R_e in region 3. But at the same point of the spacetime, the absorber also produces a half-amplitude advanced wave A_a, which reinforces R_e in region 2 and cancels A_e in region 1. An observer sees only a full-amplitude retarded wave in region 2 traveling from the emitter to the absorber.

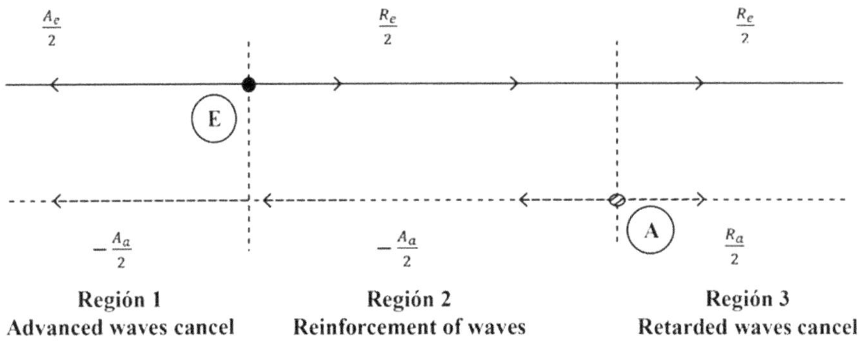

FIGURE 2.5 Scheme of the Wheeler–Feynman Absorber Theory. Dashed lines indicate a 180° phase shift (based on Ref. [29]).

Source: Adapted from Ref. [29]

It must be considered that eqs. (2.96) and (2.97) imply that emitter and absorber interact with time-symmetric fields that they themselves have produced. Therefore, they cannot change their energy or momentum, since the simultaneous emission of advanced and retarded waves do not produce such changes. This raises problems in its correspondence with observation because the emitter experiences neither recoil (i.e., momentum transfer) nor energy loss in the act of emission. However, if the absorption of the emitted retarded wave occurs sometime later, then the correspondence with observation is restored. As demonstrated by Wheeler and Feynman, if we consider a pair of electrons, the observed recoiling of each electron during emission and absorption occurs because it moves in the electromagnetic field waves (advanced or retarded) produced by the other electron.

As mentioned previously, the process described above can be thought of as the emitter sending out a probe wave in various allowed directions, seeking a transaction. An absorber, responding to one of these probe waves, sends a verifying wave back to the emitter verifying the transaction and arranging for the transfer of energy and momentum.

2.10 OPTICAL DIRAC-LIKE EQUATION AND TIME-REVERSAL SYMMETRY

In this section, we study an optical Dirac-like equation for the electromagnetic field and the symmetry under time reversal in absence of dissipation. We show that this equation can be used for the analysis of the photon states with positive and negative energy. This has been studied in detail in Ref. [30] so here we will give a summary.

2.10.1 OPTICAL DIRAC-LIKE EQUATION

As shown in Ref. [30] and in several publications, as for example in Ref. [31], photons propagating in free space satisfy a Dirac-like equation for a massless particle

$$i\hbar\frac{\partial\psi}{\partial t} = \mathcal{H}\psi \tag{2.98}$$

where $\mathcal{H}$ is a Hamiltonian operator which for a massless particle has the form

$$\mathcal{H} = c\vec{\alpha}\cdot\vec{p} \tag{2.99}$$

and $\vec{p} = -i\hbar\nabla$ is the momentum operator,

As in Ref. [30], a possible election for $\vec{\alpha}$ is a vector operator of the form

$$\vec{\alpha} = \begin{pmatrix} 0 & \vec{\tau} \\ \vec{\tau} & 0 \end{pmatrix} \quad (2.100),$$

where $\vec{\tau}$ is a vector-matrix whose components are the spin-1 matrices

$$\tau_x = \begin{pmatrix} 0 & 0 & 0 \\ 0 & 0 & -i \\ 0 & i & 0 \end{pmatrix}; \tau_y = \begin{pmatrix} 0 & 0 & i \\ 0 & 0 & 0 \\ -i & 0 & 0 \end{pmatrix}; \tau_z = \begin{pmatrix} 0 & -i & 0 \\ i & 0 & 0 \\ 0 & 0 & 0 \end{pmatrix} \quad (2.101)$$

There are several Dirac-like formulations of Maxwell's equations and alternative ways for choosing the wave function ψ. Considering that photons have only energy and no other scalar quantities such as mass or charge, it is convenient to choose ψ so that its modulus squared correspond to energy density, instead of probability density for localization as is the case of a particle with mass like the electron. Therefore, we choose as wave function the following column vector of dimension 6×1

$$\psi = \begin{pmatrix} k_1 \vec{E} \\ i k_2 \vec{B} \end{pmatrix} \quad (2.102)$$

In this last equation, the components of $\vec{E}$ and $\vec{B}$ are written as column vectors 3×1 and taking $k_1 = \sqrt{\epsilon_0 / 2}$ and $k_2 = 1/\sqrt{2\mu_0}$ we obtain

$$\psi^+ \psi = \frac{1}{2} \epsilon_0 \vec{E}^2 + \frac{\vec{B}^2}{2\mu_0} = w_{em} \quad (2.103)$$

This last expression corresponds to the density of energy in the electromagnetic field.

For an electromagnetic wave propagating in a linear magnetodielectric and non-conducting medium, there is an induced polarization and magnetization classically represented by the polarization and magnetization vectors $\vec{P}$ and $\vec{M}$, respectively. These vectors are related with the electric and magnetic fields as $\vec{D} = \epsilon_0 \vec{E} + \vec{P}$ and $\vec{B} = \mu_0 \left(\vec{H} + \vec{M} \right)$. Using these relations as shown in Example 2.11, Maxwell's equation corresponding to Ampere's law may be rewritten as

$$\nabla \times \vec{H} = \frac{\partial \vec{D}}{\partial t} + \vec{J}_{ind} \quad (2.104)$$

where $\vec{J}_{ind} = \frac{\partial \vec{P}}{\partial t} + \nabla \times \vec{M}$ is an equivalent current density associated to the polarization and magnetization of the medium. In this case, the Dirac-like equation becomes

$$i\hbar\frac{\partial}{\partial t}\psi = c\vec{\alpha}\cdot\vec{p}\psi - \frac{i\hbar}{2k_1}\tilde{J} \tag{2.105}$$

where $\tilde{J} = \begin{pmatrix} \vec{J}_{ind} \\ 0 \end{pmatrix}$ is a column matrix of dimension 6 × 1.

From eq. 2.105, a continuity equation for the wave function ψ may be derived taking the scalar product with ψ^+ on both sides, obtaining

$$\frac{1}{2}\frac{\partial}{\partial t}\left|\psi^2\right| = -c\psi^+\vec{\alpha}\cdot\nabla\psi - \frac{1}{2k_1}\psi^+\cdot\tilde{J} \tag{2.106}$$

Using the above and the definition of the column matrix $\tilde{J}$ given previously one obtains the classical equation for conservation of electromagnetic energy

$$-\nabla\cdot\left(\vec{E}\times\vec{H}\right) = \frac{\partial}{\partial t}\left\{\frac{\epsilon_o}{2}\vec{E}^2 + \frac{\mu_o}{2}\vec{H}^2\right\} + \vec{E}\cdot\vec{J}_{ind} = \frac{\partial}{\partial t}w_{em} + \vec{E}\cdot\vec{J}_{ind} \tag{2.107}$$

Using the divergence theorem, this last equation can be recast as

$$\oint\left(\vec{E}\times\vec{H}\right)\cdot d\vec{S} = \frac{\partial}{\partial t}\int w_{em}dV + \int\vec{E}\cdot\vec{J}_{ind}dV \tag{2.108}$$

Since $\vec{E}\times\vec{H}$ is the Poynting vector, the left side of this equation represents the flux of energy entering into the volume V bounded by a surface S. Part of this incoming energy is used to increase the energy stored in the electromagnetic field (first term at the right side of this equation) and the other is the energy required for changing the polarization and magnetization state. The derivation of this equation is made in more detail in Example 2.12.

EXAMPLE 2.11

Show that the effect of time and spatial variations in the polarization and magnetization vectors $\vec{P}$ and $\vec{M}$, respectively, induce a current density $\vec{J}_{ind} = \frac{\partial\vec{P}}{\partial t} + \nabla\times\vec{M}$. Discuss how this quantity transforms under time reversal.

SOLUTION

In absence of conduction, current Ampere's law is

$$\nabla\times\vec{H} = \frac{\partial\vec{D}}{\partial t} \tag{2.109}$$

where the displacement vector $\vec{D}$ is related to the electric field and polarization $\vec{P}$ by

$$\vec{D} = \varepsilon_o \vec{E} + \vec{P} \tag{2.110}$$

The magnetic field density vector $\vec{B}$ is related to the magnetic field and magnetization $\vec{M}$ by

$$\vec{B} = \mu_o (\vec{H} + \vec{M}) \tag{2.111}$$

From this last expression, we obtain

$$\vec{H} = \frac{\vec{B}}{\mu_o} - \vec{M} \tag{2.112}$$

Combining the above equations and considering that $\nabla \times \vec{B} = 0$, we get

$$\nabla \times \vec{H} = \varepsilon_o \frac{\partial \vec{E}}{\partial t} + \left(\frac{\partial \vec{P}}{\partial t} + \nabla \times \vec{M} \right) \tag{2.113}$$

In this last equation, the vectors $\vec{E}$ and $\vec{H}$ denote the electric and magnetic field, respectively, in vacuum. The first term at the right is the displacement current density vector in vacuum while the term inside brackets corresponds to a current density due to temporal variation of the polarization inside the material and the spatial variation of the magnetization. Therefore, this proves that $\vec{J}_{ind} = \frac{\partial \vec{P}}{\partial t} + \nabla \times \vec{M}$.

This current density has not time-reversal symmetry since

$$\mathcal{T}\left\{ \frac{\partial \vec{P}}{\partial t} + \nabla \times \vec{M} \right\} = \frac{\partial \vec{P}}{\partial(-t)} + \nabla \times \left(-\vec{M} \right) = -\frac{\partial \vec{P}}{\partial t} - \nabla \times \vec{M} = -\vec{J}_{ind} \tag{2.114}$$

It has been considered that vector $\vec{P}$ associated to charge distribution (more precisely distribution of induced electric dipolar moments) is even under time-reversal operation, whereas $\vec{M}$ which is associated with the circulation of charges in a classical description of magnetic dipoles is odd under this transformation. Therefore, $\vec{J}_{ind}$ is odd under time-reversal operation. We shall see that this result is compatible with time-reversal symmetry of Maxwell's equations in absence of dissipation.

EXAMPLE 2.12

Show eqs. 2.107 and 2.108. Discuss if there is time-reversal symmetry.

SOLUTION

Multiplying by ψ^+ each term of eq. 2.105 and combining with eq. 2.102, we obtain

$$\psi^+ i\hbar\frac{\partial\psi}{\partial t} = i\hbar\left(k_1\vec{E} \quad -ik_2\vec{B}\right)\frac{\partial}{\partial t}\begin{pmatrix} k_1\vec{E} \\ ik_2\vec{B}\end{pmatrix} = \psi^+ c\vec{\alpha}\cdot\vec{p}\psi - \frac{i\hbar}{2k_1}\psi^+\tilde{J}$$

$$\Rightarrow \psi^+ i\hbar\frac{\partial\psi}{\partial t} = i\hbar\left\{k_1^2\vec{E}\cdot\frac{\partial\vec{E}}{\partial t} + k_2^2\vec{B}\cdot\frac{\partial\vec{B}}{\partial t}\right\} = \psi^+ c\vec{\alpha}\cdot\vec{p}\psi - \frac{i\hbar}{2k_1}\left(k_1\vec{E} \quad -ik_2\vec{B}\right)\cdot\begin{pmatrix}\vec{J}_{ind} \\ 0\end{pmatrix}$$

$$\Rightarrow \psi^+ i\hbar\frac{\partial\psi}{\partial t} = \frac{i\hbar}{2}\left\{k_1^2\frac{\partial\vec{E}^2}{\partial t} + k_2^2\frac{\partial\vec{B}^2}{\partial t}\right\} = \psi^+ c\vec{\alpha}\cdot\vec{p}\psi - \frac{i\hbar}{2}\vec{E}\cdot\vec{J}_{ind} \tag{2.115}$$

With $k_1 = \sqrt{\frac{\varepsilon_o}{2}}$ and $k_2 = \frac{1}{\sqrt{2\mu_o}}$ eq. 2.115 can be recast as

$$\psi^+ i\hbar\frac{\partial\psi}{\partial t} = \frac{i\hbar}{2}\frac{\partial}{\partial t}\left\{\frac{1}{2}\epsilon_0\vec{E}^2 + \frac{\vec{B}^2}{2\mu_0}\right\} = \psi^+ c\vec{\alpha}\cdot\vec{p}\psi - \frac{i\hbar}{2}\vec{E}\cdot\vec{J}_{ind}$$

$$\Rightarrow \frac{i\hbar}{2}\frac{\partial}{\partial t}w_{em} = \psi^+ c\vec{\alpha}\cdot\vec{p}\psi - \frac{i\hbar}{2}\vec{E}\cdot\vec{J}_{ind} \tag{2.116}$$

For evaluating $\psi^+ c\vec{\alpha}\cdot\vec{p}\psi$, we use eqs. 2.100 and 2.101 with the operator momentum given by $\vec{p} = -i\hbar\nabla$, obtaining

$$\vec{\alpha}\cdot\vec{p}\psi = \begin{pmatrix} 0 & \vec{\tau}\cdot\vec{p} \\ \vec{\tau}\cdot\vec{p} & 0\end{pmatrix}\begin{pmatrix}k_1\vec{E} \\ ik_2\vec{B}\end{pmatrix} = \begin{pmatrix} ik_2\vec{\tau}\cdot\vec{p}\,\vec{B} \\ k_1\vec{\tau}\cdot\vec{p}\,\vec{E}\end{pmatrix} = \hbar\begin{pmatrix} ik_2\nabla\times\vec{B} \\ k_1\nabla\times\vec{E}\end{pmatrix} \tag{2.117}$$

Therefore, we have

$$\psi^+ c\vec{\alpha}\cdot\vec{p}\psi = \hbar c\left(k_1\vec{E} \quad -ik_2\vec{B}\right)\begin{pmatrix} ik_2\nabla\times\vec{B} \\ k_1\nabla\times\vec{E}\end{pmatrix} = i\hbar c k_1 k_2\left(\vec{E}\cdot\nabla\times\vec{B} - \vec{B}\cdot\nabla\times\vec{E}\right)$$

$$= -i\hbar c k_1 k_2\nabla\cdot\left(\vec{E}\times\vec{B}\right) \tag{2.118}$$

With $k_1 = \sqrt{\frac{\varepsilon_o}{2}}$, $k_2 = \frac{1}{\sqrt{2\mu_o}}$, $\vec{B} = \mu_o\vec{H}$ and $c = 1/\sqrt{\varepsilon_o\mu_o}$, eq. 2.127 can be recast as

$$\psi^+ c\vec{\alpha}\cdot\vec{p}\psi = -\frac{i\hbar}{2}\nabla\cdot\left(\vec{E}\times\vec{H}\right) \tag{2.119}$$

Combining the above equations, we obtain

$$-\nabla\cdot\left(\vec{E}\times\vec{H}\right)=\frac{\partial}{\partial t}w_{em}+\vec{E}\cdot\vec{J}_{ind}$$

The application of the divergence theorem in the previous expression leads us to eq. 2.108.

Considering that under time reversal, the electric field and the electromagnetic energy density are invariant, whereas $\vec{H}\rightarrow-\vec{H}$ and $\vec{J}_{ind}\rightarrow-\vec{J}_{ind}$, the above equation becomes

$$\nabla\cdot\left(\vec{E}\times\vec{H}\right)=-\frac{\partial}{\partial t}w_{em}-\vec{E}\cdot\vec{J}_{ind} \tag{2.120}$$

Integration of this equation leads to

$$\oint\left(\vec{E}\times\vec{H}\right)\cdot d\vec{S}=-\left(\frac{\partial}{\partial t}\int w_{em}dV+\int\vec{E}\cdot\vec{J}_{ind}dV\right) \tag{2.121}$$

Eqs. 2.120 and 2.121 are identical to the eqs. 2.107 and 2.108 expressing that in the absence of dissipation, there is time-reversal symmetry and the process of establishing the electromagnetic energy and the polarization and magnetic states inside the volume V is reversible.

2.10.2 ANALYSIS OF TIME-REVERSAL SYMMETRY OF THE OPTICAL DIRAC EQUATION

All the above analysis are based on Maxwell's equations in vacuum which as discussed in Section 2.2, are symmetrical under time reversal.

However, it is illustrative to show this symmetry directly from the optical Dirac equation given by (2.105) for an electromagnetic wave propagating in a polarizable and magnetic medium, without dissipation of energy. Applying the time-reversal operator to the left side of this equation:

$$\mathcal{T}\left\{i\hbar\frac{\partial}{\partial t}\psi\right\}=-i\hbar\mathcal{T}\left\{\frac{\partial}{\partial t}\psi\right\}=-i\hbar\frac{\partial\mathcal{T}\{\psi\}}{\partial(-t)}=i\hbar\frac{\partial\mathcal{T}\{\psi\}}{\partial t} \tag{2.122}$$

In the above equation, it has been considered that similarly to the time reversal of Schrodinger's equation, the application of the corresponding operator requires complex conjugation in addition to make $t\rightarrow-t$. Time reversal operator applied over the first term at the right side of eq. 2.105 gives

$$\mathcal{T}\{c\vec{\alpha}\cdot\vec{p}\}=c\vec{\alpha}^{*}\cdot\vec{p}^{*} \tag{2.123}$$

Considering $\vec{p} = -i\hbar\nabla$ and eq. (2.100), we obtain that $\vec{\alpha} \cdot \vec{p}$ is a real matrix of dimension 6 × 6 given by

$$\vec{\alpha} \cdot \vec{p} = \begin{pmatrix} 0 & \vec{\tau} \cdot \vec{p} \\ \vec{\tau} \cdot \vec{p} & 0 \end{pmatrix} \tag{2.124}$$

with

$$\vec{\tau} \cdot \vec{p} = \hbar \begin{pmatrix} 0 & -\dfrac{\partial}{\partial z} & \dfrac{\partial}{\partial y} \\ \dfrac{\partial}{\partial z} & 0 & -\dfrac{\partial}{\partial x} \\ -\dfrac{\partial}{\partial y} & \dfrac{\partial}{\partial x} & 0 \end{pmatrix} \tag{2.125}$$

Therefore, $\vec{\alpha}^* \cdot \vec{p}^*$ is equal to $\vec{\alpha} \cdot \vec{p}$ since the operator $\vec{\tau} \cdot \vec{p}$ does not include complex terms.

With respect to the time reversal of the second term in the right-hand member of eq. 2.105, we have:

$$\mathcal{T}\left\{\frac{i\hbar}{2k_1}\tilde{J}\right\} = \left(-\frac{i\hbar}{2k_1}\right)\mathcal{T}\left\{\tilde{J}\right\} \tag{2.126}$$

Since $\tilde{J} = \begin{pmatrix} \vec{J}_{ind} \\ 0 \end{pmatrix}$ and $\vec{J}_{ind} = \dfrac{\partial \vec{P}}{\partial t} + \nabla \times \vec{M}$ we have

$$\mathcal{T}\left\{\tilde{J}\right\} = \begin{pmatrix} \dfrac{\partial \vec{P}}{\partial(-t)} + \nabla \times \left(-\vec{M}\right) \\ 0 \end{pmatrix} = -\tilde{J} \tag{2.127}$$

In this last equation, we have considered that vectors $\vec{P}$ and $\vec{M}$ are even and odd, respectively, under time-reversal operation. Therefore, combining the above two equations we get

$$\mathcal{T}\left\{\frac{i\hbar}{2k_1}\tilde{J}\right\} = \frac{i\hbar}{2k_1}\tilde{J} \tag{2.128}$$

Collecting terms, we have that the time reversed form of eq. 2.105 is

$$i\hbar\frac{\partial}{\partial t}\mathcal{T}\{\psi\} = c\vec{\alpha} \cdot \vec{p}\mathcal{T}\{\psi\} - \frac{i\hbar}{2k_1}\tilde{J} \tag{2.129}$$

The time reversal of the wave function of eq. 2.105 is given by:

$$\mathcal{T}\{\psi\} = \mathcal{T}\begin{Bmatrix} k_1\vec{E} \\ ik_2\vec{B} \end{Bmatrix} \tag{2.130}$$

Since the electric field vectors $\vec{E}$ and $\vec{B}$, respectively, are even and odd under time reversal and considering that this operation must include complex conjugation, we get

$$\mathcal{T}\{\psi\} = \mathcal{T}\begin{Bmatrix} k_1\vec{E} \\ -ik_2\left(-\vec{B}\right) \end{Bmatrix} = \begin{Bmatrix} k_1\vec{E} \\ ik_2\vec{B} \end{Bmatrix} = \psi \tag{2.131}$$

In conclusion, from the above equations we have shown that the optical Dirac equation (2.105) is symmetric under time reversal.

All the preceding analysis is not valid if there is dissipation of electromagnetic energy, as occurs for example in a conducting medium.

EXAMPLE 2.13

The objective of this example is to analyze the propagation of an electromagnetic wave in a conductive medium and determine how the conductivity affects the symmetry of Maxwell's equations.

SOLUTION

The combination of Faraday and Ampere laws leads to the following differential equation for the electric field vector for propagation in a conducting medium with conductivity σ

$$\nabla^2\dot{E} - \mu\sigma\frac{\partial\vec{E}}{\partial t} - \mu\epsilon\frac{\partial^2\vec{E}}{\partial t^2} = 0 \tag{2.132}$$

The presence of a term proportional to the first time derivative corresponds to dissipation of electromagnetic energy and so it is expected that the propagation of an electromagnetic wave is not time-reversal symmetry in this case. This can be corroborated if we consider the case of a plane and monochromatic wave of angular frequency ω propagating in a medium with finite conductivity σ. In this case, Faraday and Ampere laws in complex notation and with time dependence given by the factor $exp[i\omega t]$ are expressed as follows:

$$\nabla \times \vec{E} = -i\omega\mu\vec{H} \tag{2.133}$$

$$\nabla \times \vec{H} = i\omega\in\vec{E} + \sigma\vec{E} \tag{2.134}$$

Combining the above two equations, we obtain

$$\nabla^2\vec{E} - i\omega\mu\left(\vec{J}_{cond} + \vec{J}_D\right) = 0 \tag{2.135}$$

where

$\vec{J}_{cond} = \sigma\vec{E}$ is the conduction current density and

$\vec{J}_D = i\omega\in\vec{E}$ is the displacement current density

In metals, the displacement current density can be neglected, since in general much lower than the conduction current density, so that eq. 2.135 may be rewritten as

$$\nabla^2\vec{E} - i\omega\mu\sigma\vec{E} = 0 \tag{2.136}$$

For simplicity, we consider the case of an electromagnetic wave linearly polarized along the x direction and propagating in the direction z so that its electric field vector is

$$\vec{E}(z,t) = \hat{x}E_x(z)e^{i\omega t} \tag{2.137}$$

Since eq. 2.136 is a second-order differential equation it has two solutions one of which is

$$\vec{E}_1(z,t) = \hat{x}Aexp\left[-\sqrt{i\omega\mu\sigma z}\right]e^{i\omega t} = \hat{x}Aexp\left[i\left(\omega t - \sqrt{\frac{\omega\mu\sigma}{2}}z\right)\right]exp\left[-\sqrt{\frac{\omega\mu\sigma}{2}}z\right] \tag{2.138}$$

According to this last equation as the wave moves along the z direction, its amplitude is exponentially damping by the factor $exp\left[-\sqrt{\frac{\omega\mu\sigma}{2}}z\right]$, meaning that this is an irreversible process where electromagnetic energy is converted into heat transferred to the medium.

The other solution of eq. 2.136 is

$$\vec{E}_2(z,t) = \hat{x}Aexp\left[+\sqrt{i\omega\mu\sigma z}\right]e^{i\omega t} = \hat{x}Aexp\left[i\left(\omega t + \sqrt{\frac{\omega\mu\sigma}{2}}z\right)\right]exp\left[+\sqrt{\frac{\omega\mu\sigma}{2}}z\right] \tag{2.139}$$

Nevertheless, this last solution is discarded since it gives an electromagnetic wave whose amplitude increases during its propagation in a conducting medium, a fact which is against all the experimental evidences.

2.10.3 PHOTON STATES WITH POSITIVE AND NEGATIVE ENERGIES

We will show that the optical Dirac-like equation in vacuum (2.98) leads to the relation $E = c|\vec{p}|$ for a photon propagating in vacuum so that a solution analog to the energy of a free fermion is obtained from Dirac's equation in the limit of zero mass. Nevertheless, this must be considered only as a formal analogy since photons have spin 1 and Dirac's equation is applicable for particles of spin ½. In addition, it is worth to relate the analysis presented in this subsection with what we can call a time Dirac-like equation as will be analyzed in Chapter 6, Section 6.6, where we will discuss a time operator for a free relativistic quantum particle, whose eigenvalues correspond to time intervals in the context of a formalism that has some similarity with the analysis here presented.

The energy eigenvalues are obtained by looking for stationary state solutions of the Dirac-like equation. A plane wave solution of this equation has the form

$$\psi_k(\vec{r},t) = A_k \exp\left[\frac{i}{\hbar}(\vec{p}\cdot\vec{r} - \hbar\omega t)\right] \tag{2.140}$$

where A_k for $k = 1$ to 6, denotes the components of the electric and magnetic field vectors, $\vec{p}$ is the momentum vector and ω is the angular frequency. Considering, a momentum vector with components along two directions, for example parallel to x and y axes, from eq. 2.101 it can be shown that the values of the energy for a photon can be obtained from the eigenvalues of the 3 × 3 matrix $(c\vec{\tau}\cdot\vec{p})^2$ for this case given by:

$$(c\vec{\tau}\cdot\vec{p})^2 = c^2\begin{bmatrix} p_y^2 & -p_x p_y & 0 \\ -p_x p_y & p_x^2 & 0 \\ 0 & 0 & p_x^2 + p_y^2 \end{bmatrix} \tag{2.141}$$

The corresponding solutions for the energy are $E = +cp$, $E = -cp$ and $E = 0$ where $p = |\vec{p}|$. A possible interpretation of the negative energy solution could be given using the Feynman's concept of antiparticles[32] associating to photon states going backward in time and postulating that the photon has its own antiparticle. Nevertheless, this is still a non-solved question. For example, in Ref. [33] it is postulated that there exist negative energy photons in the universe and with this assumption, a solution of Einstein's equation without introducing the cosmological constant is found. In Ref. [34], it is shown that although negative energy matter remains undiscovered at present, all relativistic field equations admit genuine negative energy field solutions creating and annihilating negative energy quanta. The unitary time-reversal transformation links these negative energy fields to the positive energy ones.

The solution $E = 0$ has no physical meaning since it would imply that a photon in vacuum could be at rest (zero momentum) and from the point of view of classical electrodynamics, it would result in an electromagnetic wave propagating in an unbounded medium with a field component parallel to the direction of propagation, what is not compatible with Maxwell's equations. That can be seen if one considers, for example, motion along the z direction with momentum $p_z = p$. In that case, the calculation of the eigenvalues and eigenvectors of the Hamiltonian given by eq. 2.99 shows that the positive and negative energy solution are valid because are compatible with an electromagnetic wave propagating in vacuum and satisfying the condition that the electric and magnetic field vectors must have only transversal components (that is perpendicular to the z axis in this case). On the other hand, the solutions with zero energy are not valid since they imply that the electric or magnetic field has longitudinal components (parallel to the direction of propagation). This is shown in Example 2.13.

It is important to consider that Dirac's equation for a particle like an electron only gives positive and negative energy solutions.

EXAMPLE 2.13

Show that photon states with zero energy cannot be associated to an electromagnetic wave propagating in vacuum.

SOLUTION

Considering propagation along the z direction, from eq. 2.149, the wave function can be written as:

$\psi_k(z,t) = A_k exp\left[\frac{i}{\hbar}(Pz - Et)\right]$ where P and E are the corresponding momentum and energy of the associated photons.

Considering that $\vec{p} = -i\hbar\nabla$ is the momentum operator, the eigenvalue equation for this case is obtained as

$$\mathcal{H}\psi = c\vec{\alpha}\cdot\vec{p}\psi \rightarrow c\alpha_z p_z \psi = E\psi$$

Since $p_z\psi = -i\hbar\frac{\partial}{\partial z}\psi = -i\hbar\frac{i}{\hbar}P\psi = P\psi$ we obtain the following equation for calculating eigenvalues and eigenvectors

$$c\alpha_z P\psi = E\psi,$$

where

$$\alpha_z = \begin{pmatrix} 0 & \tau_z \\ \tau_z & 0 \end{pmatrix} \text{ and } \tau_z = \begin{pmatrix} 0 & -i & 0 \\ i & 0 & 0 \\ 0 & 0 & 0 \end{pmatrix}$$

Therefore, the matrix $c\alpha_z P$ is a matrix 6 × 6 whose eigenvalues are $E = \pm cP = \pm\hbar ck$ and $E=0$. In particular, the eigenvectors corresponding to the eigenvalue E = 0 are matrices 6 × 1 proportional to the column vectors

$$\psi_1(E=0) = (0 \;\; 0 \;\; 1 \;\; 0 \;\; 0 \;\; 0)^T \text{ and } \psi_2(E=0) = (0 \;\; 0 \;\; 0 \;\; 0 \;\; 0 \;\; 1)^T$$

The positions of the elements different of zero indicate that the electromagnetic field associated to the wave has only electric or magnetic field component along the *z* direction, that is, the electromagnetic field can be described as

$$\left\{\vec{E} = \hat{z}E_z, \vec{B} = 0\right\} \text{ or } \left\{\vec{E} = 0, \vec{B} = \hat{z}B_z\right\}.$$

This result implies that there is not wave propagation explaining why the corresponding energy eigenvalue is zero corresponding to a photon at rest in vacuum, what certainly has not any physical sense.

2.11 ELECTROMAGNETIC NONRECIPROCITY AND TIME REVERSAL

2.11.1 RECIPROCITY AND TIME-REVERSAL SYMMETRY

A reciprocal system is defined as a system that exhibits the same received–transmitted field ratios when its source(s) and detector(s) are exchanged. The lack of this property is called non-reciprocity.

Non-reciprocity is essential for applications where one-way propagation is required, for example, in a radar system using a single antenna for transmitting and receiving at the same time.

For a linear system, non-reciprocity is based on time-reversal symmetry breaking by an external bias, as occurs in the Faraday effect to be discussed in this section. Time-reversal symmetry breaking is always a necessary condition for non-reciprocity.

Based on Ref. [35], let us imagine that we follow the temporal evolution of a given physical phenomenon occurring in a certain system and that we monitor the process between ports or terminals denoted as *P1* and *P2* via the state vector $\Psi(t)=\left[\psi_{P1}(t)\quad \psi_{P2}(t)\right]^{T}$ where $\psi_{P1}(t),\psi_{P2}(t)$ are the wave states at ports P1 and P2, respectively, and T is the transpose of the corresponding raw vector. Initially, the system is in an equilibrium state that is free of external excitations. At time $t = 0$, we apply an excitation through port *P1* and trace the response of the system until the time $t = T$ of complete transmission to *P2*. Then we apply the time-reversal operation which results in a system not necessarily identical to the original one, being excited at time $t = -T$ and evolving until the time $t = 0$ of complete transmission to terminal *P1*. This process is illustrated in Figure 2.6 where at time $t = 0$, the system at a state Ψ(0) receives an excitation through port P1 (for example, an input consisting in a voltage signal in the case of an electric circuit) and after a time of observation T it reaches the state $\Psi(T)=\left[\psi_{P1}(T)\quad \psi_{P2}(t)\right]^{T}$. If at this instant, one flips the sign of the time variable so that the system is excited at $t = -T$ through terminal P2 with $\psi_{P2}(-T)$ and the system evolves until the time $t = 0$, if there is time-reversal symmetry, we would expect that the system comes back to its initial state, that is $\Psi'(0) = \Psi(0)$. That means that if there is not any other excitation, the input $\psi_{P2}(-T)$ through P2 produces $\psi_{P1}(0)$ at port P1 after a time interval T.

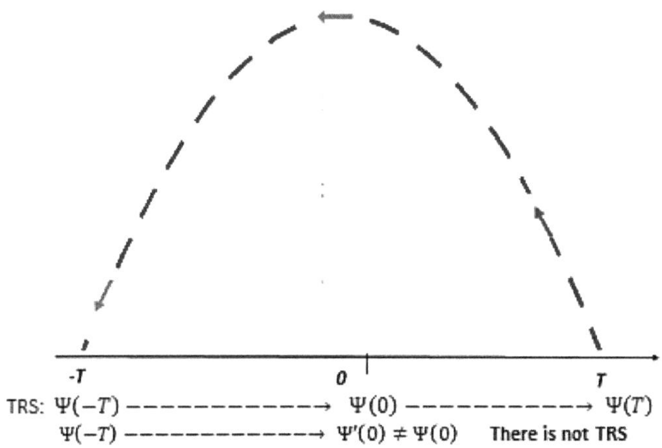

FIGURE 2.6 Time-reversal symmetry (TRS): as shown by the dotted curve at time *T*, the state of the system is time reversed and left to evolve coming back to the initial state (if there is TRS) or to a different state (if there is not TRS).

If under the time-reversal transformation sketched in Figure 2.6, the system comes back to its initial state Ψ(0) and remains the same, there is time-reversal symmetry. Since the direct and reverse parts of the process typically describe the response of the system for opposite transmission directions, time-reversal symmetry is closely related with reciprocity.

Nevertheless, it must be considered that the idea of interchanging cause and effect that underlies the reciprocity principle is different from time reversibility, since for a given position of the source, there are alternative positions for the detector. Each of these alternatives corresponds to a distinct reciprocal configuration.

The concept of electromagnetic reciprocity is closely related to that of time-reversal symmetry of Maxwell's equations. However, it must be considered that in general an external or internal bias, can destroy the symmetry of a process. This operation is called ***Time-reversal symmetry breaking*** and requires reversing/maintaining the sign of at least one of the time-reversal even/odd quantities (bias) involved in the system. Only the latter option, that is maintaining the sign of a time-reversal odd quantity, is meaningful since it is the only case which leaves the system the same. We shall see that this is the case of the Faraday effect where the sign of the magnetic field is maintained (it must be remembered that the magnetic field is odd under time reversal). Time-reversal symmetry breaking is always a necessary condition for non-reciprocity, since the field ratios cannot differ if the field levels do not. Non-reciprocity is equivalent to time-reversal symmetry breaking in systems with negligible loss. Systems with significant loss or gain are time-reversal asymmetric even when they are reciprocal, save if reciprocity is understood in the restricted sense of equal field ratios rather than field levels.

2.11.2 NON-RECIPROCITY IN FARADAY EFFECT

As an example of a process which has not reciprocity, we describe the Faraday effect showing that even assuming that there is not dissipation, time-reversal symmetry is broken by an external excitation, in this case a magnetic field.

Roughly speaking, this effect consists on the rotation of the polarization of a light beam when it passes through a region where a stationary magnetic field is applied. This effect was discovered in 1845 by Michael Faraday who found the first physical phenomenon linking light and magnetism. He was able to rotate the polarization of light when he induced a magnetic field in the same direction as the path of the light. This only occurs when light passes through transparent dielectrics. Applying a magnetic field to the dielectric causes it to exhibit circular birefringence. Birefringent materials decompose an incident ray of light into two rays, called an ordinary and an extraordinary ray. In circular birefringent materials, these rays have the additional property of being circularly polarized with opposite rotations. Different indices of refraction exist for the left and right circularly polarized rays. As a result, these rays propagate at different speeds through the medium. The two rays combine at the end, yielding a ray that is offset in polarization from the incident ray. The relationship between the rotation of the ray and the magnetic density field B is given as

$$\theta = k_V B l \tag{2.142}$$

where l is the length of the path followed by the beam subject to action of the magnetic field and k_V is known as the Verdet constant which depends on the properties of the material filling the region where the magnetic field is applied. This is shown schematically in Figure 2.7.

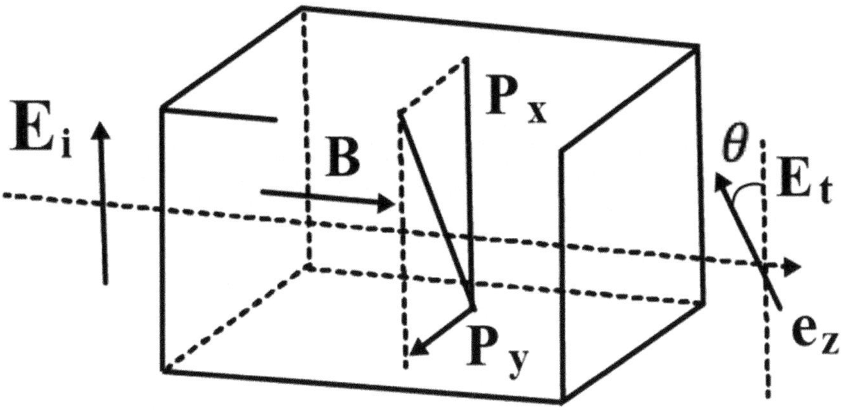

FIGURE 2.7 Light beam linearly polarized entering to a region where a static magnetic field is applied in a direction parallel to the propagation. At the output of this region, the electric field vector is rotated by an angle θ.

Let us consider the case of a linearly polarized electromagnetic wave which is reflected at the boundary separating air from a medium filled with a plasma where a uniform magnetic field is applied in a direction parallel to the direction of propagation of the incident wave, as shown in Figure 2.8.

Case 1: We consider first the case where the incident wave is linearly polarized in the x direction. As known this wave is equivalent to the superposition of two circular polarized waves rotating in opposite senses, one clockwise and the other counter clockwise:

$$\vec{E}_{inc} = E_{inc}\hat{x} = \vec{E}_{RCP} + \vec{E}_{LCP} = \underbrace{E_{inc}\left(\frac{\hat{x}+i\hat{y}}{2}\right)}_{\vec{E}_{RCP}} + \underbrace{E_{inc}\left(\frac{\hat{x}-i\hat{y}}{2}\right)}_{\vec{E}_{LCP}} \tag{2.143}$$

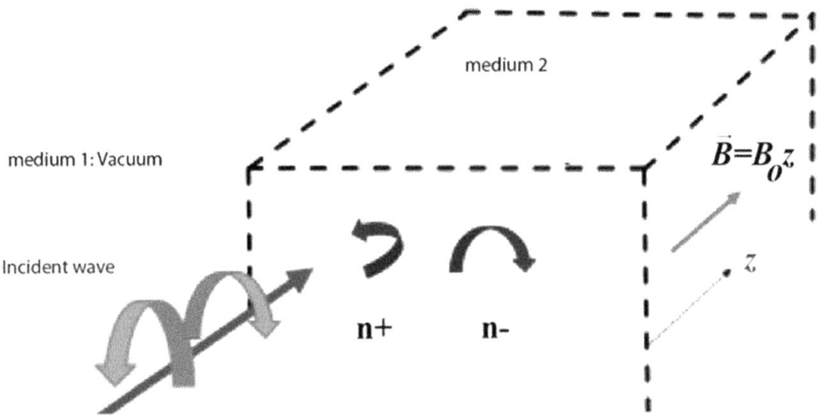

FIGURE 2.8 A linearly polarized wave enters to a region where a static magnetic field is applied as shown. This wave is decomposed into two circularly polarized waves rotating in opposite senses. Due to the magnetic field, the refraction index is different for each circularly polarized wave.

If these waves enter into a region where a static magnetic field with direction parallel to that of propagation is applied, the interaction with the particles of the medium produces different dielectric functions and refractive indexes given by

$$\varepsilon_{\pm} = 1 + \frac{\omega_p^2}{\omega_o^2 - \omega^2 \mp \omega\omega_L};\ n_{\pm} = \sqrt{\varepsilon_{\pm}} \tag{2.144}$$

where ω_o is the natural frequency of an equivalent harmonic oscillator that represents the motion of the electrons subject to an oscillating electric field, and $\omega_L = eB_o/m$ is the Larmor frequency of an electron in a static magnetic field of magnitude B_o (e is the absolute value of the charge of the electron and m its mass). These last two relations are shown in Example 2.14.

Considering Fresnel coefficients, the reflected wave is

$$\vec{E}_{reflected} = \left(\frac{1-n_{RCP}}{1+n_{RCP}}\right)\vec{E}_{RCP} + \left(\frac{1-n_{LCP}}{1+n_{LCP}}\right)\vec{E}_{LCP}$$

$$= \left(\frac{1-n_{RCP}}{1+n_{RCP}}\right)\frac{E_{inc}}{2}(\hat{x}+i\hat{y}) + \left(\frac{1-n_{LCP}}{1+n_{LCP}}\right)\frac{E_{inc}}{2}(\hat{x}-i\hat{y}) \quad (2.145)$$

The x and y components of the reflected wave are given by

$$E_{reflected\ x} = \frac{E_{inc}}{2}\left\{\frac{1-n_{RCP}}{1+n_{RCP}} + \frac{1-n_{LCP}}{1+n_{LCP}}\right\} = E_{inc}\left\{\frac{1-n_{RCP}n_{LCP}}{(1+n_{RCP})(1+n_{LCP})}\right\} \quad (2.146)$$

$$E_{reflected\ y} = i\frac{E_{inc}}{2}\left\{\frac{1-n_{RCP}}{1+n_{RCP}} - \frac{1-n_{LCP}}{1+n_{LCP}}\right\} = iE_{inc}\left\{\frac{n_{LCP}-n_{RCP}}{(1+n_{RCP})(1+n_{LCP})}\right\} \quad (2.147)$$

Therefore, the relation between the reflected components of the electric field in the x and y direction is

$$\frac{E_{reflected\ y}}{E_{reflected\ x}} = i\left(\frac{n_{LCP}-n_{RCP}}{1-n_{RCP}n_{LCP}}\right) \quad (2.148)$$

Case 2: We consider now the case where the incident wave is linearly polarized in the y direction.

$$\vec{E}_{inc} = E_{inc}\hat{y} = \vec{E}_{RCP}/i - \vec{E}_{LCP}/i = E_{inc}\left(\frac{\hat{x}+i\hat{y}}{2i}\right) - E_{inc}\left(\frac{\hat{x}-i\hat{y}}{2i}\right) \quad (2.149)$$

$$\vec{E}_{reflected} = \left(\frac{1-n_{RCP}}{1+n_{RCP}}\right)\vec{E}_{RCP}/i - \left(\frac{1-n_{LCP}}{1+n_{LCP}}\right)\vec{E}_{LCP}/i$$

$$= \left(\frac{1-n_{RCP}}{1+n_{RCP}}\right)\frac{E_{inc}}{2i}(\hat{x}+i\hat{y}) - \left(\frac{1-n_{LCP}}{1+n_{LCP}}\right)\frac{E_{inc}}{2i}(\hat{x}-i\hat{y}) \quad (2.150)$$

For this case, the x and y components of the reflected wave are given by

$$E_{reflected\ x} = \frac{E_{inc}}{2i}\left\{\frac{1-n_{RCP}}{1+n_{RCP}} - \frac{1-n_{LCP}}{1+n_{LCP}}\right\} = -iE_{inc}\left\{\frac{n_{LCP}-n_{RCP}}{(1+n_{RCP})(1+n_{LCP})}\right\} \quad (2.151)$$

$$E_{reflected\ y} = \frac{E_{inc}}{2}\left\{\frac{1-n_{RCP}}{1+n_{RCP}} + \frac{1-n_{LCP}}{1+n_{LCP}}\right\} = E_{inc}\left\{\frac{1-n_{RCP}n_{LCP}}{(1+n_{RCP})(1+n_{LCP})}\right\} \quad (2.152)$$

For this last case of linear polarization of the incident wave along the y direction, we have:

$$\frac{\vec{E}reflected_x}{\vec{E}reflected_y} = -i\frac{n_{lCP}-n_{RCP}}{1-n_{RCP}n_{LCP}} \quad (2.153)$$

Comparing eqs. (2.148) and (2.153), we observe that for case 1 (incident wave polarized in the x direction), some part is reflected with y-polarization. However, when the incident light is y-polarized, the same part of it is reflected with x-polarization but with an additional phase flip of 180° (reciprocal response would imply no phase change).

EXAMPLE 2.14

Show relations given by eq. 2.144. For that purpose consider a polarizable medium where we can consider that each electron (charge—e and mass M) is bound to the origin by a spring of constant $K = m\omega_o^2$ under the influence of a uniform external magnetic field $\vec{B} = B_o\hat{z}$ and a weak transverse electromagnetic wave with electric field vector $\vec{E}_w e^{i(kz-\omega t)}$ where $\vec{E}_w = \hat{x}E_x + \hat{y}E_y$.

SOLUTION

The equation of motion of each electron is given by

$$m\ddot{\vec{r}} + m\omega_o^2\vec{r} = -e\left(\vec{E}_w e^{i(kz-\omega t)}\right) - e\dot{\vec{r}} \times B_o\hat{z} \tag{2.154}$$

It will be assumed that the electrons remain close to its rest position, such the coordinate z in the above equation can be considered as a constant. Using the trial solution $\vec{r} = \vec{r}_o e^{i(kz-\omega t)}$, one gets

$$\left(\omega_o^2 - \omega^2\right)\vec{r}_o = -\frac{e\vec{E}_w}{m} + \frac{i\omega e B_o \vec{r}_o \times \hat{z}}{m} \tag{2.155}$$

Since both $\vec{E}_w$ and $\vec{r}_o \times \hat{z}$ are in the xy plane, the same occurs with the motion of the electron.

This equation implies that the motion of the electron is in the xy plane.

As known a linearly polarized electromagnetic wave may be expressed as equivalent to the superposition of two circularly polarized waves, one with the electric field vector rotating in the clockwise sense and the other rotating in the opposite sense (right and left circular polarized light, respectively). For that purpose, we consider that vectors $\vec{E}_w$ and $\vec{r}_o$ can be written in terms of a new basis of unitary vectors (u_+, u_-)

$$\vec{E}_w = \hat{x}E_x + \hat{y}E_y = E_x\left(\frac{\hat{x}+i\hat{y}}{2} + \frac{\hat{x}-i\hat{y}}{2}\right) - iE_y\left(\frac{\hat{x}+i\hat{y}}{2} - \frac{\hat{x}-i\hat{y}}{2}\right)$$

$$= \frac{E_x - iE_y}{\sqrt{2}} \frac{\hat{x} + i\hat{y}}{\sqrt{2}} + \frac{E_x + iE_y}{\sqrt{2}} \frac{\hat{x} - i\hat{y}}{\sqrt{2}} = E_- u_+ + E_+ u_- \tag{2.156}$$

where

$$E_\pm = \frac{E_x \pm iE_y}{\sqrt{2}}, \text{ and } u_\pm = \frac{\hat{x} \pm i\hat{y}}{\sqrt{2}} \tag{2.157}$$

Similarly, the vector $\vec{r}_o = x\hat{x} + y\hat{y}$ can be expressed in the new basis as

$$\vec{r}_o = r_{o-} u_+ + r_{o+} u_-, \text{ where } r_{o\pm} = \frac{x \pm iy}{\sqrt{2}} \tag{2.158}$$

With these transformations, we get a system of equations for (r_{o+}, r_{o-}) where each of these coordinates are not mixed:

$$m\left(\omega_o^2 - \omega^2\right) r_{o-} = -\omega e B_o r_{o-} - eE_- \tag{2.159}$$

$$m\left(\omega_o^2 - \omega^2\right) r_{o+} = \omega e B_o r_{o+} - eE_+ \tag{2.160}$$

From these equations, we obtain

$$r_{o\pm} = -\frac{e}{m} \frac{E_\pm}{\omega_o^2 - \omega^2 \mp \omega\omega_L} \tag{2.161}$$

where $\omega_L = eB_o/m$ is the Larmor frequency of an electron in a static magnetic field. From this result, one can see that electric dipoles are induced in the medium and a corresponding polarization vector $\vec{P}$ which can be expressed as:

$\vec{P} = -Ne\vec{r} = -Ne\left(r_- u_+ + r_+ u_-\right) = P_- u_+ + P_+ u_-$ where N is the average number density of electrons and

$$r_+ = r_{o+} e^{i(k_+ z - \omega t)} = -\frac{e}{m} \frac{E_+}{\omega_o^2 - \omega^2 - \omega\omega_L} e^{i(k_+ z - \omega t)} \tag{2.162}$$

$$r_- = r_{o-} e^{i(k_- z - \omega t)} = -\frac{e}{m} \frac{E_+}{\omega_o^2 - \omega^2 + \omega\omega_L} e^{i(k_- z - \omega t)} \tag{2.163}$$

In these last two equations, it is considered that due to the external magnetic field applied over the medium, the two circularly polarized waves move with different velocities so that the corresponding wave vector have different magnitudes given by $k_\pm = \frac{\omega}{c} n_\pm = \frac{\omega}{c} \sqrt{\varepsilon_\pm}$ where $\varepsilon_\pm$ are the correponding dielectric functions to be determined as follows. First, we determine the corresponding polarization vectors which represent the volumetric densities of induced electric dipole moments:

$$\left.\begin{aligned} P_+ &= -Ner_{o+}e^{i(k_+z-\omega t)} = \frac{\varepsilon_o\omega_p^2}{\omega_o^2-\omega^2-\omega\omega_L}E_+e^{i(k_+z-\omega t)} = \varepsilon_o\chi_+E_+e^{i(k_+z-\omega t)} \\ P_- &= -Ner_{o-}e^{i(k_-z-\omega t)} = \frac{\omega_p^2}{\omega_o^2-\omega^2+\omega\omega_L}E_-e^{i(k_-z-\omega t)} = \varepsilon_o\chi_-E_-e^{i(k_-z-\omega t)} \end{aligned}\right\} \quad (2.164)$$

where $\omega_p = \sqrt{\frac{Ne^2}{m\varepsilon_o}}$ is the plasma frequency and $\chi_\pm$ is the corresponding electric susceptibilities given by

$$\chi_\pm = \frac{\omega_p^2}{\omega_o^2-\omega^2\mp\omega\omega_L} \quad (2.165)$$

The displacement vector $\vec{D} = \varepsilon_o\vec{E} + \vec{P}$ is expressed in terms of the basis of unitary vectors (u_+, u_-) as

$$\vec{D} = D_-u_+ + D_+u_- = \varepsilon_o\left(1+\chi_-\right)E_-e^{i(k_-z-\omega t)}u_+ + \varepsilon_o\left(1+\chi_+\right)E_+e^{i(k_+z-\omega t)}u_- \quad (2.166)$$

From these expressions, we prove that the dielectric function corresponding to the right- and left-handed circular waves is given by eq. 2.144, that is

$$\varepsilon_\pm = 1+\chi_\pm = 1+\frac{\omega_p^2}{\omega_o^2-\omega^2\mp\omega\omega_L}$$

2.11.3 LORENTZ RECIPROCITY THEOREM

As previously discussed in this book, on a microscopic scale most physical laws are time-reversal symmetry. Although the traditional definition of a time-reversal operation which satisfies the microscopic initial conditions is very important for many branches of physics, especially in quantum field theory, it is in practice not easily applicable for classical electrodynamics. Indeed, the main subjects of study in classical electrodynamics are macroscopic systems and processes. The reversal of such processes is usually understood in the macroscopic sense where only the macroscopic initial conditions are satisfied. In this framework, lossy materials remain lossy for the reversed process, and wave processes in lossy materials appear irreversible. For that reason, in Ref. [36], an alternative notion of *restricted-time reversal* is considered. Under this transformation, it is assumed that one reverses the time in the Maxwell's equations of the considered system, but the time in equations governing all other processes which are coupled to the electromagnetic system under study (such as equations of motion of atoms in materials) is not reversed. Moreover, the external bias fields are not reversed. In this scenario, considering time-reversed processes, only macroscopic initial conditions are considered and properly reversed. The electromagnetic processes remain dissipative under the restricted time reversal (loss is not transformed into equivalent gain). Most importantly, since all the laws of classical physics are time-symmetric, all the dissipation processes will be governed by exactly the same laws after restricted time reversal, including the formulas for calculating dissipated power.

The concept of reciprocity is closely related with that of restricted time reversal. For time-invariant systems (whose properties do not change with time), the condition of reciprocity at each point holds if the restricted time reversal does not change the system, and vice versa. However, while the time-inversion concept is intrinsically of theoretical nature and implies process inversion with correct macroscopic initial conditions, the reciprocity principle can be easily applied to real systems and requires only the interchange of the source and detector locations. Reciprocity is an important concept in antennas because it has implications when the role of emitting and receiving antennas is reversed.[37]

Let us consider two sets of sources $\vec{J}_1$ and $\vec{J}_2$ which each produce electromagnetic fields $\vec{E}_1$, $\vec{H}_1$ and $\vec{E}_2$, $\vec{H}_2$, respectively.

From the quantity $\nabla\cdot(\vec{E}_1 \times \vec{H}_2 - \vec{E}_2 \times \vec{H}_1)$ and Maxwell's rotational equations, it can be proved the following relation (see Example 2.15):

$$\oint d\vec{S}'\cdot\left(\vec{E}_1\times\vec{H}_2-\vec{E}_2\times\vec{H}_1\right)=\int dV'\left(\vec{J}_1\cdot\vec{E}_2-\vec{J}_2\cdot\vec{E}_1\right) \tag{2.167}$$

EXAMPLE 2.15

Prove eq. 2.167.

SOLUTION

From the divergence theorem

$$\oint d\vec{S}'\cdot\left(\vec{E}_1\times\vec{H}_2-\vec{E}_2\times\vec{H}_1\right)=\int\nabla\cdot\left(\vec{E}_1\times\vec{H}_2-\vec{E}_2\times\vec{H}_1\right)dV' \tag{2.168}$$

Use of the following identity of vector calculus $\nabla\cdot\left(\vec{a}\times\vec{b}\right)=\left(\nabla\times\vec{a}\right)\cdot\vec{b}-\left(\nabla\times\vec{b}\right)\cdot\vec{a}$, it allows to rewrite the term inside the volume integral of eq. 2.168 as

$$\nabla\cdot\left(\vec{E}_1\times\vec{H}_2-\vec{E}_2\times\vec{H}_1\right)$$

$$= \left(\nabla\times\vec{E}_1\right)\cdot\vec{H}_2 - \left(\nabla\times\vec{H}_2\right)\cdot\vec{E}_1 - \left(\nabla\times\vec{E}_2\right)\cdot\vec{H}_1 + \left(\nabla\times\vec{H}_1\right)\cdot\vec{E}_2 \quad (2.169)$$

For a harmonically time varying electromagnetic field with angular frequency ω, Faraday and Ampere laws without magnetic sources can be written as

$$\nabla\times\vec{E}_1 = i\omega\mu\vec{H}_1,\ \nabla\times\vec{E}_2 = i\omega\mu\vec{H}_2 \quad (2.170)$$

$$\nabla\times\vec{H}_1 = \vec{J}_1 - i\omega\in\vec{E}_1,\ \nabla\times\vec{H}_2 = \vec{J}_2 - i\omega\in\vec{E}_2 \quad (2.171)$$

Combining eqs. (2.170) and (2.171) with eq. 2.169

$$\nabla\cdot\left(\vec{E}_1\times\vec{H}_2 - \vec{E}_2\times\vec{H}_1\right) = i\omega\mu\vec{H}_1\cdot\vec{H}_2 - \vec{J}_2\cdot\vec{E}_1 + i\omega\in\vec{E}_2\cdot\vec{E}_1 - i\omega\mu\vec{H}_2\cdot\vec{H}_1$$

$$+\vec{J}_1\cdot\vec{E}_2 - i\omega\in\vec{E}_1\cdot\vec{E}_2$$

$$= \vec{J}_1\cdot\vec{E}_2 - \vec{J}_2\cdot\vec{E}_1 \quad (2.172)$$

Using again the divergence theorem, we obtain

$$\oint d\vec{S}'\cdot\left(\vec{E}_1\times\vec{H}_2 - \vec{E}_2\times\vec{H}_1\right) = \int\left(\vec{J}_1\cdot\vec{E}_2 - \vec{J}_2\cdot\vec{E}_1\right)dV' \quad (2.173)$$

The right side of equation (2.173) is an integral over a volume containing the sources $\vec{J}_1$ and $\vec{J}_2$ and the left side corresponds to an integral over its corresponding bounding surface.

A more useful form of this theorem, applicable to antennas, is found by noticing that for electric and magnetic fields observed at a large distance from a source (e.g., a sphere of infinite radius surrounding an antenna), we must consider that:

- $\vec{E}\times\vec{H}$ which is the Poynting vector points in the direction normal to the sphere, that is parallel to the unitary radial vector $\hat{r}$
- The vectors $\vec{E}$ and $\vec{H}$ are related as $\vec{H} = \vec{E}\times\hat{r}/Z_{medium}$ where $Z_{medium} = \sqrt{\mu/\in}$ is the intrinsic impedance of the medium

In this case, the following relations is obtained:

$$\int dV'\,\vec{E}_1\cdot\vec{J}_2 = \int dV'\,\vec{E}_2\cdot\vec{J}_1 \quad (2.174)$$

The proof of this equation is as follows:

At large distances from the sources, the magnetic field vector produced by each one is related with the corresponding electric field as $\vec{H}_j = \hat{r}\times\vec{E}_j/Z_{medium}$ (for $j=1,\ 2$), where $\hat{r}$ is the unitary radial vector and Z_{medium} is the wave impedance of the medium.

Using the vector identity $\vec{a}\times\vec{b}\cdot\vec{c} = \vec{c}\times\vec{a}\cdot\vec{b}$, the term $\vec{E}_1\times\vec{H}_2 - \vec{E}_2\times\vec{H}_1$ in the surface integral of eq. 2.173 can be written as

$$\left(\vec{E}_1 \times \vec{H}_2 - \vec{E}_2 \times \vec{H}_1\right) \cdot d\vec{S}' = \left(\vec{E}_1 \times \vec{H}_2 - \vec{E}_2 \times \vec{H}_1\right) \cdot \hat{r} dS'$$

$$= \left(\hat{r} \times \vec{E}_1 \cdot \vec{H}_2 - \hat{r} \times \vec{E}_2 \cdot \vec{H}_1\right) dS' = Z_{medium} \vec{H}_1 \cdot \vec{H}_2 - Z_{medium} \vec{H}_2 \cdot \vec{H}_1 = 0$$

This last equality considers that at large distances from the sources $\vec{E}_j = Z_{medium}$ $\hat{r} \times \vec{H}_j$ (for $j=1, 2$).

Use of this last result in eq. 2.173 leads to

$$\int \left(\vec{J}_1 \cdot \vec{E}_2 - \vec{J}_2 \cdot \vec{E}_1\right) dV' = 0 \text{ or } \int \vec{J}_1 \cdot \vec{E}_2 \, dV' = \int \vec{J}_2 \cdot \vec{E}_1 \, dV'$$

The left side of eq. 2.174 is defined as the *reaction* of the electric field $\vec{E}_1$ to the source $\vec{J}_2$. Similarly, the right side of this equation is defined as the *reaction* of the electric field $\vec{E}_2$ to the source $\vec{J}_1$. It must be noted that these quantities have the dimension of power and represent the effect of each source current at one point on the corresponding electric field that produces at another point. Equation (2.174) corresponds to the condition of reciprocity previously discussed at the beginning of this section and in the following example, we will illustrate the close relation with the reciprocity of electrical circuits.

EXAMPLE 2.16

Let us consider two antennas operating as a transmitter and a receiver. We can regard the terminals of the two antennas as the terminal of a two-port obeying the following relation between the currents and voltages at each terminal:

$$V_1 = Z_{11} I_1 + Z_{12} I_2$$

$$V_2 = Z_{21} I_1 + Z_{22} I_2$$

In these relations, (V_1, I_1) and (V_2, I_2) are the current and voltage corresponding to terminals 1 and 2, respectively. The coefficients (Z_{11}, Z_{22}) and (Z_{12}, Z_{21}) are associated to self and mutual impedances, respectively. This shows that the condition of reciprocity expressed by eq. 2.174 implies that (Z_{12}, Z_{21}).

SOLUTION

Assuming that antennas 1 and 2 are excited with terminal currents I_1 and I_2 and that no magnetic sources are present. From eq. 2.174, it can be readily obtained that the condition of reciprocity for this case can be written as

$$V_1^{oc} I_1 = V_2^{oc} I_2$$

where $V_1^{oc} = -\int \vec{E}_2 \cdot d\vec{l}$ is the open circuit voltage at the terminal corresponding to antenna 1 due to the electric field produced by the antenna 2, and similarly V_2^{oc} is the open circuit voltage at the terminal corresponding to antenna 2 due to the electric field produced by the antenna 1.

Since $V_1^{oc} = Z_{12} I_2$ and $V_2^{oc} = Z_{21} I_1$ we conclude that the condition of reciprocity leads to the equality Z_{12}, Z_{21}.

2.12 APPLICATION OF TIME REVERSAL FOR LOCATING LIGHTNING DISCHARGES

In this section, we show how time-reversal invariance of electromagnetic waves can be applied to specific situations of the electrical engineering. One of these situations corresponds to the location of a lightning discharge which is a typical issue of interest for studies and projects related with the design and operation of electrical transmission lines.

2.12.1 A BRIEF DESCRIPTION OF THE LIGHTNING DISCHARGE PHENOMENON

We give here, a brief description about the mechanism of production of electric discharges as can be found in several publications. For example, a clear and brief description can be found in Ref. [38].

The lightning is a natural phenomenon generated during thunderstorm by electrostatic discharge which produces an electric current passing through the lightning channel.

Lightning is a large-scale natural spark discharge that can occur between the atmosphere and the earth's surface. On discharge, a highly electrically conductive plasma channel is created within the air, and when current flows within this channel, it rapidly heats the surrounding air. Lightning occurs when there is formation and separation of positive and negative electric charges within the atmosphere that creates the highly intensive electric field needed to support this natural spark discharge.

The formation of electric charges in the atmosphere is due mainly to the ionization of air molecules by cosmic rays. Cosmic rays are high-energy particles such as protons that originate from outside the solar system. On colliding with air molecules, they produce a shower of lighter particles, some of which are charged.

Within a thundercloud, the rapid upward and downward movement of water droplets and ice crystals can separate and concentrate these charges. The negative charges accumulate at the bottom part of the cloud and the positive charges toward the top.

As the area of negative charge at the base of the thundercloud builds up, it induces a region of positive charge on the ground below. As a result of this, a potential difference or voltage is created across the cloud-to-ground gap. Once the voltage reaches a certain strength, the air between the base of the cloud and the ground develops an electrical conductivity. At first a channel, known as a stepped leader, is formed. Although invisible to the naked eye, this allows electrons to move from the cloud to the ground. The stepped leader travels with a slight pause in between, to the ground and as it nears the ground, a positively charged streamer fires upward from the ground to connect with it. Streamers are most often initiated from tall objects on the ground. A stepped leader grows down from the cloud until it contacts a leader moving up from the ground. A return stroke is then generated.

Once connected, electrons from the cloud can flow to the ground and positive charges can flow from the ground to the cloud. It is this flow of charge that is the visible lightning stroke.

After the first discharge, it is possible for another leader to form down the channel. Once again, a visible lightning stoke is seen. This can happen 3–4 times in quick succession. All of this happen in a time interval of about 200 milliseconds.

2.12.2 LOCATING LIGHTNING DISCHARGES BASED ON TIME-REVERSAL SYMMETRY OF ELECTROMAGNETIC FIELDS

An application of time-reversal symmetry of electromagnetic fields in the ideal case where dissipation effects can be neglected in a first approach, is the electromagnetic time reversal (EMTR) method to locate lightning strikes. This method is presented in several references, for example, Ref. [39].

In the EMTR method, the fields recorded at each sensor are flipped over in time, retransmitted using the sensors as emitters and propagated back by simulation. After having recorded the field waveform, each sensor injects the time-reversed field back into the same medium. Therefore, the back propagation will be characterized by the same geometric ($1/r$) dependence as in the direct propagation. As a result, the contribution from each sensor might become prohibitively small beyond some critical distances, leading to numerical errors. To prevent this effect and also to remove the singularities arising at the positions of the sensors (during back propagation), it is convenient to keep the field amplitude constant while the field back propagates, as explained in Ref. [40].

As a result, the back propagation and the propagation models are not exactly the same. Hence, the contribution of each sensor n to the time-reversed field in this first model is

$$\vec{E}_{TRn}(\vec{r},t) = \vec{E}_n\left(T - t - \frac{|\vec{r} - \vec{r}_n|}{c}\right) \tag{2.175}$$

where T is the duration of the signal. This means that only the interval $0 < t < T$ is considered.

Let us consider the case of a lightning stroke impacting on a point $\vec{r}_s$ as sketched in Figure 2.9. Several sensors are located at points designated as $\vec{r}_n$ with $n = 1, 2, \ldots\ldots, N$.

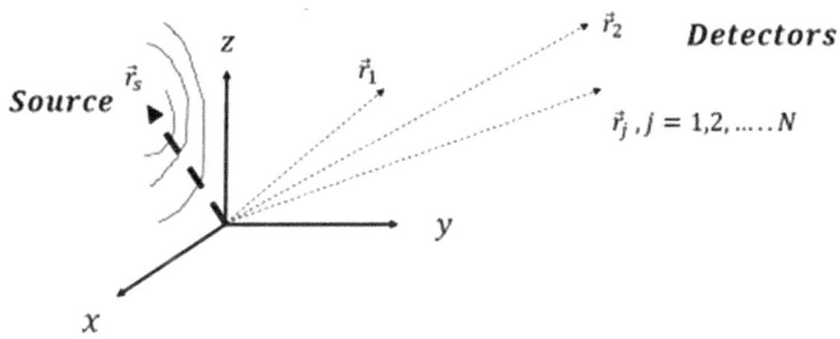

FIGURE 2.9 Position of the impact point of a lightning stroke and N detectors of the corresponding electromagnetic wave that it is produced.

Therefore, the total back propagated field is equal to

$$\vec{E}_{TR}(\vec{r},t) = \sum_{n=1}^{N} \vec{E}_{TRn}(\vec{r},t) = \sum_{n=1}^{N} \frac{\vec{\varphi}\left(T - t + \frac{|\vec{r}_n - \vec{r}_s|}{c} - \frac{|\vec{r} - \vec{r}_n|}{c}\right)}{|\vec{r}_n - \vec{r}_s|} \tag{2.176}$$

where $\vec{\varphi}$ is a vector function depending on the electric field at the source point $\vec{r}_s$. For this case, $\vec{\varphi}$ corresponds to the radiation component of the vertical electric field normalized to a distance of 1 m and its magnitude is given by

$$|\vec{\varphi}(t)| = -\frac{\upsilon}{2\pi\varepsilon_o c^2} i_s(t) \tag{2.177}$$

where

$i_s(t)$ is the channel-base lightning current
υ is the return stroke speed
c is the light velocity
ε_o is the dielectric permittivity of vacuum

According to Refs. [41 and 42], a typical lightning discharge current has the form shown in Figure 2.11 and an analytical expression for this current is of the form:

$$i_s(t) = \frac{i_{max}}{\eta}\left[\frac{f(t)+(t/T)^n}{1+(t/T)^n}\right]y(t) \tag{2.178}$$

where $f(t) = sin^2\omega t$ and $y(t) = e^{-t/\tau}$, with η, τ, n adjustable parameters chosen to adapt to the registered current which can have the form shown in Figure 2.10.

The current waveform can be described by the following equation[41,42]]

$$i(t) = \frac{i_{max}}{\eta}\frac{(t/T_f)^{10}}{1+(t/T_f)^{10}}exp(-t/\tau) \tag{2.179}$$

The parameters of this equation for the first and subsequent strokes are given in the following table taken from Ref. [42]:

TABLE 2.1 Parameters for Lightning Current Waveform.

Parameter	Symbol	First stroke	Subsequent strokes
Front time coefficient (μs)	*Tf*	19.0	0.454
Decay time coefficicient (μs)	τ	485	143
Correction factor	η	0.93	0.993

From eq. 2.179 and using the parameters for the first stroke given in Table 1, Figure 2.10 shows the waveform of the lightning strike current considering a peak value of 100 kA.

At the strike point, $\vec{r} = \vec{r}_s$ and eq. 2.176 leads to

$$\vec{E}_{TR}(\vec{r},t) = \varphi(T-t)\sum_{n=1}^{N}\frac{1}{|\vec{r}_n - \vec{r}_s|} \tag{2.180}$$

It can be seen that at the point source, the back-propagated field has the maximum value if the contributions of the different sensors are in phase, meaning that they interfere constructively. This is the condition used to locate the fault and implies that the phase term in eq. 2.176 should be equal for all the sensors. This condition can be expressed as:

$$T + \frac{|\vec{r}_n - \vec{r}_s|}{c} - \frac{|\vec{r} - \vec{r}_n|}{c} = K \qquad (2.181)$$

where K is a constant.

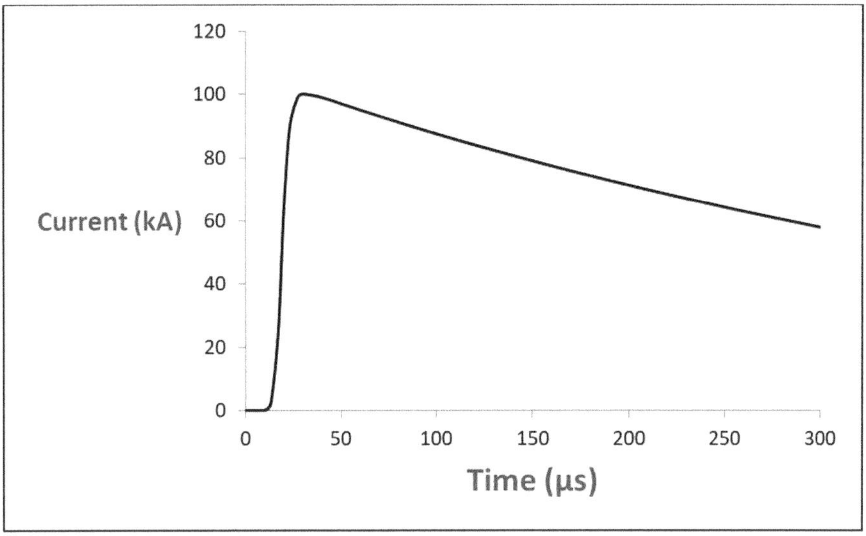

FIGURE 2.10 Waveform of the current of a lightning strike with a peak of 100 kA, according to the model proposed in standard IEC 61312-1.[42]

Rearranging eq. 2.181 and writing it for two sensors locates at sites $\vec{r}_i$ and $\vec{r}_j$ we obtain

$$\frac{|\vec{r}_i - \vec{r}_s|}{c} - \frac{|\vec{r}_j - \vec{r}_s|}{c} = \frac{|\vec{r} - \vec{r}_i|}{c} - \frac{|\vec{r} - \vec{r}_j|}{c} \qquad (2.182)$$

The left side of this last equation can be recognized as the difference in time of arrival (ToA) of the signal from the strike point to the sensors i and j, which can be determined from the waveforms recorded at those two sensors. The right-hand side represents the difference in ToA of a signal radiated from a general point $\vec{r}$ toward the same two sensors. This equation can be used to locate a disturbance by means of the ToA technique, as shown in the following illustrative example:

Let us consider the case of a fault occurring in a transmission line as shown in Figure 2.11. This fault originates electromagnetic waves traveling toward both terminals.

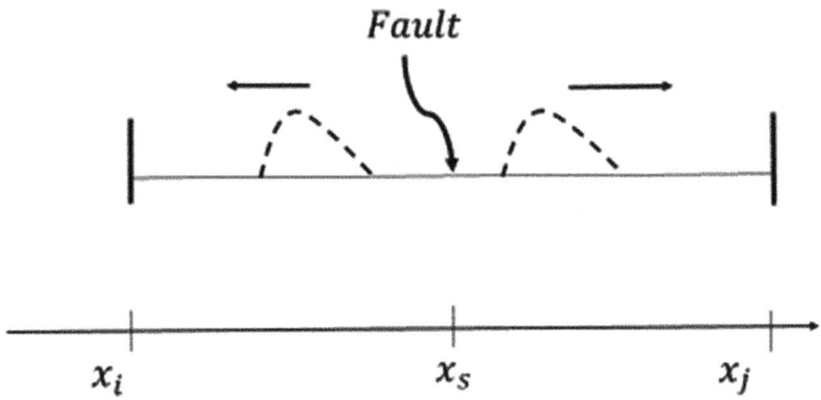

FIGURE 2.11 Fault occurring in a transmission line between two points where detectors are placed at $x = xi$ and $x = xj$.

Figure 2.11 shows a disturbance occurring at a point placed at position x_s which originates two pulses traveling toward detectors placed at positions xi and xj. In this case, we can make a one dimensional analysis and the corresponding arrival times to these detectors are:

$$t_i(x_s) = \frac{|\vec{r}_i - \vec{r}_s|}{c} = \frac{x_s - x_i}{c} \tag{2.183}$$

$$t_j(x_s) = \frac{|\vec{r}_j - \vec{r}_s|}{c} = \frac{x_j - x_s}{c} \tag{2.184}$$

Combining these last two equations, we get that the difference in the arrival times of the signal emitted at a source located at position x to the detectors placed at positions x_i and x_i is given by

$$\Delta t_{ij}(x_s) = t_i(x_s) - t_j(x_s) = \frac{2x_s}{c} - \frac{2x_i + L}{c} \tag{2.185}$$

Therefore, the location of the source is given by

$$x_s = \frac{c\Delta t_{ij}(x_s)}{2} + x_i + \frac{L}{2} \tag{2.186}$$

EXAMPLE 2.17

Consider a lightning strike with a peak value of 100 kA falling between two points P1 and P2 separated by a distance L = 300 Km. At each of these points, there is a detector and a time-reversal system.

a) If the difference between arrival times to the detectors is 100 μs, determine the impact point of the lightning stroke.
b) Plot the waveforms of the electric field measured by the detectors before the corresponding time reversal.
c) Plot the waveform of the time-reversed electric field returned to the source point.

SOLUTION

a) The impact point can be found directly from eq. 2.186 making $x_1 = 0$ and assuming that L is the distance in straight line between the sources, so that:

$$x_s = \frac{c \cdot t_{ij}(x_s)}{2} + x_1 + \frac{L}{2} = \frac{3 \cdot 10^5 \cdot 100 \cdot 10^{-6}}{2} + 0 + \frac{300}{2} = 165\, Km$$

Therefore, the impact point is at 165 Km from detector 1 and at 135 Km from the detector 2.

b) The electric field measured by the detectors is obtained considering that at time t, the detected signal corresponds to that produced by the source at the previous instant $t - \frac{|\vec{r}_j - \vec{r}_s|}{c}$ where $\vec{r}_j$ denotes the vector position of each detector (j = 1, 2 in this case). The magnitudes of these fields are obtained as

$$\left|\vec{E}_{Dj}(t)\right| = \frac{\phi_s\left(t - \frac{|\vec{r}_j - \vec{r}_s|}{c}\right)}{|\vec{r}_j - \vec{r}_s|} u\left(t - \frac{|\vec{r}_j - \vec{r}_s|}{c}\right) \tag{2.187}$$

where

$$\phi_s\left(t - \frac{|\vec{r}_j - \vec{r}_s|}{c}\right) = \frac{v}{2\pi\varepsilon_o c^2} i_s\left(t - \frac{|\vec{r}_j - \vec{r}_s|}{c}\right) \tag{2.188}$$

The step function in eq. 2.187 considers that the electric field cannot be recorded in each detector before the corresponding arrival time. Using eq. 2.179 for representing the wave form of the current at instant $t-\frac{|\vec{r}_j-\vec{r}_s|}{c}$, we have

$$i_s\left(t-\frac{|\vec{r}_j-\vec{r}_s|}{c}\right)=\frac{I_s}{\eta}\frac{\left(t-\frac{|\vec{r}_j-\vec{r}_s|}{c}\right)^{10}}{\left(t-\frac{|\vec{r}_j-\vec{r}_s|}{c}\right)^{10}+T^{10}} \tag{2.189}$$

where I_s is the peak value of the current corresponding to the lightning strike.

The waveform of the electric fields at each detector has a waveform given by the combination of eqs. 2.187–2.189. The corresponding plot is shown in Figure 2.12 and it can be seen that the waveforms are similar but the intensity of the electric field recorded by detector 1 is lower than that corresponding to detector 2 considering the difference in distances from the source.

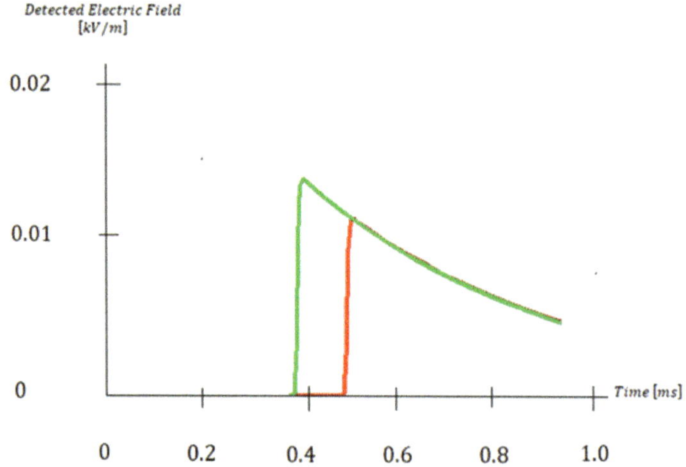

FIGURE 2.12 Waveforms of the electric fields measured by detector 1 (red curve) and detector 2 (green curve). The electric field is in kV/m and time is in seconds. The parameters considered are: $I_s = 100\ kA$, $T_f = 19\ \mu s$, $\tau = 485\ \mu s$, and $\eta = 0.93$.

c) The waveform of the time-reversed electric field returned to the source point is shown in Figure 2.13 (blue curve). For comparison, the curves of Figure 2.13 are shown again. This field is obtained from eq. 2.180 which for this case leads to

$$\vec{E}_{TR}(\vec{r}_s) = \phi_s(T-t)\{1-u(t-T_{observ})\}\sum_{j=1}^{2}\frac{u\left(t-\frac{|\vec{r}_j-\vec{r}_s|}{c}\right)}{|\vec{r}_j-\vec{r}_s|} \tag{2.190}$$

where

$$\phi_s(T-t) = \frac{\nu}{2\pi\varepsilon_o c^2} I_s exp\left[-(T-t)/\tau\right]\frac{(T-t)^{10}}{T^{10}+(T-t)^{10}} \tag{2.191}$$

In eq. 2.190, the factor $1 - u(t - T_{observ})$ allows to represent the vanishing of the electric field for a time of observation T_{observ} long enough so that the perturbation has decayed exponentially.

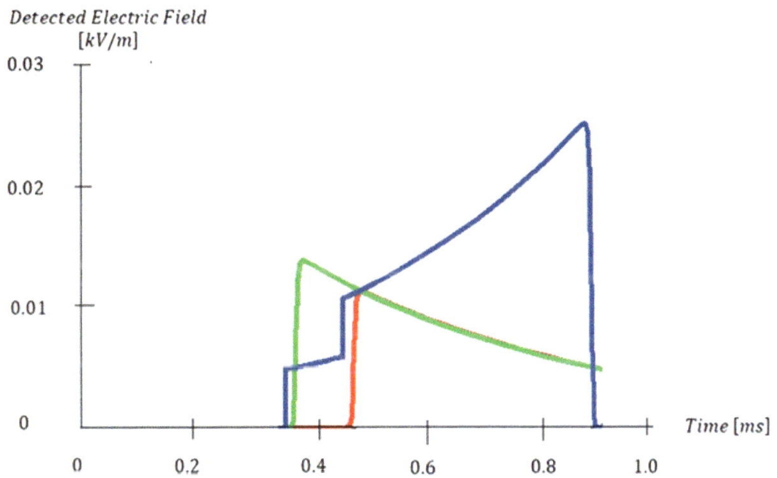

FIGURE 2.13 The blue curve shows the time reversed electric field (kV/m) at the source point. The red and green curves show the electric field measured, respectively, by detectors 1 and 2 before being time-reversed.

2.13 CONCLUSIONS

In this chapter, we have analyzed the effect of time reversing the equations ruling the behavior of electromagnetic fields. The main conclusions can be summarized as follows:

- The equations describing the electromagnetic field in empty space (without charges and current external sources) are symmetric under time reversal if in addition to make the change $t \to -t$, the magnetic field vectors $\vec{B}$, $\vec{H}$ change to $-\vec{B}$, $-\vec{H}$, respectively.
- In the absence of dissipation, the propagation of an electromagnetic wave is symmetric under time reversal and Maxwell's equations which allow the propagation of a retarded and an advanced solution.
- For a medium where dissipation can be considered weak, the advanced solution of Maxwell's equations can be used in engineering techniques for locating sources of electromagnetic waves.
- Considering dissipation, the advanced solution of Maxwell's equation represents a wave propagating toward the source with increasing amplitude. This solution is mathematically possible but its probability of occurrence makes it physically impossible, due to the enormous number of microscopic conditions to be satisfied.
- Equations of *macroscopic* electrodynamics are not time-reversal invariant when the system has dissipative elements. It must be considered that a description based on a model which only includes the electromagnetic degrees of freedom is incomplete.
- In contrast, in a microscopic formalismwith all the light and matter degrees of freedom included in the analysis—the system dynamics is time-reversal symmetric. Thus, in some senses, macroscopic dissipative systems (e.g., lossy dielectrics) have a *hidden time-reversal symmetry*.

KEYWORDS

- **electromagnetic field**
- **electromagnetic wave**
- **dielectric absorption**
- **dispersion**
- **radiation**
- **superluminal propagation**
- **causality**

REFERENCES

1. Lewis, G. N. The Symmetry of Time in Physics. Science **1930,** 71 (1849), 569–577.
2. Frisch, M.; Pietsch, W. Reassessing the Ritz–Einstein Debate on the Radiation Asymmetry in Classical Electrodynamics. *Stud. Hist. Phil. Modern Phys.* **2016,** 55, 13–23.
3. Ritz, W.; Einstein, A. Zum gegenwärtigen Stand des Strahlung-Problems. *Physikalische Zeitschrift* **1909,** *10* (9), 323–324.
4. Popper, K. Time's Arrow and Entropy. *Nature* **1965,** *2007*, 233–234.
5. Penrose, O.; Percival, I. C. The Direction of Time. *Proc. Phys. Soc.* **1962,** *79*, 605–616.
6. Wheeler, J. A.; Feynman, R. P. Interaction with the Absorber as the Mechanism of Radiation. *Rev.Mod. Phys*. **1945,** *17*, 157–181.
7. Hogarth, J. E. Cosmological Considerations of the Absorber Theory of Radiation. *Proc. R. Soc. Lond. Ser A Math. Phys. Sci.* **1962,** *267* (1330), 365–383.
8. Dirac, P. A. M. Classical Theory of Radiating Electrons. *Proc. R. Soc. Lond. Ser. A Math. Phys. Sci.* **1938,** *167*, *929*, 148–169.
9. Earman, J. *Sharpening the Electromagnetic Arrow(s) of Time*, edited by Craig Callender; The Oxford Handbook of Philosophy of Time, 2011.
10. Jackson, J. D. *Classical Electrodynamics*; John Wiley and Sons, Inc., 3rd edition, 1999.
11. Nunes, F. D.; Borges, B. V.; Weiner, J. Analysis of Dispersive and Dissipative Media with Optical Resonances. *Optics Express* **2012,** *14*, 15679.
12. Deák, L.; Fülöp, T. Reciprocity in Quantum, Electromagnetic and Other Wave Scattering. *Ann. Phys. 327*, 1050–1077.
13. Caloz, C.; Alù, A.; Tretyakov, S.; Sounas, D.; Achouri, K.; Deck-Léger, Z-L. Electromagnetic Nonreciprocity. *Phys. Rev. Appl.* **2018,** *10*, 047001.
14. Silveirinha, M. G. Hidden Time-Reversal Symmetry in Dissipative Reciprocal Systems. *Phys. Rev. Appl.* **2018,** *10*, 047001.
15. Claro, F.; Robles,P.; Rojas, R. Laser Induced Dynamics of Interacting Small Particles. *J. Appl. Phys.* **2009,** *106*, 084311.
16. Fuchs, R.; Claro, F. Enhanced Nonconservative Forces Between Polarizable Nanoparticles in a Time-Dependent Electric Field. *Appl. Phys. Lett.* **2004,** *85*, 3280–3282.
17. Claro, F.; Fuchs, R.; Robles, P.; Rojas, R. The Effect of Dissipation on the Torque and Force Experienced by Nanoparticles in an AC Field. *Int. J. Modern Phys. B* **2014,** 28, 1450240.
18. Robles, P.; Claro, F.; Rojas, R. Dynamical Response of Polarizable Nanoparticles to a Rotating Electric Field. *Am. J. Phys.* **2011,** *79*, 945–949.
19. Veselago, V. The Electrodynamics of Substances with Simultaneously Negative Values of $\in$ and μ. *Soviet Phys. Uspekhi* **1968,** *10*, 509–514.
20. Chen, J.; Wang, Y.; Jia, B.; Geng, T.; Li, X.; Feng, L.; Qian, W.; Liang, B.; Zhang, X.; Gu, M.; Zhuang, S. Observation of the Inverse Doppler Effect in Negative-Index Materials at Optical Frequencies. www.nature.com/naturephotonics,2011
21. Veselago, V.; Braginsky, L.; Shklover, V.; Hafner, C. Negative Refractive Index Materials. *J. Comput. Theor. Nanosci.* **2006,** *3*, 1–30.
22. Lee, S. H.; Park, C. M.; Seo, Y. M.; Wang, Z. G.; Kim, C. K. Reverse Doppler Effect of Sound. *Phys. Rev. Lett.* **2010,** *104*, 054301.
23. Robles, P. Exploring Superluminal Transmission of Electromagnetic Waves Through Dielectric Barriers and Causality: A Pedagogical Insight. *Eur. J. Phys.* **2014,** 35, 065022.

24. Young, H. D.; Freedman, R. A.; Ford, L. *University Physics*, Vol. 2; Pearson Addison-Wesley: Reading, 2007.
25. Winful, H. G. The Meaning of Group Delay in Barrier Tunneling: A Re-Examination of Superluminal Group Velocities. *New J. Phys.* **2006,** *8* 101.
26. Cramer, J. G. The Arrow of Electromagnetic Time and the Generalized Absorber Theory. *Found. Phys.* **1983,** *13*, 887.
27. Griffiths, D. J. *Introduction to Electrodynamics*, 3rd ed.; Prentice Hall: Upper Saddle River, 1999; p 425.
28. Wheeler, J. A.; Feynman, R. Interaction with the Absorber as the Mechanism of Radiation. *Rev. Modern Phys.* **1945,** *17*, 157–181.
29. Cramer, J. G. Generalized Absorber Theory and the Einstein-Podolsky-Rosen Paradox. *Phys. Rev. D* **1980,** *22*, 362–376.
30. Robles, P.; Pizarro, F. *Photon Propagation Through Dispersive Media, Chap. 4 of Book Wave Propagation Concepts for Near-Future Telecommunication Systems*; Intech, 2017.
31. Kobe, D. H. A Relativistic Schrödinger-like Equation for a Photon and Its Second Quantization. *Found. Phys.* **1999,** *29*, 1203–1231. DOI: 10.1023/A:1018855630724.
32. Feynman, R. P. The Theory of Positrons. *Phys. Rev.* **1949,** *76*, 749–759.
33. Qin, Y-P. A Cosmological Model with Negative Energy Photons. *Chin. Phys. B* **2010,** 19, 019803.
34. Henry-Couannier, F. Negative Energies and Time Reversal in Quantum Field Theory. *Glob. J. Sci. Front. Res. Math. Decision Sci.* **2012,** *12* (13) Version 1.0.
35. Caloz, C.; Alù, A.; Tretyakov, S.; Sounas, D.; Achouri, K.; Deck-Léger, Z-L. Electromagnetic Nonreciprocity. *Phys. Rev. Appl.* **2018,** *10*, 047001.
36. Asadchy, V. S.; Mirmoosa, M. S.; Díaz-Rubio, A.; Fan, S.; Tretyakov, S. Tutorial on Electromagnetic Nonreciprocity and Its Origins. https://arxiv.org/abs/2001.04848
37. Orfanidis, S. J. Electromagnetic Waves and Antennas, 2009. http://www.ece.rutgers.edu/~orfanidi/ewa
38. Dvyer, J. R.; Uman, M. The Physics of Lightning. *Phys. Rep.* **2014,** 534, 147–241, https://www.sciencelearn.org.nz/resources/239-lightning-explained
39. Mora, N.; Rachidi, F.; Rubinstein, M. Locating Lightning Using Time Reversal of Electromagnetic Fields. 30th International Conference on Lightning Protection—ICLP, 2010. Cagliari, Italy, September 13th–17th, 2010.
40. Lugrin, G.; Mora, N.; Rachidi, F.; Rubinstein, M.; Diendorfer, G. On the Use of the Time Reversal of Electromagnetic Fields to Locate Lightning Discharges. In *Proc. Int. Conf. Lightning Protection* 2012; pp 1–4.
41. Heidler, F.; Cveti`c, J. A Class of Analytical Functions to Study the Lightning Effects Associated with the Current Front. *ETEP* **2002,** *12* (2).
42. IEC 62305-1, International @ Protection Against Lightning—Part 1: General Principles, Edition 2.0; 2010.

CHAPTER 3

Time in Non-Relativistic Quantum Mechanics

ABSTRACT

Time is used in non-relativistic quantum theory as a classical concept. On the one hand, time is the parameter entering Schrödinger's equation and measured by an external laboratory clock. But it could be considered that time also plays the role of an observable when a given event occurs (e.g., when a particle crosses through a potential barrier), and one would expect that like every observable, it should be represented in the theory by an operator whose properties are predictors of the outcome of time measurements made on physical systems. Several time operators have been proposed in different publications, but still, there is not a widely accepted consensus.

That subject, among others concerned with time in quantum mechanics, is analyzed in the present chapter, where we focus on the analysis of time-reversal symmetry in non-relativistic quantum systems. The search for symmetries in Physics helps to study the structure of theories. Briefly speaking, a symmetry is a transformation that does not change the structure of the theory.

We start with the analysis of Schrödinger's Equation verifying that for obtaining a time-symmetric description of the dynamics of a particle, the operation of time-reversal must include conjugation represented by the action of an antiunitary operator acting over a wave function. In particular, we analyze quantum systems whose evolution is determined by a set of external parameters varying cyclically, discussing how even though these parameters return to their initial values, the state of the system acquires a geometric phase, which depends on the path followed by the external parameters in the parameter space and which in general can be associated with the lack of the time-reversal symmetry of the system.

Looking at Time from a Physics Perspective. Patricio Robles (Author)

Another of the issues covered in this chapter is that related to the claim to have a Hermitian Hamiltonian in a quantum mechanical system, which can be relaxed if the Hamiltonian is invariant under the combined *PT*-symmetry, where *P* denotes the parity operation and *T* denotes time reversal.

Although the topics included in this chapter are covered in several specialized papers, our objective is to present them in a form as simple as possible and complemented with several illustrative examples. One appendix with additional examples is included and also other appendixes which provide more details about the concepts of quantum mechanics, quantization of the electromagnetic field, and analysis of the application of time-reversal operation to the phenomenon known as revival of Rabi oscillations in a quantum two-level system.

3.1 INTRODUCTION

In Chapters 1 and 2, we have discussed the effect of time reversal on the description of a physical system if its dynamics is ruled by Newton's equation and Maxwell's equations. In doing so, we have considered that the dimensions of the particles and the volume they occupy is such that a macroscopic description is adequate so that the equations of classical mechanics and electrodynamics are valid. We have shown that in the presence of dissipation, the classical description is not symmetric under time-reversal but the absence of dissipation does not necessarily secure time-reversal symmetry, as, for instance, occurs in the case of a charged particle subjected to the action of an externally applied magnetic field.

Nevertheless, as the involved dimensions and distances between particles become smaller than some nanometers, the relevance of quantum effects increases and cannot be neglected. If the velocities with which the considered particles move are very small compared with light velocity, the dynamics of each particle can be adequately described by Schrödinger's equation and the laws of non-relativistic quantum mechanics (NRQM). Appendix 3.1 contains a summarized explanation of the principles of NRQM and the corresponding formalism. Figure 3.1 shows a qualitative scheme of the range of validity of classical, quantum, and relativistic physics according to the dimensions and magnitude of the involved velocities. For dimensions lower than a few nanometers and velocities much lower than light velocity *c*, predictions of classical physics are not valid and the behavior of a system must be described by the laws of quantum mechanics. For the case of a particle like an electron, if its velocity is not negligible with respect to *c*, it is necessary to use a description based on quantum field theory, which is out of the scope of this book. On the other hand, the dynamics of objects with macroscopic dimensions, but moving with a velocity close to light velocity must be described according to special relativity theory. In addition, if gravitational effects are not negligible, the analysis must be made using the general relativity theory.

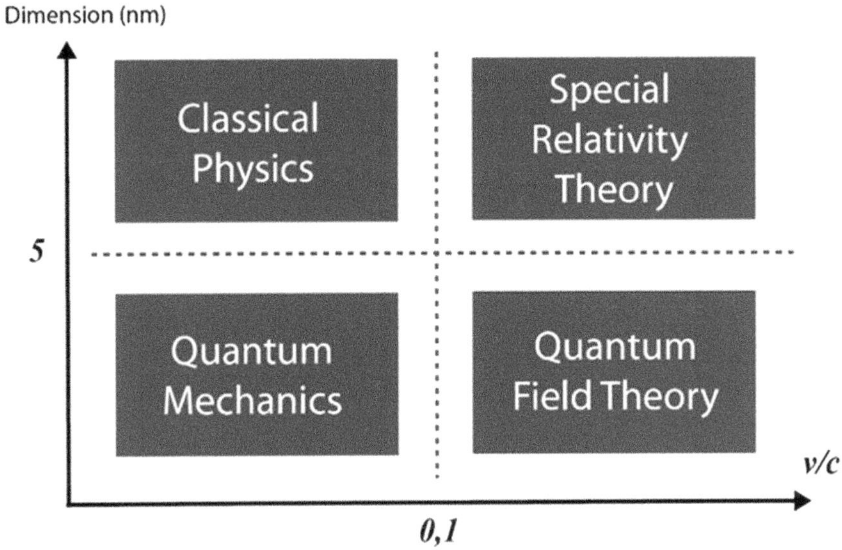

FIGURE 3.1 Range of validity of the current physical theories considering the dimensions and the velocity of the components of a physical system with respect to the light speed *c*.

In this chapter, we analyze the symmetry of a quantum system if it is subject to a time-reversal transformation. As we have said, symmetries are valuable tools since they can be used as guides to look into the structure of the physical theories. In simple terms, symmetry can be regarded as a transformation that leaves invariant the structure of the theory. The question if a theory considers a fundamental direction of time, is equivalent to asking whether such a physical theory provides solutions in both directions of time, independently of the particular initial or boundary conditions. The fact that a physical theory has a fundamental, privileged temporal direction may be formally represented by means of a symmetry transformation. There are two notions which are relevant to understand the directed nature of time (or the nonexistence of directness) expressed in terms of physical symmetries: the first one is the "time reversal," as the symmetry transformation that changes the direction of time, and that of "time-reversal invariance," as the symmetry which results from the former. Roughly speaking, time-reversal symmetry means that the evolution and properties of a system do not change if in the corresponding dynamical equations, the parameter t ordering the sequence of occurrence of the events is changed by $-t$. Time-reversal invariance of a process means that the properties of a system are independent of the instant of observation.

Can we say that our fundamental physical laws are invariant under time reversal? According to the profuse quantity of publications related to this question, it is widely accepted that the predictions of our fundamental physical laws are actually invariant under time reversal, that is, it is not possible to identify in a fundamental sense the difference between both direction of time: from past to future and from future to past.

The objective of this chapter is to provide a pedagogical review of this question in the context of the non-relativistic quantum mechanical theory. The concepts of quantum mechanics here used are based on well-known books used in the first semesters of postgraduate programs in Physics.[1–3]

We start with the analysis of Schrödinger's equation verifying that for obtaining a time-symmetric description of the dynamics of a particle, the operation of time-reversal must include conjugation represented by the action of an antiunitary operator acting over a wave function (Sections 3.2.2 and 3.2.3). The meaning of antiunitary operators is explained in Section 3.2.3. The invariance of one of the basic relations of quantum mechanics under time reversal is explained in Section 3.2.4, as is the case of the uncertainty principle momentum-position. In Section 3.2.5, it is explained that for systems with half-integer total angular momentum, time-reversal symmetry implies that a given eigenstate and its reversal are different states but sharing the same energy, what is called Kramers degeneracy. The breaking of time-reversal symmetry due to the application of a magnetic field to a quantum system is analyzed in Section 3.3.

In Section 3.4, we analyze the time reversal of the photoelectric effect which was a crucial step toward the theoretical formulation of quantum mechanics and that gave rise to the concept of photon or "quantum of light" introduced by Albert Einstein. As an analogy, we discuss the time reversal of the dynamics of a mechanical system which presents a formal analogous behavior.

In Section 3.5, we present the time-reversal operator for systems with spin, which belongs to a class of operators that as a subset includes those that when repeated, restore the original state. We analyze the result of the consecutive application of the time-reversal operator for systems containing an even or an odd number of particles with spin ½.

In Section 3.6, we analyze quantum systems whose evolution is dependent on external parameters changing cyclically. In that case, in addition to the dynamic phase, the state of the system acquires a geometric phase, which depends on the path followed by the parameters in the parameter space. Even though the parameters return to the same initial values, the wave function that describes the system acquires a geometric phase which depends on the path followed by the external parameters in parameter space. The dynamic phase contains information about how long time the evolution took, while the geometric phase indicates the parametric path followed by the system. In case of a slow cyclic change of the external parameters, the acquired geometric phase is called the Berry phase.

Section 3.7 contains the analysis of dissipation in quantum systems showing how the effect of damping can be considered, in particular for the case of a quantum-damped harmonic oscillator. Considering that any microscopic system is always embedded in some macroscopic environment, so that it is not isolated, the dynamics of a damped oscillator can be analyzed as an effective isolated system which includes the environment represented as a bath that receives the energy that is dissipated in the interaction of the oscillator with the medium. In this equivalent system, there is an energy exchange between the two subsystems, so that at a scale of dimensions such that a quantum description is required, the irreversibility of time evolution can be understood as transitions between the set of states in which the system splits, due to multiple interactions between the subsystems.

Section 3.8 includes an analysis of the entanglement effect resulting from the non-locality implicit in NRQM and which consists that measurements made in two quantum systems, let us say particles like electrons or photons, are correlated independently of the separation between these particles. This effect was discussed by the famous physicists A. Einstein, B. Podolsky, and N. Rosen in a famous paper published in 1935 where arguments were given in order to question the reality of quantum mechanics. The analysis of this effect has been included in the present chapter since apparently, it seems that superluminal transmission of signals would be possible, what as seen in Chapter 2 is not allowed by the special relativity theory and would be against the principle of causality.

In Section 3.9, an analysis of time-reversal symmetry in quantum optics is made giving an example where in the case of interaction of a coherent state of light with a two-level quantum system, there is a collapse and revival of oscillations which have the form of a breakdown of time-reversal symmetry in comparison with the predictions based on a semiclassic model where Rabi oscillations show a symmetry with respect to time.

Section 3.10 contains an analysis of quantum systems with a Hamiltonian containing a complex potential, but giving real values of energy if they are symmetric under a combined application of time-reversal and parity transformations. If this *PT*-symmetry is not broken, the Hamiltonian exhibits all properties of a quantum theory described by a Hermitian Hamiltonian.

Section 3.11 includes the conclusions about the main concepts explained in this chapter.

3.2 TIME-REVERSAL SYMMETRY

3.2.1 SCHRÖDINGER EQUATION

Let us consider one spinless particle of mass m undergoing an interaction with the medium represented by a potential energy $V(\vec{r})$ dependent only on the position. The corresponding Schrödinger's equation is:

$$i\hbar\frac{\partial\psi}{\partial t} = -\frac{1}{2m}\nabla^2\psi + V(\vec{r})\psi \tag{3.1}$$

If time t is replaced by $-t$, it can be verified that $\psi(\vec{r}, -t)$ is not a solution due to the first-order derivative with respect to time. Nevertheless $\psi^*(\vec{r}, t)$ is a solution. This is shown as follows:

Let us assume that the solution of eq. 3.1 is $\psi^*(\vec{r}, t) = \varphi(\vec{r})e^{-\frac{i}{\hbar}Et}$ considering a stationary state with constant energy E. In this case, if we only change t by $-t$ we would have:

$$\psi(\vec{r}, -t) = \varphi(\vec{r})e^{-\frac{i}{\hbar}E(-t)} = \varphi(\vec{r})e^{\frac{i}{\hbar}Et} \tag{3.2}$$

Inserting eq. 3.2 into 3.1 gives

$$-E\varphi(\vec{r}) = -\frac{\hbar^2}{2m}\nabla^2\varphi(\vec{r}) + V(x)\varphi(\vec{r})\varphi(\vec{r}) \tag{3.3}$$

The minus sign at the left side of eq. 3.3 does not agree with Schrödinger's equation since, for instance, for the case of a free particle propagating with momentum $+p$ in the x direction, so that

$$\varphi(x) = e^{\frac{i}{\hbar}px} \tag{3.4}$$

Equation 3.3 would lead to a negative energy = $-p^2/2m$, an incorrect result according to NRQM which predicts only positive energies.

On the contrary $\psi^*(x,-t)$ is a solution as can be verified since for the same case of a free particle propagating along the x direction we would have:

$$\psi(x,t) = e^{\frac{i}{\hbar}(px-Et)} \Rightarrow \psi^*(x,-t) = e^{-\frac{i}{\hbar}px}\, e^{+\frac{i}{\hbar}E(-t)} = e^{\frac{i}{\hbar}(-px-Et)} = e^{-\frac{i}{\hbar}(px+Et)} \tag{3.5}$$

Equation (3.5) corresponds to a wave propagating along the x-axis with negative momentum $-p$ and inserted in eq. 3.1 allows us to obtain the correct expression for the energy of a free particle, as can be readily verified:

$$E = p^2/2m \tag{3.6}$$

Coming back to eq. 3.1, it can be verified that if as expected, the potential energy $V(\vec{r})$ is real, the complex conjugation at both sides of that equation and the substitution $t \rightarrow -t$ lead to

$$-i\hbar\frac{\partial\psi^*(\vec{r},-t)}{\partial(-t)} = -\frac{1}{2m}\nabla^2\psi^*(\vec{r},-t) + V(\vec{r})\psi^*(\vec{r},-t) \Rightarrow$$

$$i\hbar\frac{\partial\psi^*(\vec{r},-t)}{\partial t} = -\frac{1}{2m}\nabla^2\psi^*(\vec{r},-t) + V(\vec{r})\psi^*(\vec{r},-t)$$

Therefore, if $\psi^*(\vec{r},t)$ is a solution of eq. 3.1, $\psi^*(\vec{r},-t)$ also satisfies that equation.

These results suggest that the time-reversal operation should include the conjugation of a complex wave function as analyzed in the next section.

3.2.2 TIME-REVERSAL OPERATOR

In quantum mechanics the time operator $\hat{T}$ is defined so that if it is applied over an state defined by the wave function $\phi(\vec{r},t)$ gives the function $\phi(\vec{r},-t)$. If these functions are associated with kets $|\phi\rangle$ and $|\tilde{\phi}\rangle$, respectively, the action of this operator is expressed as

$$\hat{T}|\phi\rangle = |\tilde{\phi}\rangle \tag{3.7}$$

If $\hat{T}$ acts over a state resulting from the action of another operator $\hat{O}$ over the ket $|\phi\rangle$, that is $\hat{T}(\hat{O}|\phi\rangle)$, this can be expressed as

$$\hat{T}\hat{O}|\phi\rangle = (\hat{T}\hat{O}\hat{T}^{-1})(\hat{T}|\phi\rangle = \tilde{O}|\tilde{\phi}\rangle \tag{3.8}$$

From eq. 3.8 the time reversal of the operator $\hat{O}$ is defined as $\tilde{O} = \hat{T}\hat{O}\hat{T}^{-1}$.

If the operator $\hat{O}$ has time-reversal symmetry $\tilde{O} = \hat{O}$ and therefore the following relation is fulfilled:

$$\hat{T}\hat{O} = \hat{O}\hat{T} \tag{3.9}$$

Equation (3.9) implies that if the operator $\hat{O}$ is symmetric under time-reversal operation, it must commute with the time-reversal operator, that is $\left[T, \hat{O}\right] = 0$.

3.2.3 ANTIUNITARITY OF THE TIME-REVERSAL OPERATOR

In quantum mechanics the way a physical system evolves is determined by the Hamiltonian $\mathcal{H}$. If this evolution is symmetric under time- reversal, we will show that the time-reversal operator T must have the form of the product of two operators:

$$T = UK \tag{3.10}$$

It must be remembered that in quantum mechanics, operators can be represented by matrices whose elements depend on the representation used.

U is an unitary operator meaning that $U^{\dagger}U = I$, where $U^{\dagger}$ is the matrix resulting from conjugating and transposing the elements of U.

K is the complex conjugate operator that results in the complex conjugate of any coefficient that multiplies a given ket.

In this way, the time-reversal operator T is an antiunitary operator considering the following definition:

The transformation T acting over kets $|\alpha\rangle$ and $|\beta\rangle$ such as:

$$|\alpha\rangle \rightarrow |\tilde{\alpha}\rangle = T\,|\,\alpha\rangle, \qquad |\beta\rangle \rightarrow |\tilde{\beta}\rangle = T\,|\,\beta\rangle \tag{3.11}$$

is defined as antiunitary if the following conditions are fulfilled

$$\langle\tilde{\beta}\,|\tilde{\alpha}\,\rangle = \langle\beta|\alpha\rangle^{*} \tag{3.12a}$$

$$T\left(c_1\left|\alpha\right\rangle + c_2\left|\beta\right\rangle\right) = c_1^{*}T\left|\alpha\right\rangle + c_2^{*}T\left|\beta\right. \tag{3.12b}$$

where c_1 and c_2 are in general complex coefficients corresponding to the expansion of a given state as a superposition of two base states $|\alpha\rangle$ and $|\beta\rangle$.

The requirement that T must be an antiunitary operator can be better understood considering the following arguments:

Consider a quantum system whose state at $t = 0$ is $|\alpha\rangle$. At a slightly later time, $t = \delta t$, the system has evolved to the following state[3]:

$$\left|\alpha, t_o = 0; t = \delta t\right\rangle = \left(1 - \frac{i}{\hbar}\mathcal{H}\delta t\right)\left|\alpha\right\rangle \tag{3.13}$$

In the above equation, $\mathcal{H}$ is the corresponding Hamiltonian operator characterizing the evolution of the system. In eq. 3.13, it has been considered the approximation of the evolution operator for a small time interval. As known, the evolution operator U_{ev} for a interval of time between t_o and t is given by

$$U_{ev}\left(t,t_o\right) = exp\left[-\frac{i}{\hbar}\int_{t_o}^{t}\mathcal{H}\left(t\right)dt\right] \tag{3.14}$$

For the evolution operator, we will use here the symbol U_{ev} for avoiding confusion with the antiunitarity operator U associated with the time-reversal operator, given by eq. 3.10.

If the time interval $t - t_o = \in$ is small so that $\frac{\in}{t} \ll 1$ the evolution operator may be expressed as

$$U_{ev}\left(\in\right) = 1 - \frac{i}{\hbar}\mathcal{H}\in \tag{3.15}$$

Let us suppose that instead of equation (3.13), the time-reversal operator is applied at time $t = 0$ and the system is left to evolve under the action of the Hamiltonian $\mathcal{H}$. At time $t = \delta t$, we would have the state given by the ket

$$\left(1 - \frac{i}{\hbar}\mathcal{H}\delta t\right)T\left|\alpha\right\rangle \tag{3.16}$$

If the motion has time-reversal symmetry, we expect that the preceding ket be the same as:

$$T\left|\alpha, t_o = 0; t = -\delta t\right\rangle = T\left(1 - \frac{i}{\hbar}\mathcal{H}\left(-\delta t\right)\right) \tag{3.17}$$

The state given by eq. 3.17 is obtained by considering the ket $|\alpha\rangle$ at instant $t - \delta t$ and then changing the sign of the corresponding momentum $\vec{p}$ as illustrated in Figure 3.2 which for purposes of this section has been adapted from a book frequently used in the first courses of quantum mechanics, as is the case of the one corresponding to reference.[3] This is mathematically expressed as

$$\left(1 - \frac{i}{\hbar}\mathcal{H}\delta t\right)T\left|\alpha\right\rangle = T\left(1 - \frac{i}{\hbar}\mathcal{H}\left(-\delta t\right)\right)\left|\alpha\right\rangle \tag{3.18}$$

If eq. 3.18 is valid for any ket denoted as $|⬚\rangle$ we should have

$$-i\mathcal{H}T\left|⬚\right\rangle = \mathrm{T}i\mathcal{H}\left|⬚\right\rangle \tag{3.19}$$

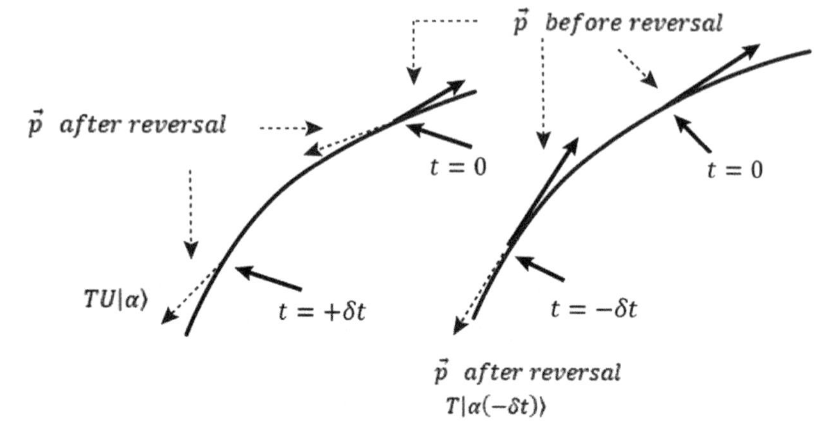

(a) Time reversal at $t = 0$ (b) Time reversal at $t = -\delta t$

FIGURE 3.2 Time reversal and evolution of the state of a particle: (a) Time reversal at $t = 0$ and evolution up to $t = +\delta t$; (b) Time reversal at $t = -\delta t$.

Source: Adapted from Ref. [3].

We could believe that it is possible to cancel the i's in each member of this last equation and so we would obtain

$$\mathcal{H}T = -T\mathcal{H} \tag{3.20}$$

Is this a correct result? For answering this question, let us consider an energy eigenket $|n\rangle$ with energy eigenvalue E_n. From eq. 3.20 we would have

$$\mathcal{H}T\,|n = -T\mathcal{H}\rangle|n\rangle = -E_nT\,|n\rangle \tag{3.21}$$

This equation indicates that $T|n\rangle$ is an eigenket of a Hamiltonian with energy eigenvalues $-E_n$. But this is contradiction with the fact that in NRQM, the energy spectrum is positive semidefinite from 0 to ∞.

We will now show that if the operator T is antiunitary a time-reversal operation gives a result compatible with the postulates of NRQM, specifically that the corresponding eigenvalues of the Hamiltonian $\mathcal{H}$ correspond to positive energies.

If T is an antiunitary operator, it must be considered that its action over any complex number gives its complex conjugate, meaning that i must be changed by $-i$, so that equations (3.18) and (3.19) lead to

$$-i\mathcal{H}T\,|\square\rangle = Ti\mathcal{H}|\square\rangle = -\,iT\mathcal{H}|\square\rangle \tag{3.22}$$

Therefore, the operators $\mathcal{H}$ and T must satisfy the condition

$$\mathcal{H}T = T\mathcal{H} \text{ or } [T,\mathcal{H}] = 0 \tag{3.23}$$

This result implies that a quantum system has time-reversal symmetry if its Hamiltonian commutes with the time-reversal operator T.

It must be noted that eq. 3.23 does not imply that the time-reversal operator commutes with the evolution operator U, that is

$$[T, U_{ev}] \neq 0 \tag{3.24}$$

Equation (3.24) can be proved as follows:

Applying first the evolution operator over a ket $|\square\rangle$ for a small time interval $\in$ and then the time-reversal operator T we obtain:

$$TU_{ev}(\in)|\square\rangle = T\left(1-\frac{i}{\hbar}\in\mathcal{H}\right)|\square\rangle = \left(T+\frac{i}{\hbar}\in T\mathcal{H}\right)|\square\rangle \tag{3.25}$$

On the other hand, if we change the sequence of operations of T and U_{ev} we obtain:

$$U_{ev}(\in)T|\square\rangle = \left(1-\frac{i}{\hbar}\mathcal{H}\in\right)T|\square\rangle = \left(T-\frac{i}{\hbar}\in\mathcal{H}T\right)|\square\rangle \tag{3.26}$$

Comparing equations (3.25) and (3.26), we see that even if T and H commute, $TU_{ev}(\in) \neq U_{ev}(\in)T$.

EXAMPLE 3.1

Verify and discuss eq. 3.24 for the case of a free particle with motion in one dimension described by the wave function:

$$\psi(x,t) = e^{\frac{i}{\hbar}(px-Et)} \tag{3.27}$$

SOLUTION

For $t = 0$: $\psi(x,0) = \psi_o = e^{\frac{i}{\hbar}px}$

If we apply to ψ_o, the combined transformation $U_{ev}T$, where U_{ev} is the evolution operator we get:

$$(U_{ev}T)\psi_o = U_{ev}T\left(e^{\frac{i}{\hbar}px}\right) = U_{ev}\left(e^{\frac{i}{\hbar}px}\right)^* = U_{ev}e^{-\frac{i}{\hbar}px} =$$

$$= e^{-\frac{i}{\hbar}Et} e^{-\frac{i}{\hbar}px} = e^{-\frac{i}{\hbar}(px+Et)} = e^{\frac{i}{\hbar}(-px+E(-t))} \equiv \phi_1(x,t) \tag{3.28}$$

Note that this last function is a plane wave with time running backwards and with the corresponding momentum reversed so that the associated wave moves along the negative x direction.

On the other hand, applying to ψ_o first the evolution operator and then the time-reversal operation, we get

$$(TU_{ev})\psi_o = T\left(e^{\frac{i}{\hbar}(px-Et)}\right) = e^{\frac{i}{\hbar}(-px-E(-t))} \equiv \phi_2(x,t) \tag{3.29}$$

where it has been considered that the action of the time-reversal operator includes conjugation and the change $t \rightarrow -t$. Taking the complex conjugate of the wave function described by eq. 3.27 is equivalent to change the sign of the momentum.

Since $\phi_1(x,t) \neq \phi_2(x,t)$ we verify that $TU_{ev} \neq U_{ev}T$.

Let us analyze if $\phi_1(x,t)$ and $\phi_2(x,t)$ satisfy Schrödinger's equation:

If we insert $\phi_1(x,t) = e^{-\frac{i}{\hbar}(px+Et)}$ and then $\phi_2(x,t) = e^{-\frac{i}{\hbar}(px-Et))}$ in the Schrödinger's equation (3.1) we obtain:

$$i\hbar\frac{\partial\phi_1}{\partial t} = i\hbar\left(-\frac{i}{\hbar}E\right)\phi_1 = E\phi_1$$

$$i\hbar\frac{\partial\phi_2}{\partial t} = i\hbar\left(+\frac{i}{\hbar}E\right)\phi_2 = -E\phi_2$$

Therefore, with $\phi_1(x,t) = (U_{ev}T)\psi_o$, we obtain $E = \frac{p^2}{2m} + V(x)$ which clearly is in agreement with Schrödinger's eq. 3.1, while $\phi_2(x,t)$ is not a correct solution of this equation. This is explained considering that only $\phi_1(x,t)$ corresponds to a correct time-reversed wave function since, as we have seen, starting from the solution $\psi(x,t) = exp\left[\frac{i}{\hbar}(px - Et)\right]$ time-reversal operation implies complex conjugation and the change $t \rightarrow -t$, so that

$$T\{\psi(x,t)\} = exp\left[-\frac{i}{\hbar}(px - E(-t))\right] = exp\left[\frac{i}{\hbar}(-px + E(-t))\right] = \phi_1(x,t)$$

3.2.4 INVARIANCE OF THE UNCERTAINTY PRINCIPLE UNDER TIME REVERSAL

In this subsection, we show that the Uncertainty Principle expressed as $[x_i,p_j] = i\hbar\delta_{ij}$ is invariant under time-reversal operation. As kown, this is a basic principle of quantum mechanics implying that in the specific case of the simultaneous measurement of the momentum and position of a particle, there are always small but finite uncertainties in the results, so that if for example we are making simultaneous measurements of the position of a particle along the x-axis and the component of the momentum parallel to this direction, the product of the uncertainties cannot be less than a minimum value proportional to the Plank's constant. If we can measure with absolute precision one of these quantities, the other will be totally undetermined.

For any ket $|\square\rangle$ we have that the action of the operator $[x_i,p_j]$ gives

$$[x_i,p_j]\,|\square\rangle = i\hbar\delta_{ij}\,|\square\rangle \tag{3.30}$$

where subscripts i and j correspond to components along the directions corresponding to the x, y, and z axes.

Applying the time-reversal operator T to both sides of this last equation and considering that this operation implies change in the sign in the momentum we have

$$\begin{array}{cc} \underline{T\,[x_i,p_j]\,|\square\rangle} & \underline{T i\hbar\delta_{ij}\,|\square\rangle} \\ =[x_i,-p_j]\,T|\square\rangle & =-i\hbar\delta_{ij}\,T|\square\rangle \end{array} \tag{3.31}$$

This last equation leads to

$$[x_i,-p_j]\,T|\square\rangle - -i\hbar\delta_{ij}\,T|\square\rangle \tag{3.32}$$

showing that for the time reversed of any state given by the ket $T|\square\rangle$ the commutation relation $[x_i,-p_j] = i\hbar\delta_{ij}$ is preserved if T is an antiunitary operator.

3.2.5 KRAMERS DEGENERACY AND TIME-REVERSAL SYMMETRY

Suppose a system with a Hamiltonian symmetric under time reversal, that is: $[T,\mathcal{H}] = 0$. If $|n\rangle$ and $T|n\rangle$ are the corresponding eigenkets of the operators $\mathcal{H}$ and T, they correspond to the same energy E_n. This can be shown as follows:

From $\mathcal{H}|n\rangle = E_n|n\rangle$ and applying the time-reversal operator T we have

$$T(\mathcal{H}|n\rangle) = T(E_n|n\rangle) = E_n T|n\rangle \tag{3.33}$$

Since $[T, \mathcal{H}] = 0$ the left side of this last equation is equal to $H(T|n\rangle)$ while the right side is equal to $E_n T|n\rangle$ if the eigenvalues of the Hamiltonian operator $\mathcal{H}$ are real. Therefore, $T(\mathcal{H}|n\rangle) = E_n T|n\rangle$ proving that the eigenkets $|n\rangle$ and $T|n\rangle$ have the same energy E_n.

Do $|n\rangle$ and $T|n\rangle$ represent the same state? If so, the time-reversed state $|n\rangle$ should be of the form $T|n\rangle = e^{i\delta}|n\rangle$ with δ being a phase.

But this would lead to $\mathrm{T}(T|n\rangle) = e^{i\delta} e^{-i\delta}|n\rangle = |n\rangle$ implying $T^2 = I$

Considering that for systems with half-integer spin $T^2 = -1$ (Ref. [3]), the eigenkets $|n\rangle$ and $T|n\rangle$ must correspond to different states, implying a degeneracy called Kramers degeneracy.

3.3 BREAKING OF TIME-REVERSAL SYMMETRY BY AN EXTERNAL MAGNETIC FIELD

If a charged particle with charge q is subject to an external electromagnetic field represented by a vector potential $\vec{A}(\vec{r}, t)$ and a scalar potential $\phi(\vec{r}, t)$, the Hamiltonian operator is given by the following expression (considering MKS units, so that both $q\vec{A}$ and $\vec{P}$ have units of momentum):

$$\mathcal{H} = \frac{1}{2m}\left(\vec{P} - q\vec{A}(\vec{r}, t)\right)^2 + q\phi(\vec{r}, t) \tag{3.34}$$

Expanding the first term of this equation we get

$$\mathcal{H} = \frac{\vec{P}^2}{2m} + \frac{q^2\vec{A}^2}{2m} - \frac{q}{2m}\left(\vec{P}\cdot\vec{A} + \vec{A}\cdot\vec{P}\right) + q\phi(\vec{r}, t) \tag{3.35}$$

Since the momentum operator $\vec{P}$ is odd under time reversal and the electromagnetic field is externally imposed over the particle, the term containing $\vec{P}\cdot\vec{A} + \vec{A}\cdot\vec{P}$ ibreaks the time-reversal symmetry since it changes sign when applying the time-reversal operation.

EXAMPLE 3.2

Going deeper in the above statement about breaking of time-reversal symmetry by an external magnetic field, analyze the result of time-reversal operation of the motion of a charged particle subject to an external magnetic field. Compare your result with a prediction based on classical electrodynamics.

SOLUTION

If one deals with a magnetic field externally applied and which can only be modified acting over the corresponding source, the time-reversal symmetry is broken.

In fact, the Schrödinger's equation for this case is

$$i\hbar\frac{\partial\psi(\vec{r},t)}{\partial t}=\frac{1}{2m}\left\{-i\hbar\nabla-\text{q}\vec{A}(\vec{r},t)\right\}^2\psi(\vec{r},t) \tag{3.36}$$

We can verify that $\psi^*(\vec{r},-t)$ is not a solution of eq. 3.36. In fact, if we apply the time-reversal operator T to both sides of this equation (meaning the replacements $i\rightarrow -i$ and $t\rightarrow -t$), we obtain

$$-i\hbar\frac{\partial\psi^*(\vec{r},-t)}{\partial(-t)}=\frac{1}{2m}\left\{+i\hbar\nabla-\text{q}\vec{A}(\vec{r},-t)\right\}^2\psi^*(\vec{r},-t)\Rightarrow$$

$$i\hbar\frac{\partial\psi^*(\vec{r},-t)}{\partial t}=\frac{1}{2m}\left\{+i\hbar\nabla-\frac{q}{c}\vec{A}(\vec{r},t)\right\}^2\psi^*(\vec{r},-t) \tag{3.37}$$

In the last equation, we have considered that since the magnetic field is imposed from an external source and cannot be modified, $\vec{A}(\vec{r},-t)=\vec{A}(\vec{r},t)$. Therefore, by comparison of eqs. 3.36 and 3.37, we conclude that $\psi^*(\vec{r},t)$ does not satisfy eq. 3.36, and consequently, the presence of an external magnetic field breaks the time-reversal symmetry.

We can corroborate this statement using Heisenberg's equation of motion[2,3] for obtaining a dynamical equation similar to that corresponding to the classical description of the motion of a charge in a magnetic field, as follows:

From eq. 3.36 the Hamiltonian is

$$\mathcal{H}=\frac{\left(\vec{P}-q\vec{A}\right)^2}{2m} \tag{3.38}$$

where $\vec{P}$ is the canonical momentum obeying the following relation with the momentum operator $\vec{p}$:

$$\vec{P}=\vec{p}+q\vec{A}=m\vec{\upsilon}+q\vec{A} \tag{3.39}$$

Using Heisenberg's equation of motion we have that

$$\frac{d\vec{P}}{dt}=\frac{1}{i\hbar}\left[\vec{P},\mathcal{H}\right]=-\nabla\mathcal{H} \tag{3.40}$$

For the i-th component of the momentum, we have

$$\frac{dP_i}{dt}=\dot{P}_i=-\frac{\partial\mathcal{H}}{\partial x_i}=\frac{\left(\vec{P}-q\vec{A}(\vec{r},t)\right)}{m}\cdot q\frac{\partial\vec{A}}{\partial x_i}=q\upsilon_j\nabla_i A_j \tag{3.41}$$

where υ_j is the j-th component of the velocity operator $\vec{\upsilon}=\frac{\vec{P}-q\vec{A}}{m}=\frac{\vec{p}}{m}=\frac{d\vec{r}}{dt}$ and we are using the rule of repeated indexes for summation.

From the relation $\vec{P}=\vec{p}+q\vec{A}$, we have that for the i-th component of the canonical momentum $\vec{p}$:

$$\dot{P}_i=m\ddot{x}_i+q\dot{A}_i=m\ddot{x}_i+q\left(\frac{\partial A_i}{\partial t}+\upsilon_j\nabla_j A_i\right) \tag{3.42}$$

Combining eqs. 3.41 and 3.42, we get that

$$m\ddot{x}_i=q\left(-\frac{\partial A_i}{\partial t}\right)+q\left(\upsilon_j\nabla_i A_j-\upsilon_j\nabla_j A_i\right)=qE_i+q\left(\nabla_i\upsilon_j A_j-\upsilon_j\nabla_j A_i\right) \tag{3.43}$$

We have considered that the charge is subject only to a magnetic field, so that $\vec{E}=-\frac{\partial\vec{A}}{\partial t}$, together with the vectorial identity $\vec{\upsilon}\times\left(\nabla\times\vec{A}\right)=\nabla\left(\vec{\upsilon}\cdot\vec{A}\right)-(\vec{\upsilon}\cdot\nabla)\vec{A}$ and with $\vec{B}=\nabla\times\vec{A}$. Therefore, eq. 3.43 leads to the expression for the Lorentz force $\vec{F}=q\left(\vec{E}+\vec{\upsilon}\times\vec{B}\right)$ which is not invariant under time reversal since $\vec{\upsilon}$ changes to $-\vec{\upsilon}$ while the other quantities do not vary, confirming that there is not time-reversal symmetry if the external magnetic field $\vec{B}$ is fixed by an external source.

This example shows that both for a classical system as for a quantum system, the presence of an external magnetic field breaks the time-reversal symmetry in their evolution. In addition, it can be seen that this loss of time-reversal symmetry occurs even there is not dissipation of energy, so that we are considering a conservative system. Therefore, as we have seen in Chapter 1, this example confirms that the absence of dissipation does not necessarily imply time-reversal symmetry.

3.4 TIME REVERSAL OF THE PHOTOELECTRIC EFFECT

In this section, we analyze the time reversal of the photoelectric effect whose theory was formulated by Albert Einstein during the first years of the 20th century, introducing the concept of photon which was a crucial step toward the theoretical formulation of quantum mechanics some years later.

We first discuss the motion of an electron emitted by means of the photoelectric effect and subject to a stopping potential as occurs in typical experiments analyzing in particular the time reversal of the corresponding dynamics determined by Schrödinger's equation once the electron has been emitted. Then as a pedagogical analogy, we analyze the quantum description of a particle in a gravitational field and discuss the dynamics of a mechanical system which presents a formal analogous behavior.

3.4.1 SCHRÖDINGER EQUATION FOR AN ELECTRON EMITTED BY PHOTOELECTRIC EFFECT

The microscopic theory of emission of electrons from a cold metal surface subjected to an electric field has been developed by Fowler and Nordhein,[4] considering and idealized situation for which the electrons in the conduction band are treated as free independent particles. Their energies are described by a Fermi distribution with a maximum energy which corresponds to the Fermi energy $E_F = \hbar^2 k_F^2 / 2m$, with a zero-temperature distribution, an acceptable approach for normal room temperatures.

In the absence of an external field, the electrons are confined by a potential of magnitude $U = E_F + W$ caused by the positive ions, where W is the work function, that is, the minimum energy required to extract an electron from the metal.

We consider that the electrons are emitted in a direction perpendicular to a flat surface at $x = 0$, corresponding to the surface of metal which receives the incident radiation. If the emitted electrons are subject to a stopping potential as occurs in typical experiments of photoelectric effect, so that they move along the x-axis against an external electric field $\vec{E} = -\hat{x}E_o$, the corresponding one-dimensional Schrödinger equation is

$$i\hbar\frac{\partial\psi(x,t)}{\partial t} = \left\{-\frac{\hbar^2}{2m}\frac{\partial^2}{\partial x^2} + V(x)\right\}\psi(x,t) = \mathcal{H}(x)\psi(x,t) \tag{3.44}$$

$$\text{with } V(x) = \begin{cases} 0 & x < 0 \\ U + |e|E_o x & x > 0 \end{cases} \tag{3.45}$$

where the symbol $|e|$ denotes the absolute value of the charge of the electron and m its mass. In this case, the potential energy $V(x)$ increases along the positive x-axis since the emitted electron moves against an electric field

$\vec{E} = -\hat{x}E_o$. A classsical analogy corresponds to the case of a mass sliding upwards through an inclined plane and against the gravity force, as will be discussed in Section 3.4.2.

The variation of the potential energy along the x-axis is sketched in Figure 3.3.

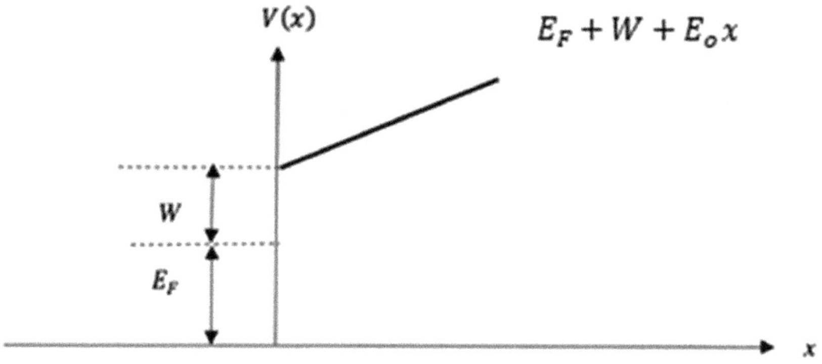

FIGURE 3.3 Potential energy acquired by an electron emitted by the photoelectric effect.

This situation is analogous with the case of a quantum particle of mass m subject to gravity with a potential $\Phi_{grav} = gy$ where the corresponding Schrödinger's equation for a one-dimensional analysis with motion along the y direction is

$$i\hbar\frac{\partial\psi(y,t)}{\partial t} = \left\{-\frac{1}{2m}\frac{\partial^2}{\partial y^2} + m\Phi_{grav}\right\}\psi(y,t) \tag{3.46}$$

From eq. 3.46 and using the Ehrenfest theorem,[3] an equation quite similar to that obtained from classical mechanics can be obtained:

$$\frac{d^2\langle y\rangle}{dt^2} = -g \tag{3.47}$$

where $\langle y\rangle$ is the expected value of the position of the particle along the y-axis. This last equation clearly indicates that in the absence of dissipation, we have a conservative systems with time-reversal symmetry. Nevertheless, it must be considered that the left-hand side of eq. 3.47 cannot be considered as an acceleration, but rather the result of calculate the second derivative with respect to time of the expectation value of the position. Quantities such as velocity and acceleration which are familiar for us in classical mechanics cannot be considered in an analogous form in quantum mechanics, since if a particle does not have a determinate position (prior to a measurement), neither does it have a well-defined velocity nor acceleration. It is only possible to obtain the probability of finding a particular value.

Figure 3.4 shows a typical arrangement for observing experimentally the photoelectric effect. The battery shown in this figure establishes an electric field between both plaques so that the emitted electrons from the anode are subject to a force acting against their motion. Only the fastest electrons with kinetic energy $K_{max} = h\nu - W = |e|V_s$ can reach the cathode. For each value of the frequency ν of the incident radiation, maintaining the polarity of the battery, we can adjust the voltage V_s so that there will not be electrons detected by the galvanometer G. In this case, the photoemitted electrons will follow a path like that represented as abc in Figure 3.4.

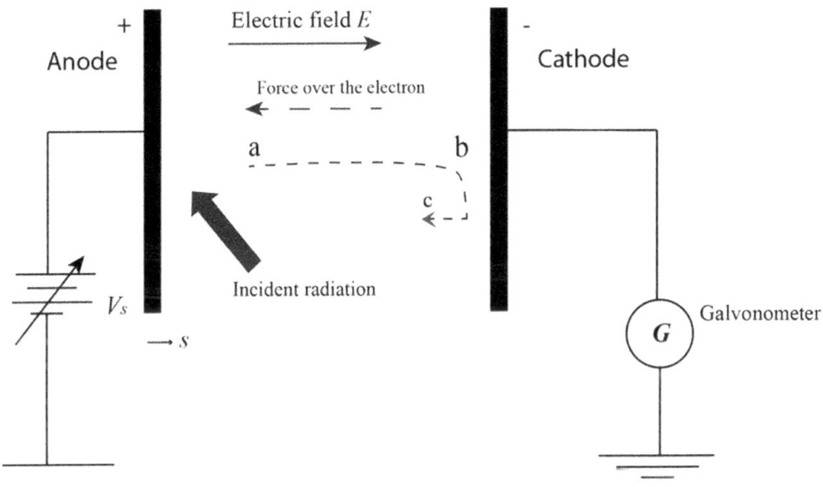

FIGURE 3.4 Typical experimental set for detecting the photoelectric effect.

3.4.2 MECHANICAL SYSTEM ANALOGOUS WITH THE PHOTOELECTRIC EFFECT

An extension of the analogy between an electron emitted by photoelectric effect and subject to an electric field and a mass subject to a gravitational field can be illustrated considering as made in Ref. [5], where the authors present a gravitational analogy to explain the role of the voltage source in the photoelectric effect and how the kinetic energy of the fastest electrons can be determined. In the example analyzed in this reference and sketched in

Figure 3.5, a mass m receives an amount E of external energy at point A from an external source not shown.

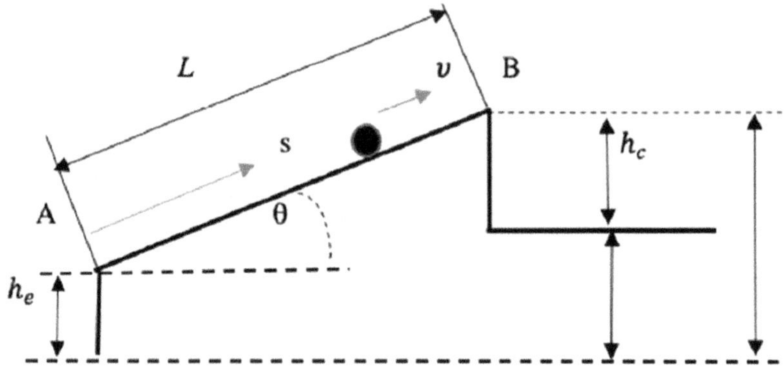

FIGURE 3.5 Mechanical system with a behavior analogous to that of the photoelectric effect.

Let υ_o be the initial velocity of mass m at point A, so that for arriving at point B, the energy E must be as minimum equal to the sum of kinetic energy $m\upsilon_o^2/2$ and the corresponding potential energy $mg(H-h_e)$ at B.

The dynamics of the mass during its ascent toward point B over the inclined plane is described by Newton's second law

$$-m\frac{d\upsilon}{dt}=mgsin\theta \tag{3.48}$$

Equation 3.48 can be recast as

$$-m\frac{d\upsilon}{ds}\frac{ds}{dt}=-m\upsilon\frac{d\upsilon}{ds}=mgsin\theta \Rightarrow \upsilon^2(s)=\upsilon_o^2-(2gsin\theta)s \tag{3.49}$$

From eq. 3.49, we can obtain that the minimum value of the initial velocity that the particle must have for arriving at point B with zero-velocity is

$$\upsilon_{omin}=\sqrt{L(2gsin\theta)} \tag{3.50}$$

Therefore, the minimum kinetic energy which the particle must have at point A for making possible to reach point B with zero-velocity is

$$K_{min}=\frac{1}{2}m\upsilon_{omin}^2=mgLsin\theta \tag{3.51}$$

From Figure 3.4, we have $Lsin\theta = H - h_e$, so that from eq. 3.51 we get the following expression for the minimum energy required from the external source so that a marble can go from A to B through the inclined plane

$$E_{min} = mgH = K_{min} + mgh_e \tag{3.52}$$

We can see that the dynamics of this process is time-reversal symmetric. In fact, the effect of time reversing the process can be seen directly from eq. 3.48, making the replacements $\upsilon \to -\upsilon$ and $t \to -t$. The result is that the dynamics of this process does not change as expected for an ideal case where mechanical energy is conserved. This conclusion is reinforced if we imagine that we film this process and when the particle reaches point B, we let the film to run backwards, that is, we let the mass m fall from point B with zero initial velocity through the inclined plane. At point A, it will have the same velocity υ_o it had as initial value for the upward motion. In this case, an amount of energy mgH will be returned to the external source.

The analogy between the mechanical system of Figure 3.5 and the photoelectric effect can be further analyzed as made in Ref. [5], if we imagine that at point ***A***, an unknown agent raises the marbles from the bottom of the lower step and throws them up along the inclined track, with different speeds. By trial and error, a critical height $h = h_s$ is found such that just below it only the fastest marbles reach point B. Let K_{max} be the kinetic energy of the fastest marbles at the beginning A of the inclined plane. All this energy is converted in potential energy mgh_s when a marble reach a point at a height h_s and at this point, it comes back to A. Therefore, the mechanical energy provided to the fastest marbles by the external source must be

$$E = mgh_e + K_{max} = mgh_e + mgh_s \tag{3.53}$$

The analogy with the photoelectric effect discussed in the previous section can be established making the energy E equivalent to that of an incident photon of frequency v impacting over a cold metal surface with energy $E = hv$ as shown in Figure 3.4 and the work function W analog to the potential energy of mass m at point A of Figure 3.5, that is $\triangleq mgh_e$. Eq. (3.53) can be directly compared with that corresponding to the photoelectric effect given by

$$hv = K_{max} + W \tag{3.54}$$

with $K_{max} = |e| V_s$ for the photoelectric effect and equivalent to mgh_s for the analogous mechanical system of Figure 3.5.

3.4.3 TIME REVERSAL OF THE PHOTOELECTRIC EFFECT

In the case of the electron emitted by the photoelectric effect, the time-reversed process corresponds to the emission of radiation when electrons impact over a metallic surface as occurs in the emission of X rays (Bremsstrahlung). In fact, the basis for the models used to describe the photoemission consider this process as a result of radiative deexcitation of electrons incident on a solid.[6] The most usual ways for describing photoemission an inverse photoemission processes are the so-called "three-step" models, which are briefly described here as follows:

For photoemission the basic steps are:

i. Excitation of electrons in the solid by incident photons (photo-excitation).
ii. Motion of the energetic electrons through the crystal to the surface.
iii. Passage of the electrons towards the vacuum crossing the barrier imposed by the surface.

The three-step model for inverse photoemission is based on an inversion of Spicer's model for photoemission[7] and considers the following steps:

i. Penetration of the electrons through the surface barrier.
ii. Competition between the effects produced by radiative and non-radiative decay.
iii. Optical deexcitation.

A schematic representation for these processes is shown in Figure 3.6a for photoemission and in Figure 3.6b for inverse photoemission.

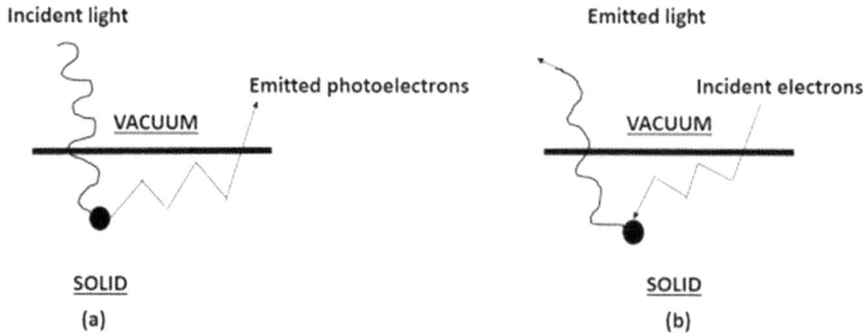

FIGURE 3.6 Sketch of the photoelectric effect (a) and inverse photoelectric effect (b).

3.4.4 TIME-REVERSAL ANALYSIS OF THE PHOTOELECTRIC EFFECT USING EHRENFEST THEOREM

In order to illustrate the concepts presented in this section, we discuss the time-reversal symmetry for the systems shown in Figures 3.4 and 3.5 using the Ehrenfest theorem.

First, we consider that the Hamiltonian for a particle moving under the effect of a potential $V(s)$ dependent on the position is

$$\mathcal{H} = \frac{P^2}{2m} + V(s) \tag{3.55}$$

where P is the momentum operator and s is the position operator. For the case of an electron emitted by photoelectric effect s is associated with the position of this particle with respect to the anode once it has left it, as shown in Figure 3.4, while for the analog mechanical system of Figure 3.5, it corresponds to the displacement from point A.

The potential $V(s)$ is

$$V(s) = \begin{cases} E_F + W + |e|\, E_o s \text{ for the photoelectron of Figure 3.3} \\ mgh_e + mgs\, sin\theta \text{ for the mass } m \text{ of Figure 3.5} \end{cases} \tag{3.56}$$

The momentum operator is related to the position operator s as $P = m\frac{ds}{dt}$. From Heisenberg's equation of motion we obtain that

$$\frac{d^2 s}{dt^2} = \frac{d}{dt}\left(\frac{ds}{dt}\right) = \frac{1}{i\hbar}\left[\frac{ds}{dt}, \mathcal{H}\right] = \frac{1}{i\hbar m}[P, \mathcal{H}] = \frac{1}{i\hbar m}\left[P, V(s)\right] = -\frac{1}{m}\frac{\partial V}{\partial s} \tag{3.57}$$

By taking the expectation values at both sides of this last equation with respect to a Heisenberg state ket that does not vary with time, we obtain

$$\frac{d^2}{dt^2}\langle s\rangle = -\frac{1}{m}\langle\frac{\partial V}{\partial s}\rangle \tag{3.58}$$

This last equation is second Newton's law but applied to the expectation value of the position operator s. It can be seen that this equation does not change if we make the replacement $t \rightarrow -t$, and therefore, the quantum description for this case has also time-reversal symmetry as in the classical description.

This last equation leads that for the electron emitted by photoelectric effect and moving against a stopping potential as shown in Figure 3.4, the expectation value for its position obeys the following equation which in the absence of any dissipation has time-reversal symmetry:

$$\frac{d^2}{dt^2}x = -\frac{eE_o}{m} \tag{3.59}$$

3.5 TIME-REVERSAL OPERATOR FOR SYSTEMS WITH SPIN

We have seen in Section 3.2.3 that the time-reversal operator T must be anti-unitary in order that a time-reversal operation gives a result compatible with the postulates of NRQM, specifically that the eigenvalues of the Hamiltonian $\mathcal{H}$ correspond to positive energies.

The antiunitary time-reversal operator was first obtained on the basis of theoretical group analysis by E. Wigner in 1932, starting from the invariance of the Schrödinger equation under time-reversal transformation.[9] To this requirement, it must be added that this operator must be anti-linear.[10] An operator A is anti-linear if satisfies the following relations:

$$A(|\varphi_1\rangle + |\varphi_2\rangle) = A\,|\varphi_1\rangle + A|\varphi_2\rangle \quad (3.60)$$

$$A(c|\psi\rangle) = c^*\,A|\psi\rangle \quad (3.61)$$

where c is a complex number.

For the case of a spinless particle, the time-reversal operator T is equal to the complex conjugation operator K of eq. 3.10, so that the inner product of two time-reversed states $T(|\varphi_1\rangle$ and $T(|\varphi_2\rangle$ is transformed as

$$\langle T\varphi_1|T\varphi_2\rangle = \langle K\varphi_1|K\varphi_2\rangle = \int\left(K\varphi_1(\vec{r})\right)^* K\varphi_2(\vec{r})\,d^3\vec{r} = \int\left(\varphi_2(\vec{r})\right)^* \varphi_1(\vec{r})\,d^3\vec{r} = \langle\varphi_2|\varphi_1\rangle \quad (3.62)$$

For the case of a particle with spin ½, the conventional time-reversal operator is $T = UK$ with U an unitary operator given by[3,4,10]

$$U = e^{-i\pi\sigma_y/2} = -i\sigma_y = \begin{pmatrix} 0 & -1 \\ 1 & 0 \end{pmatrix} \quad (3.63)$$

where σ_y is the Pauli matrix associated with rotation about the y-axis which is given by

$$\sigma_y = \begin{pmatrix} 0 & -i \\ i & 0 \end{pmatrix} \quad (3.64)$$

It can be readily verified that the inverse of the operator U given by eq. 3.63 is:

$$U^{-1} = U^{\dagger} = \begin{pmatrix} 0 & 1 \\ -1 & 0 \end{pmatrix} \quad (3.65),$$

where the symbol † means the transposition and conjugation operation applied over a matrix.

The matrix form of the time-reversal operator U given by eq. 3.63 allows that the spin operator $\vec{S} = \frac{\hbar}{2}\vec{\sigma}$ be odd under time reversal, that is

$$T\vec{S}T^{-1} = -\vec{S} \tag{3.66}$$

where $\vec{\sigma}$ is the spin vector matrix whose components are the Pauli matrices $\sigma_x, \sigma_y, \sigma_z$, of dimension 2 × 2 with

$$\sigma_x = \begin{pmatrix} 0 & 1 \\ 1 & 0 \end{pmatrix};\ \sigma_y = \begin{pmatrix} 0 & -i \\ i & 0 \end{pmatrix};\ \sigma_z = \begin{pmatrix} 1 & 0 \\ 0 & -1 \end{pmatrix}. \tag{3.67}$$

It is worth to mention that the time-reversal operator T belongs to a class of operators, which as a subset includes those operators that when repeated, restore the original state. If we consider a state $|\psi\rangle$, the corresponding time-reversed state is $T|\psi\rangle$. Multiplication of $T|\psi\rangle$ by a phase factor is the only transformations that does not change this state, and in general, we have that $T^2 = \eta\mathbb{I}$, where $\mathbb{I}$ represents the identity and $\eta = e^{i\phi}$ is a constant phase factor. In Ref. [11], Wigner has shown that $\eta = \pm 1$. A proof of this result can be found in Ref. [10] and it shows that with respect to time reversal, there are two classes of quantum mechanical quantities, "even" and "odd" ones. The coordinates, total energy, and kinetic energy belong to the first class and they are not connected with time, or have an even-order dependency on time, such as t^2, t^4, and so on. The operators corresponding to quantities of this class commute with the time-reversal operator. On the other hand, quantities, such as velocity, linear, and angular momentum belong to the second class and their corresponding operators have the property of anticommuting with the time-reversal operator. For example, for the momentum operator P, we have $TP = -PT$.

For spinless particles, $T = K$ and $T^2 = \mathbb{I}$ while for a system containing n particles with spin ½ (e.g., n electrons), $T^2 = +\mathbb{I}$ if n is an even number and $T^2 = -\mathbb{I}$ if n is an odd number.

EXAMPLE 3.3

Give arguments for interpreting eq. 3.66 and that the time-reversal operator for a system with spin ½ is $T = UK$ with U given by eq. 3.63.

SOLUTION

Equation 3.66 means that the spin operator is odd under time-reversal operation. This can be seen considering that if $|\alpha\rangle$ is a ket representing a given state, $|\tilde{\alpha}\rangle = T|\alpha\rangle$ is the corresponding time-reversed state. Considering that under time reversal the spin of a particle like its angular momentum changes the sign, the expectation value of the spin operator $\vec{S}$ should satisfy the following condition

$$\langle\tilde{\alpha}|\vec{S}|\tilde{\alpha}\rangle = -\vec{S}$$

With $\langle\tilde{\alpha}| = \langle\alpha|T^{-1}$, we have

$$\langle\tilde{\alpha}|\vec{S}|\tilde{\alpha}\rangle = -\vec{S} \Rightarrow \langle\alpha|T^{-1}\vec{S}\,T|\alpha\rangle = -\vec{S}$$

This implies that

$$T^{-1}\vec{S}\,T = -\vec{S} \tag{3.68}$$

Multiplying this last equation from the left by T and by the right by T^{-1}, we get eq. 3.66.

EXAMPLE 3.4

Show that the time-reversal operator T with U given by eq. 3.63 allows that the spin operator $\vec{S} = \frac{\hbar}{2}\vec{\sigma}$ be odd under time reversal according to eq. 3.66.

SOLUTION

Let us verify that the transformation given by eq. 3.66 is valid for the components S_x, S_y, S_z of the spin given by $\vec{S} = \frac{\hbar}{2}\vec{\sigma}$:

$$TS_xT^{-1} = \vec{S} = \frac{\hbar}{2}(UK)\sigma_x(UK)^{-1} = \frac{\hbar}{2}UK\sigma_xK^{-1}U^{-1}$$

$$= \frac{\hbar}{2}\begin{pmatrix} 0 & -1 \\ 1 & 0 \end{pmatrix}\sigma_x^*\begin{pmatrix} 0 & 1 \\ -1 & 0 \end{pmatrix}$$

$$= \frac{\hbar}{2}\begin{pmatrix} 0 & -1 \\ 1 & 0 \end{pmatrix}\sigma_x\begin{pmatrix} 0 & 1 \\ -1 & 0 \end{pmatrix} = \frac{\hbar}{2}\begin{pmatrix} 0 & -1 \\ 1 & 0 \end{pmatrix}\begin{pmatrix} 0 & 1 \\ 1 & 0 \end{pmatrix}\begin{pmatrix} 0 & 1 \\ -1 & 0 \end{pmatrix} == \frac{\hbar}{2}\begin{pmatrix} 0 & -1 \\ 1 & 0 \end{pmatrix}\begin{pmatrix} -1 & 0 \\ 0 & 1 \end{pmatrix}$$

$$= \frac{\hbar}{2}\begin{pmatrix} 0 & -1 \\ -1 & 0 \end{pmatrix} = -\frac{\hbar}{2}\sigma_x = -S_x$$

Similarly, it can be shown that $TS_yT^{-1} = -S_y$ and $S_zT^{-1} = -S_z$.

Therefore, the spin operator is odd under time reversal with $T = UK$ with U given by eq. 3.63.

EXAMPLE 3.5

Prove and interpret the result that if the state of a particle with spin ½ is subject to two successive applications of the time-reversal operator, it does not return to its original state and specifically that $T^2 = -\mathbb{I}$, where $\mathbb{I}$ represents the identity.

SOLUTION

With the time-reversal operator given by $T = UK$, we have that T^2 acting over any ket $|\alpha\rangle$ gives

$$T^2 \mid \alpha\rangle = (UK)(UK) \mid \alpha\rangle = (UK)(KU) \mid \alpha\rangle = UK^2U \mid \alpha\rangle = UU \mid \alpha\rangle = U^2 \mid \alpha\rangle \quad (3.69)$$

We have considered that $K^2 = \mathbb{I}$. Therefore, from eq. 3.63:

$$U^2 = \begin{pmatrix} 0 & -1 \\ 1 & 0 \end{pmatrix}\begin{pmatrix} 0 & -1 \\ 1 & 0 \end{pmatrix} = \begin{pmatrix} -1 & 0 \\ 0 & -1 \end{pmatrix} = -\,\mathbb{I}$$

and $T^2|\alpha\rangle = -|\alpha\rangle$.

This can be verified by applying the time-reversal operator twice to the states with "spin up" and "spin down" written respectively as:

$$|+\rangle = \begin{pmatrix} 1 \\ 0 \end{pmatrix}, |-\rangle = \begin{pmatrix} 0 \\ 1 \end{pmatrix}$$

Therefore, we obtain:

$$T|+\rangle = UK|+\rangle = U\begin{pmatrix} 1 \\ 0 \end{pmatrix} = \begin{pmatrix} 0 & -1 \\ 1 & 0 \end{pmatrix}\begin{pmatrix} 1 \\ 0 \end{pmatrix} = \begin{pmatrix} 0 \\ 1 \end{pmatrix} = +(1)|-\rangle \quad (3.70)$$

Similarly,

$$T|-\rangle = \begin{pmatrix} 0 & -1 \\ 1 & 0 \end{pmatrix}\begin{pmatrix} 0 \\ 1 \end{pmatrix} = -\begin{pmatrix} 1 \\ 0 \end{pmatrix} = -(1)\mid +\rangle \quad (3.71)$$

Applying again the operator T over each one of these states we get

$$T^2|+\rangle = \begin{pmatrix} 0 & -1 \\ 1 & 0 \end{pmatrix}\begin{pmatrix} 0 \\ 1 \end{pmatrix} = -\begin{pmatrix} 1 \\ 0 \end{pmatrix} = -\mid +\rangle \quad (3.72)$$

$$T^2|-\rangle = -\begin{pmatrix} 0 & -1 \\ 1 & 0 \end{pmatrix}\begin{pmatrix} 1 \\ 0 \end{pmatrix} = -\begin{pmatrix} 0 \\ 1 \end{pmatrix} = -\mid -\rangle \quad (3.73)$$

These results prove that $T^2 = -\mathbb{I}$ and this can be related to Kramer's degeneracy discussed in Section 3.2.5. In fact, we can see that in particular, for the case of an electron subject to a magnetic field $\vec{B}$ in the direction z, the kets $|\pm\rangle$ and $|-\rangle$ are eigenvectors of the Hamiltonian $\mathcal{H} = -\vec{S} \cdot \vec{B}$. The corresponding eigenvalues are $-B_z\hbar/2$ and $+B_z\hbar/2$, respectively. From eqs. 3.70 and 3.71, clearly, the states $T|\pm\rangle$ are also eigenkets of the Hamiltonian $\mathcal{H}$ different from $|\pm\rangle$ but keeping the same magnitude $B_z\hbar/2$ of the corresponding eigenvalues.

EXAMPLE 3.6

Consider a spin state $|s\rangle = c_+ |+\rangle + c_- |-\rangle$ with c_+, c_- complex coefficients such that $|c_+|^2 + |c_-|^2 = 1$. Obtain the time-reversed state. Is there time-reversal symmetry for this system?

SOLUTION

The time-reversal operator $T = -i\sigma_y K$ acting over the spin state $|s\rangle$ gives

$$T\,|\,s\rangle = -i\sigma_y K\,|\,s\rangle = -i\begin{pmatrix} 0 & -i \\ i & 0 \end{pmatrix}\{c_+^*\,|+\rangle + c_-^*\} = \begin{pmatrix} 0 & -1 \\ 1 & 0 \end{pmatrix}\left\{c_+^*\begin{pmatrix} 1 \\ 0 \end{pmatrix} + c_-^*\begin{pmatrix} 0 \\ 1 \end{pmatrix}\right\}$$

$$\therefore T\,|\,s\rangle = T\{c_+\,|+\rangle + c_-\,|-\rangle\} = \left\{c_+^*\begin{pmatrix} 0 \\ 1 \end{pmatrix} - c_-^*\begin{pmatrix} 1 \\ 0 \end{pmatrix}\right\} = \{c_+^*\,|-\rangle - c_-^*\,|+\rangle\} \tag{3.74}$$

Clearly, $|s\rangle$ and $T|s\rangle$ are different states. If we apply again the time-reversal operator we obtain

$$T^2\,|\,s\rangle = T\,\{c_+^*\,|-\rangle - c_-^*\,|+\rangle\} = \begin{pmatrix} 0 & -1 \\ 1 & 0 \end{pmatrix}\left\{c_+\begin{pmatrix} 0 \\ 1 \end{pmatrix} - c_-\begin{pmatrix} 1 \\ 0 \end{pmatrix}\right\}$$

$$= -c_+\begin{pmatrix} 1 \\ 0 \end{pmatrix} - c_-\begin{pmatrix} 0 \\ 1 \end{pmatrix} = -\,|\,s\rangle \tag{3.75}$$

Therefore, $T^2 = -\mathbb{I}$ as in the above example and there is not time-reversal symmetry because the result of two successive time-reversal operations do not lead the system to the initial state. The final state acquires a phase which in this case is $\pm\pi$.

3.6 TIME REVERSAL AND GEOMETRIC PHASES

3.6.1 INTRODUCTION

A topic closely related to the time reversal of the dynamics of a quantum system is the so-called geometric phase. This can be explained by means of an illustrative example. Let us imagine a vector with a certain direction and put it over the surface of a sphere, for example, on the north pole, and with a direction and arrow pointing toward a certain meridian. Then we move the object keeping it always parallel to its initial direction down the meridian until we reach the equator and then it is moved parallel along the equator till another meridian which keeps an angle of θ with the original one. Next, one moves the vector back to the north pole along the path corresponding to the new meridian, and keeps it always parallel. When the north pole is reached we discover that the vector is not in the same initial direction. It has turned around an angle θ (see Figure 3.7). This phenomenon which is called a holonomy can be described by the so-called Hannay angles.[12]

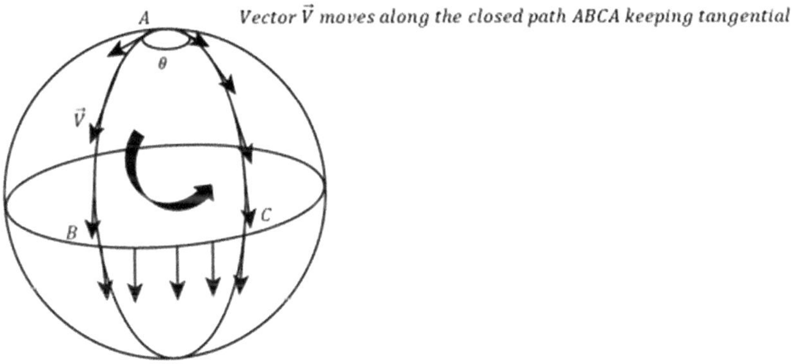

FIGURE 3.7 The holonomy under the parallel transport of a vector. A vector transported on a closed path on the surface of a sphere and kept tangential to that path, does not retain its initial orientation.

3.6.2 THE BERRY PHASE

In a paper published in 1984, Michael V. Berry showed that a quantum system having a Hamiltonian $\mathcal{H}$ depending on a set of parameters denoted as a set of values $\mathcal{R}$ and undergoing a cyclic and slow change acquires a geometric phase factor in addition to the dynamical phase factor.[13] According to this paper, "If the environment, and hence $\mathcal{H}$, is slowly altered, it follows from the adiabatic theorem (Refs. [14, 15]) that at any instant, the system will be in an eigenstate of the instantaneous $\mathcal{H}$. In particular, if the Hamiltonian is returned to its original form, the system will return to its original state, apart from a phase factor."

Let us consider a quantum system whose Hamiltonian is time-dependent through a slowly time-varying set of parameters $X_i(t)$ which are cyclic, so that in the corresponding space of parameters, they follow a closed path C. Any eigenstate of the Hamiltonian acquires besides an expected dynamical phase, a geometrical or Berry's phase, whose value depends on the closed path C. Aharonov and Anandan[16] generalized Berry's result showing that such a phase might appear for any state cyclic with respect to some evolution.

In what follows we give an explanation of the geometrical phase in quantum mechanics, based on Refs. [2] and [13].

If the Hamiltonian is time-dependent, the eigenvalues and eigenfunctions must also change with time as expressed in the following equation:

$$\mathcal{H}(t)\,\varphi_n(x,t) = E_n\varphi_n(x,t) \tag{3.76}$$

where $\varphi_n(x,t)$ is the eigenfunction for the *nth* state and E_n the corresponding eigenvalue.

According to the adiabatic theorem, if the Hamiltonian $\mathcal{H}$ changes very gradually so that a particle starting in the *nth* state will remain in this state, save a phase factor corresponding to a dynamical phase. Therefore, we would expect an evolution given by

$$\psi_n(x,t) = \varphi_n(x,t)exp\left[-\frac{i}{\hbar}\int_0^t E_n(t')dt'\right] = \varphi_n(x,t)e^{i\theta_n(t)} \tag{3.77}$$

where $\theta_n(t)$ is a dynamical phase given by

$$\theta_n(t) = -\frac{1}{\hbar}\int_0^t E_n(t')dt' \approx -\frac{E_n}{\hbar}t \tag{3.78}$$

Plugging these last two equations in the time-dependent Schrödinger's equation

$$i\hbar\frac{\partial\psi_n}{\partial t} = \mathcal{H}(t)\psi_n(x,t) = E_n\psi_n(x,t) \tag{3.79}$$

we obtain

$$i\hbar\left\{e^{i\theta_n(t)}\frac{\partial\varphi_n}{\partial t} + \varphi_n\left(-\frac{i}{\hbar}E_n\right)e^{i\theta_n(t)}\right\} = E_n e^{i\theta_n(t)}\varphi_n(x,t) \tag{3.80}$$

After some algebra, we rewrite the above equation as:

$$i\hbar\frac{\partial\varphi_n}{\partial t}+E_n(t)\varphi_n(x,t)=E_n(t)\varphi_n(x,t) \tag{3.81}$$

But this last equation implies that $\frac{\partial\varphi_n}{\partial t}=0$ which is a contradictory result since according to eq. 3.77, the function φ_n must depend on time. *Therefore, something is not working well !!*. Let us prove with some additional phase $\gamma_n(t)$ so that

$$\psi_n(x,t)=\varphi_n(x,t)e^{i\theta_n(t)}e^{i\gamma_n(t)} \tag{3.82}$$

Inserting eqs. 3.82 into 3.79 we get

$$i\hbar\left\{e^{i\theta_n(t)}e^{i\gamma_n(t)}\frac{\partial\varphi_n}{\partial t}+\left(-\frac{i}{\hbar}E_n\right)\varphi_n(t)e^{i\theta_n(t)}e^{i\gamma_n(t)}+ie^{i\theta_n(t)}e^{i\gamma_n(t)}\varphi_n(t)\frac{d\gamma_n(t)}{dt}\right\}$$
$$=E_ne^{i\theta_n(t)}e^{i\gamma_n(t)}\varphi_n(x,t) \tag{3.83}$$

From this last equation, we obtain

$$\frac{\partial\varphi_n}{\partial t}+i\varphi_n\frac{d\gamma_n}{dt}=0 \tag{3.84}$$

Taking the inner product with $\langle\varphi_n|$

$$\left\langle\varphi_n\left|\frac{\partial\varphi_n}{\partial t}\right.\right\rangle=-i\left\langle\varphi_n\left|\varphi_n\frac{d\gamma_n}{dt}\right.\right\rangle=-i\langle\varphi_n\,|\,\varphi_n\rangle\frac{d\gamma_n}{dt} \tag{3.85}$$

If the function φ_n is normalized, the above equation leads to

$$\frac{d\gamma_n}{dt}=i\left\langle\varphi_n\left|\frac{\partial\varphi_n}{\partial t}\right.\right\rangle \tag{3.86}$$

If the function φ_n depends on a set of time-dependent parameters $\mathcal{R}=\{R_1(t), R_2(t), \ldots\ldots R_L(t)\}\equiv\vec{R}$ we can write

$$\frac{\partial\varphi_n}{\partial t}=\frac{\partial\varphi_n}{\partial R_1}\frac{\partial R_1}{\partial t}+\frac{\partial\varphi_n}{\partial R_2}\frac{\partial R_2}{\partial t}+\ldots\frac{\partial\varphi_n}{\partial R_L}\frac{\partial R_L}{\partial t}=\nabla_{\vec{R}}\varphi_n\cdot\frac{d\vec{R}}{dt} \tag{3.87}$$

Therefore, from the above equations we obtain

$$\left\langle\varphi_n\left|\frac{\partial\varphi_n}{\partial t}\right.\right\rangle=\langle\varphi_n\,|\,\nabla_{\vec{R}}\varphi_n\rangle\cdot\frac{d\vec{R}}{dt} \tag{3.88}$$

$$\frac{d\gamma_n}{dt}=i\left\langle\varphi_n\left|\frac{\partial\varphi_n}{\partial t}\right.\right\rangle=i\langle\varphi_n\,|\,\nabla_{\vec{R}}\varphi_n\rangle\cdot\frac{d\vec{R}}{dt} \tag{3.89}$$

If the parameters R_j have a cyclic variation with period τ, so that for $j=1,2,\ldots\ldots L$

$$R_j(t+\tau)=R_j \tag{3.90}$$

we obtain

$$\gamma_n = i\int_0^\tau \langle\varphi_n|\nabla_{\vec{R}}\varphi_n\rangle\cdot\frac{d\vec{R}}{dt}dt = i\oint\langle\varphi_n|\nabla_{\vec{R}}\varphi_n\rangle\cdot d\vec{R} \tag{3.91}$$

This result corresponds to Berry's phase for a closed path followed by a vector $\vec{R}$ in parameter space. In a paper published in 1984, Michael V. Berry showed that a quantum system having a Hamiltonian $\mathcal{H}$ depending on a set of parameters denoted as a set of values $\mathcal{R}$ and undergoing a cyclic and slow change, acquires a geometric phase factor in addition to the dynamical phase factor.[13]. According to this paper, "If the environment, and hence $\mathcal{H}$, is slowly altered, it follows from the adiabatic theorem (Refs. [14, 15]) that at any instant, the system will be in an eigenstate of the instantaneous $\mathcal{H}$. In particular, if the Hamiltonian is returned to its original form, the system will return to its original state, apart from a phase factor."

A relevant issue is: *Is it possible to measure the Berry's phase?* In first courses of Quantum Physics one learns that the phase angle corresponding to a wave function ψ is arbitrary and that for obtaining, for instance, the probability of finding a particle in some region of space what matters is the square of the module of this wave function, that is $|\psi|^2 = \psi \cdot \psi^*$ so that if $\psi = |\psi|e^{i\alpha}$, where α is a phase that is canceled out in the calculation of $|\psi|^2$. But the geometrical phase can be measured, if for example, we take a beam of particles, let us say electrons, all will be in the same state given by ψ and split it in two. One of the beams is to pass through a region where there is an interaction with a magnetic field changing adiabatically, while the other beam does not undergo some interaction. When the two beams are recombined, the total wave function is of the form $\psi_R = |\psi| + |\psi|e^{i\Phi geom}$.

In this case, we obtain

$$|\psi_R|^2 = |\psi|^2\left(1+e^{i\,\Phi_{geom}}\right)\left(1+e^{-i\,\Phi_{geom}}\right) = 4|\psi|^2\cos^2\frac{\Phi_{geom}}{2} \tag{3.92}$$

The geometric phase can be measured by observing points of constructive and destructive interferences (where Φ_{geom} is an even or odd multiple of π, respectively).

3.6.3 EXAMPLE OF BERRY PHASE ACQUIRED BY A PARTICLE WITH SPIN ½

Let us consider a particle of spin ½ at rest in a uniform magnetic field, which points in the z direction:

$$\vec{B} = B_o \hat{z} \tag{3.93}$$

The spin vector $\vec{S}$ is given by $\vec{S} = \frac{\hbar}{2}\vec{\sigma}$ with $\hat{\sigma} = (\sigma_x, \sigma_y, \sigma_z)$ and $\sigma_x, \sigma_y, \sigma_z$ corresponding to the Pauli spin matrices:

$$\sigma_x = \begin{pmatrix} 0 & 1 \\ 1 & 0 \end{pmatrix}; \ \sigma_y = \begin{pmatrix} 0 & -i \\ i & 0 \end{pmatrix}; \sigma_z = \begin{pmatrix} 1 & 0 \\ 0 & -1 \end{pmatrix} \tag{3.94}$$

The Hamiltonian $\mathcal{H}$ corresponding to the interaction between the spin of the particle and the externally applied magnetic field is

$$\mathcal{H} = -\gamma \vec{S} \cdot \vec{B} \tag{3.95}$$

$\gamma = \frac{e}{m}$ is the gyromagnetic ratio.[2] For the specific case that the magnetic field points in the z direction, the interaction between the particle and the magnetic field is given by the following Hamiltonian:

$$\mathcal{H} = -\gamma \vec{S} \cdot \vec{B} = -\gamma B_o S_z \tag{3.96}$$

Is there time-reversal symmetry for this case?

The answer is Not, because as we have seen in Section 3.5, Example 3.4, the spin operator $\vec{S}$ is odd under time reversal, and since the magnetic field $\vec{B}$ is controlled by an external source, from eq. 3.96, it can be seen that the Hamiltonian given is not invariant under time-reversal since:

$$T\mathcal{H}T^{-1} = -\gamma B_o T S_z T^{-1} = -\gamma B_o (-S_z) = +\gamma B_o S_z = -\mathcal{H} \tag{3.97}$$

Therefore, we expect a Berry phase for this case as shown below.

Let us consider that our spin ½ particle interacts with an external magnetic field which rotates adiabatically (slowly) around the z-axis with angular velocity ω as shown in Figure 3.8:

The components of the magnetic field vector along the x, y, and z direction are:

$$B_x(t) = B_o (cos\omega t)\, sin\theta \tag{3.98}$$

$$B_y(t) = B_o\,(\sin\omega t)\,\sin\theta \tag{3.99}$$

$$B_z = B_o\,\cos\theta \tag{3.100}$$

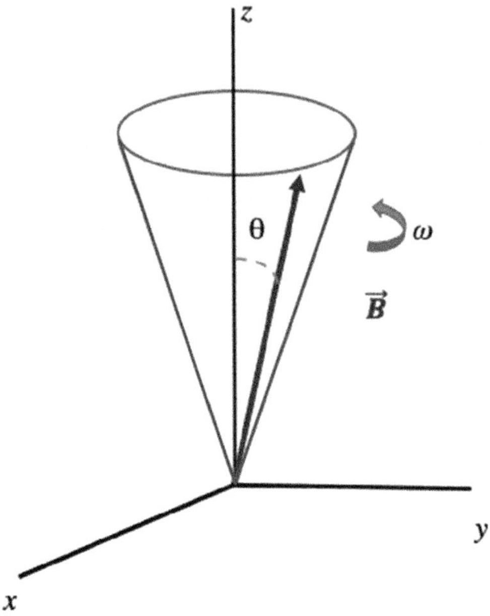

FIGURE 3.8 Magnetic field vector rotating with respect to z-axis with angular velocity ω.

Using eqs. 3.94 and 3.95, we obtain that the resulting Hamiltonian for this case is

$$\mathcal{H} = -\gamma\frac{\hbar}{2}\begin{pmatrix} \cos\theta & e^{-i\omega t}\sin\theta \\ e^{i\omega t}\sin\theta & -\cos\theta \end{pmatrix} \tag{3.101}$$

The normalized eigenstates and the corresponding energies are:

$$|n_+(t)\rangle = \begin{pmatrix} \cos\frac{\theta}{2} \\ e^{i\omega t}\sin\frac{\theta}{2} \end{pmatrix} \text{ with energy } E_+ = -\frac{\gamma B_o\hbar}{2} \tag{3.102}$$

$$|n_-(t)\rangle = \begin{pmatrix} -\sin\frac{\theta}{2} \\ e^{i\omega t}\cos\frac{\theta}{2} \end{pmatrix} \text{ with energy } E_- = +\frac{\gamma B_o\hbar}{2} \tag{3.103}$$

The Hamiltonian depends on the magnetic vector $\vec{B}(t)$ with components described by $\theta, \phi(t) = \omega t$ and with an equivalent radius r defined as $r \equiv B_o$. This vector traces out the curve $\mathbb{C}$ as shown in Figure 3.9.

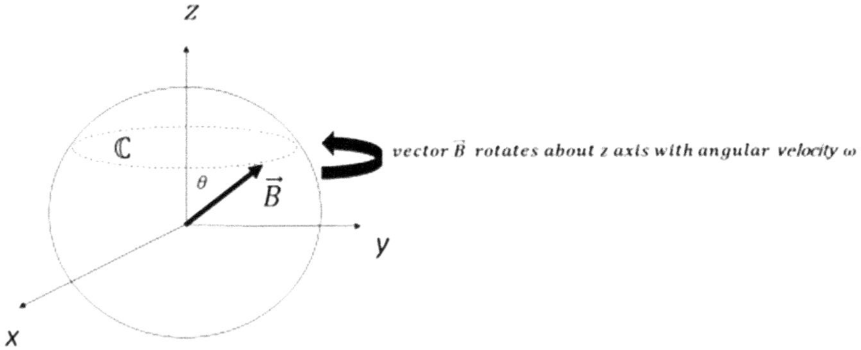

FIGURE 3.9 Curve $\mathbb{C}$ described by the tip of the magnetic field vector during a cycle.

The gradient in the parameter space spanned by $\vec{B}(t)$ is

$$\nabla \left| n_\pm(t) \right\rangle = \frac{\partial}{\partial r} \left| n_\pm(t) \right\rangle \hat{r} + \frac{1}{r} \frac{\partial}{\partial \theta} \left| n_\pm(t) \right\rangle \hat{\theta} + \frac{1}{r sin\theta} \frac{\partial}{\partial \phi} \left| n_\pm(t) \right\rangle \hat{\phi} \tag{3.104}$$

Therefore, for the eigenstates of the Hamiltonian $\mathcal{H}$, we get

$$\nabla \left| n_+(t) \right\rangle = \frac{1}{r} \begin{pmatrix} -\frac{1}{2} sin\frac{\theta}{2} \\ \frac{e^{i\omega t}}{2} cos\frac{\theta}{2} \end{pmatrix} \hat{\theta} + \frac{1}{r sin\theta} \begin{pmatrix} 0 \\ i\frac{e^{i\omega t}}{2} sin\frac{\theta}{2} \end{pmatrix} \hat{\phi} \tag{3.105}$$

$$\nabla \left| n_-(t) \right\rangle = \frac{1}{r} \begin{pmatrix} -\frac{1}{2} cos\frac{\theta}{2} \\ \frac{-e^{i\omega t}}{2} sin\frac{\theta}{2} \end{pmatrix} \hat{\theta} + \frac{1}{r sin\theta} \begin{pmatrix} 0 \\ i\frac{e^{i\omega t}}{2} cos\frac{\theta}{2} \end{pmatrix} \hat{\phi} \tag{3.106}$$

Taking the scalar product with $\langle n|$

$$\left\langle n_+ \middle| \nabla \middle| n_+ \right\rangle = i \frac{sin^2 \frac{\theta}{2}}{r sin\theta} \hat{\phi} \tag{3.107}$$

$$\left\langle n_- \middle| \nabla \middle| n_- \right\rangle = i \frac{cos^2 \frac{\theta}{2}}{r sin\theta} \hat{\phi} \tag{3.108}$$

Integration along the curve $\mathbb{C}$ of Figure 3.10 gives

$$\oint \langle n_{\pm} | \nabla | n_{\pm} \rangle r sin\theta d\theta d\phi \hat{\phi} = i\pi \left(1 \mp cos\theta\right) = i\Phi_{geom} \tag{3.109}$$

Therefore, the state of this system after a rotation of the tip of the magnetic field vector $\vec{B}$ so that $\vec{B}(\tau) = \vec{B}(0)$ is given by

$$\left| n_{\pm}(\tau) \right\rangle = exp\left[-\frac{i}{\hbar} E_{\pm}\, \tau \right] exp\left[i\pi \left(1 \mp cos\theta\right)\right] \left| n_{\pm}(0) \right\rangle \tag{3.110}$$

We see that the dynamical phase depends on the period τ of the rotation but the geometrical phase depends only on the special geometry of the problem – in this case, the opening angle θ of the cone that the magnetic field traces out (see Figure 3.9).

3.6.4 BERRY PHASE FOR PHOTONS

A similar result to that obtained in the above section has been observed when circularly polarized light propagates down a helically wound optical fiber.[17] A summary and discussion of this work is as follows:

In the quoted reference, the authors assumed that the light propagates inside the twisting waveguide in a single mode and is parametrized by s which is the optical length (change of phase for a given distance traveled along the waveguide). From the adiabatic invariance of the helicity of the photon we have that at each point s, the photon's spin state $|\vec{k}(s),\sigma\rangle$ satisfies the following equation:

$$\vec{S} \cdot \vec{k}(s) \left| \vec{k}(s), \sigma \right\rangle = \eta \left| \vec{k}(s), \sigma \right\rangle \tag{3.111}$$

where $\vec{S}$ is the spin vector of the photon, $\vec{k}(s)$ is the propagation vector of the photon and $\sigma = \pm 1$ its helicity. We must remember that a photon has an associated spin whose magnitude can be $\pm\hbar$ and its ***helicity*** is defined as the component of its spin vector along a direction parallel to that of its linear momentum. It is possible to keep the helicity quantum number σ of a photon as an adiabatic invariant during the passage through an optical fiber when no reflections occur.

Formally, this equation is similar to that considered by Berry for a spin in an adiabatically changing magnetic field. As we have seen, as the magnetic field is adiabatically changed, the components B_x and B_y of this field trace out a closed curve C lying over the surface of a sphere of radius $B = \sqrt{B_x^2 + B_y^2 + B_z^2}$ in parameter space (B_x, B_y, B_z). After one round trip in parameter space, the system must come back to its initial state, acquiring a geometrical phase factor $exp[i\Phi_{geom}]$, where Φ_{geom} is the Berry's phase.

A similar effect occurs in the case of a photon that propagates smoothly down a helical waveguide. The vector $\vec{k}$ is constrained to remain parallel to the local axis of this waveguide, since the momentum carried by the photon is in this direction. Since its helicity is adiabatically conserved, the spin $\vec{S}$ vector must also remain parallel to the local axis of the waveguide. Hence, the geometry of a helical path that in this case has the considered waveguide constraints $\vec{k}$ and hence $\vec{S}$ to trace out a closed loop C on the surface of a sphere of radius $k = \sqrt{k_x^2 + k_y^2 + k_z^2}$ in parameter space (k_x, k_y, k_z). In the special case of a uniform helical waveguide with N turns, C is a circle and the Berry's phase is

$$\Phi_{geom} = 2\pi N(1 - cos\theta) \tag{3.112}$$

In the above equation, θ is the angle between the local axis of the waveguide axis and the axis of the helix.

The phase given by eq. 3.112 can be measured in the following interference experiment: A circularly polarized laser beam enters a single-input optical fiber (SIOF). This fiber, in turn, couples an equal amount of light into two helically optical fibers, each having N turns, but wound in *contrary* senses. These last two fibers are adjusted to have equal optical path lengths. These two helices form the two arms of the interferometer. The ends of the helical fibers are then brought together and coupled into a single-output optical fiber (SOOF), in which the interference occurs. The experiment arrangement is shown in Figure 3.10 where a circularly polarized laser beam is injected into a SIOF, coupling an equal amount of light into two helically wound optical fibers (OFa and OFb), each having N turns but in a contrary sense, adjusted so that they have the same optical length. The output intensity is measured after the SOOF.

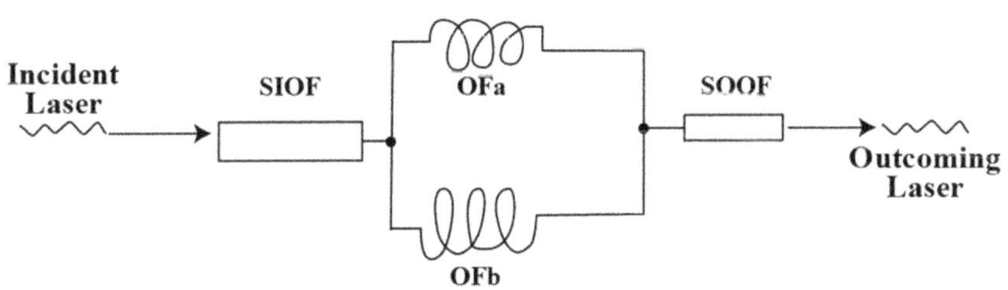

FIGURE 3.10 Outline of an experiment for measuring Berry's phase for photons propagating along helically wound optical fibers.

The predicted interference pattern gives an intensity proportional to $cos^2[2\pi N](1 - cos\theta)]$.

3.6.5 RELATION OF AHARONOV-BOHM EFFECT WITH BERRY PHASE

In his paper of 1984 (Ref. [13]), Berry showed that the Aharonov–Bohm effect can be understood as a phase factor that occurs in the geometrical space instead of in the parameter space as is the case of Berry's phase. For understanding this interpretation, we will first revisit this effect.

The Aharonov–Bohm effect is a quantum mechanical phenomenon occurring when an electrically charged particle interacts with an electromagnetic potential, despite its motion is confined to a region in which both the magnetic and electric fields are zero. The basic mechanism of this effect is the influence that the electromagnetic potential has on the phase of a charged particle's wave function, and can be detected by interference experiments.

First, we must take into account that it is possible to have a magnetic field confined to some bounded region $\mathbb{V}$ of space such that $\vec{B} \neq 0$ for each point inside this volumen, while outside this región $\vec{B} = 0$ but the magnetic potential vector $\vec{A}$ is not zero. This is sketched in Figures 3.11 and 3.12 where a particle of charge q emitted by a source A can follow a path above or below a very long impenetrable cylinder of radius a subject to a magnetic field $\vec{B} \neq 0$ only inside this cylinder and with a direction parallel to its axis.

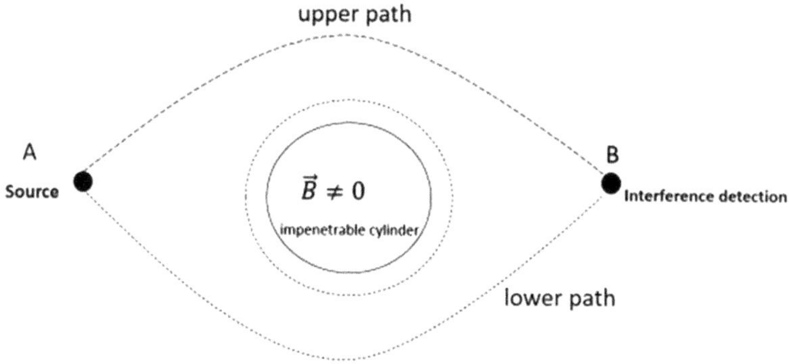

FIGURE 3.11 Sketch of the setup for observing the Aharonov–Bohm effect.

Source: Adapted from Ref. [3]

Let us imagine a circular ring of radius $r > a$ shown by the dashed contour in Figure 3.11. The magnetic flux through the circular surface S_r bounded by this imaginary ring is given by the following expression if we use the theorem of Stokes from vector calculus

$$\Phi_{mag} = B\pi a^2 = \int \left(d\vec{S} \cdot \vec{B} \right)_{S_r} = \int \left(d\vec{S} \cdot \nabla \times \vec{A} \right)_{S_r} = \oint \left(\vec{A} \cdot d\vec{l} \right)_{C_r} = A_\phi 2\pi r \quad (3.113)$$

where A_ϕ is the tangential component of the vector potential.

Therefore, the magnetic vector potential $\vec{A}$ at any point over the ring has only tangential component A_ϕ given by

$$A_\phi(b) = \frac{B\pi a^2}{2\pi r} = \frac{Ba^2}{2r} \tag{3.114}$$

This analysis shows that even though the magnetic field $\vec{B}$ is confined to the interior of the solenoid, the vector potential $\vec{A}$ at external points is not zero.

Now let us imagine a particle of charge q constrained to move only through the outside of the solenoid. As shown by Aharonov and Bohm, the vector potential can have influence on the quantum behavior of a charged particle that never encounters an electromagnetic field.[18] In fact, Schrödinger's equation in a region where $\vec{A} \neq 0$ but with zero magnetic field is:

$$i\hbar\frac{\partial\psi}{\partial t} = \frac{1}{2m}\left(-i\hbar\nabla - q\vec{A}\right)^2\psi \tag{3.115}$$

It can be verified that the solution of the above equation is[2,3]

$$\psi(\vec{r},t) = \psi_o(t)e^{ig(\vec{r})} \tag{3.116}$$

where

$$g(\vec{r}) = \frac{q}{\hbar}\int_{\vec{r}_o}^{\vec{r}} \vec{A}(\vec{s})\cdot d\vec{s} \tag{3.117}$$

and $\psi_o(t)$ is the solution for a region in which $\vec{A} = 0$. It has been assumed here that this potential is static, although the method can be generalized to time-dependent potentials.[2] The vector $\vec{r}_o$ in eq. 3.117 represents the position of some arbitrarily chosen reference point.

Figure 3.12 shows an experiment where a beam of electrons, is split into two and each beam follows a path surrounding a solenoid on the different side of it.

From eqs. 3.116 and 3.117, we can see that the presence of $\vec{A}$ results in a phase difference β for the paths indicated in Figure 3.12, where $\vec{A}_1(\vec{s})$ and $\vec{A}_2(\vec{s})$ are the corresponding values of $\vec{A}$ in points along these paths. This phase difference is:

$$\begin{aligned}\beta &= \frac{q}{\hbar}\left\{\int_{\vec{r}_1}^{\vec{r}_2} \vec{A}_2(\vec{s})\cdot d\vec{s} - \int_{\vec{r}_1}^{\vec{r}_2} \vec{A}_1(\vec{s})\cdot d\vec{s}\right\} \\ &= \frac{q}{\hbar}\left\{\int_{\vec{r}_1}^{\vec{r}_2} \vec{A}_2(\vec{s})\cdot d\vec{s} + \int_{\vec{r}_2}^{\vec{r}_1} \vec{A}_1(\vec{s})\cdot d\vec{s}\right\} = \frac{q}{\hbar}\oint \vec{A}\cdot d\vec{s} \\ &= \frac{q}{\hbar}\Phi_{mag}\end{aligned} \tag{3.118}$$

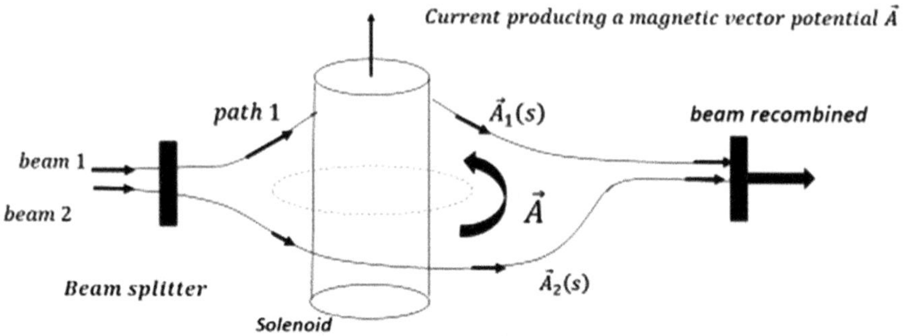

FIGURE 3.12 Illustration of the magnetic Aharonov-Bohm effect. A beam of electrons, is split into two beams, each one following a path surrounding a solenoid as shown. Due to the presence of the magnetic vector potential, there is a phase difference between the paths.

There is a similarity between Berry phase and the geometric phase of the Aharonov–Bohm effect, as can be seen comparing eqs. 3.91 and 3.118. If we analyze eq. 3.91 for the case when the parameter space is three-dimensional, that is the parameters of the set $\mathcal{R}$ considered for obtaining this equation are analogous to the component of a vector $\vec{S} = \{R_1, R_2, R_3\}$ in this space, we can write eq. 3.91 as

$$\gamma_n = i\oint \langle \varphi_n | \nabla_{\vec{R}} \varphi_n \rangle \cdot d\vec{R} = \oint \vec{\mathcal{A}}_n \cdot d\vec{s} \tag{3.119}$$

where $\vec{\mathcal{A}}_n \equiv i\langle \varphi_n | \nabla_{\vec{R}} \varphi_n \rangle$ is called the Berry connection and plays formally a role analogous to the vector potential $\vec{A}$ in eq. 3.118.

This similarity can be further appreciated if we consider the case of an atom slowly moving in a monochromatic laser field where the so-called Berry's connection $\vec{\mathcal{A}}_n \equiv \left\langle \psi_n\left(\vec{R}\right) \middle| \nabla \psi_n\left(\vec{R}\right) \right\rangle$ appears in the Hamiltonian in a form similar to the magnetic potential as given by eq. 3.119. As explained in Ref. [19], two types of degrees-of-freedom come into play. First, the center-of-mass motion of the atom can be described in terms of its position operator $\hat{r}$ and momentum operator $\hat{p} = -i\hbar\nabla_{\vec{r}}$. Second the internal dynamics of the atom correspond to transitions between electronic states induced by the laser field. Within the rotating-wave approximation, this internal dynamics is described by a time-independent Hamiltonian $\hat{H}_{int}(\vec{r})$, so that the total Hamiltonian is given by:

$$\hat{H}_{total}\left(\vec{r}\right) = \frac{\hat{p}^2}{2M} + \hat{H}_{int}\left(\vec{r}\right) \tag{3.120}$$

Considering the vector position of the atom $\vec{r}$ as an external parameter, the dressed states are defined as the eigenstates of the internal Hamiltonian

$$\hat{H}_{int}(\hat{r})\lfloor\psi_n(\hat{r})\rangle = E_n(\hat{r})\lfloor\psi_n(\hat{r})\rangle \tag{3.121}$$

The amplitude $\phi_n(\vec{r},t)$ of finding the atom at point $\vec{r}$ in the internal state $\lfloor\psi_n(\hat{r})\rangle$ satisfies the following equation:

$$i\hbar\frac{\partial\phi_n}{\partial t} = \left[\frac{\left(\hat{p}-\vec{A}_n(\vec{r})\right)^2}{2M} + E_n(\vec{r}) + \mathcal{V}_n(\vec{r})\right]\phi_n(\vec{r},t) \tag{3.122}$$

This last equation looks similar to the time-dependent Schrödinger's equation for a charged particle with unitary charge moving in the magnetic field associated with the vector potential $\vec{A}_n \equiv i\langle\varphi_n|\nabla_{\vec{R}}\varphi_n\rangle$. In addition, the particle feels a potential that is the sum of two terms. The first one is the energy $E_n(\vec{r})$ of the occupied dressed state, whereas, the second one is $\mathcal{V}_n(\vec{r})$ which represents the kinetic energy associated with the micromotion of the atom, as it makes virtual transitions between the effectively occupied dressed state ψ_n and all others are in dressed state ψ_l with $l \neq n$.[22]

As shown in Ref. [20] when a neutral atom moves in a laser field, the motion of its center-of-mass may imitate the dynamics of a charged particle in a magnetic field, appearing a Lorentz-type force and an Aharonov-Bohm phase. The latter is related to the Berry's phase that emerges when the atom adiabatically follows one of the dressed states corresponding to the atom–laser interaction.

3.7 EFFECT OF DISSIPATION IN A QUANTUM SYSTEM

Any microscopic system is always surrounded by some macroscopic environment and therefore there are interactions implying an irreversible energy transfer to the medium, so that for the system we are considering this occurs as a dissipative process with an associated arrow of time.

Dissipation effects play a relevant role in several areas of Physics, as for example, high energy physics (e.g., in quark-gluon physics) and in early universe physics, among others.

One of the main difficulties that must be considered in the study of dissipative quantum systems is that the canonical commutation relations (CCR) are not preserved due to damping terms. One alternative for solving this problem is to introduce fluctuating forces for preserving the canonical structure. Another way to handle the problem is to start with a Hamiltonian that describes the system, the bath and the system–bath interaction.

A full analysis of these issues requires a careful study at the level of specialists in quantum mechanics, that is, out of the scope and objectives of this book whose emphasis is about time in physical processes and systems. For that reason in this section, only an insight will be given focusing on the effect of dissipation on the time-reversal symmetry of a quantum system and making an analogy with a classical system. For that purpose, we will consider the case of the quantum-damped oscillator taking as reference the paper of E. Celeghini et al. (Ref. [21]).

The classical equation of motion of a damped oscillator is

$$m_{\ddot{x}} + \gamma_{\dot{x}} + K_{x} = 0 \tag{3.123}$$

The solution for this equation is of the form

$$x(t) = e^{-\Gamma t}\left(C_1 e^{i\text{Ù}t} + C_2 e^{-i\text{Ù}t}\right) \tag{3.123}$$

where $\Gamma = \gamma / 2m, \Omega = \sqrt{\frac{1}{m}\left(K - \frac{\gamma^2}{4m}\right)}$ and C_1, C_2 are constants depending on the initial conditions. If $x(0) = 0$ and $\dot{x}(0) = v_o$, the solution is of the form:

$$x(t) = \frac{v_o}{\Omega} e^{-\Gamma t} sin\Omega t \tag{3.125}$$

The presence of dissipation indicates that this process is irreversible and thus produces an asymmetry with respect to time reversal. In fact, making $t \to -t$ and $v_o \to -v_o$ in eq. 3.125 leads to

$$x(-t) = \frac{(-v_o)}{\Omega} e^{-\Gamma(-t)} sin\Omega(-t) = \frac{v_o}{\Omega} e^{\Gamma t} sin\Omega t \tag{3.126}$$

Equation 3.126 corresponds to an oscillation whose amplitude is increasing in time, which is in contradiction with the widely accepted evidence fact that in the presence of dissipation and without an external source of energy, a damped oscillator will become stopped.

To apply a canonical quantization scheme to the system described by eq. 3.123, the phase–space dimension is doubled introducing new degrees-of-freedom, to simplify the analysis considering an equivalent isolated system.

It can be assumed that these new degrees-of-freedom represent a single equivalent (collective) degree of freedom for the bath, which absorbs the energy dissipated by the oscillator. A procedure of doubling the phase–space dimension is necessary and the Lagrangian is written as

$$\mathcal{L} = m\dot{x}\dot{y} + \frac{1}{2}\gamma\left(x\dot{y} - \dot{x}y\right) - Kxy \tag{3.127}$$

The new variable y introduced in this Lagrangian grows as rapidly as the x solution decays; in this sense, y may be interpreted as describing an effective degree of freedom associated with the interaction with the heat bath to which the system described by eq. 3.123 is coupled. The dynamical equations for both oscillators can be obtained from this Lagrangian as follows:

$$\frac{\partial}{\partial t}\frac{\partial \mathcal{L}}{\partial \dot{y}} - \frac{\partial \mathcal{L}}{\partial y} = 0 \Rightarrow m\ddot{x} + \frac{\gamma}{2}\dot{x} + \mathcal{K}x = 0 \tag{3.128}$$

$$\frac{\partial}{\partial t}\frac{\partial \mathcal{L}}{\partial \dot{x}} - \frac{\partial \mathcal{L}}{\partial x} = 0 \Rightarrow m\dot{y} - \frac{\gamma}{2}\dot{y} + \mathcal{K}y = 0 \tag{3.129}$$

Equation 3.129 appears to be the time-reversed ($t \rightarrow -t$) of eq. 3.128. The canonical momenta are given by

$$p_x \equiv \frac{\partial L}{\partial \dot{x}} = m\dot{y} - \frac{1}{2}\gamma y \tag{3.130}$$

$$p_y \equiv \frac{\partial L}{\partial \dot{y}} = m\dot{x} + \frac{1}{2}\gamma x \tag{3.131}$$

From these momenta, the Hamiltonian is given by

$$\mathcal{H} = p_x\dot{x} + p_y\dot{y} - \mathcal{L} = \left(m\dot{y} - \frac{\gamma}{2}y\right)\dot{x} + \left(m\dot{x} + \frac{\gamma}{2}x\right)\dot{y} - m\dot{x}\dot{y} - \frac{\gamma}{2}\left(x\dot{y} - \dot{x}y\right) + \mathcal{K}xy$$

$$= \frac{\left(m\dot{y}\right)\left(m\dot{x}\right)}{m} + \frac{\gamma}{2}\left(\dot{x}y - y\dot{x}\right) + \mathcal{K}xy$$

$$= \frac{\left(p_x + \frac{\gamma}{2}y\right)\left(p_y - \frac{\gamma}{2}x\right)}{m} + \frac{\gamma}{2m}\left[\left(p_y - \frac{\gamma}{2}x\right)y - y\left(p_y - \frac{\gamma}{2}x\right)\right] + \mathcal{K}xy$$

$$= \frac{p_x p_y}{m} + \frac{\gamma}{2m}\left(p_y y - p_x x\right) + \left(\mathcal{K} - \frac{\gamma^2}{4m}\right)xy \tag{3.132}$$

Canonical quantization can be performed by introducing the commutators

$$[x, p_x] = [y, p_y] = i\hbar \tag{3.133}$$

$$[x, y] = [p_x, p_y] = 0 \tag{3.134}$$

We define creation and annihilation operators as

$$a = \alpha\left(p_x - i\beta x\right);\ a^{\dagger} = \alpha\left(p_x + i\beta x\right) \tag{3.135}$$

$$b = \alpha\left(p_y - i\beta y\right);\ b^{\dagger} = \alpha\left(p_y + i\beta y\right) \tag{3.136}$$

where α, β are reals defined as:

$$\alpha = \frac{1}{2\sqrt{\hbar m \Omega}} \tag{3.137}$$

$$\beta = 2\Omega m \tag{3.138}$$

and

$$\Omega = \sqrt{\frac{1}{4m}\left(K - \frac{\gamma^2}{m}\right)} \tag{3.139}$$

From eqs. 3.133–3.136, it can be verified that operators a and b satisfy the following commutation relations

$$[a, a^{\dagger}] = [b, b^{\dagger}] = 1 \tag{3.140}$$

$$[a, b] = [a, b^{\dagger}] = 0 \tag{3.141}$$

From eqs. 3.135 and 3.136:

$$p_x = \frac{a + a^{\dagger}}{2\alpha};\ p_y = \frac{b + b^{\dagger}}{2\alpha} \tag{3.142}$$

$$x = i\left(\frac{a - a^{\dagger}}{2\alpha\beta}\right);\ y = i\left(\frac{b - b^{\dagger}}{2\alpha\beta}\right) \tag{3.143}$$

Inserting these expressions in the Hamiltonian of eq. 3.132, we obtain

$$\begin{aligned}\mathcal{H}_P = {} & \frac{\left(a + a^{\dagger}\right)\left(b + b^{\dagger}\right)}{4\alpha^2 m} + \frac{i\gamma}{4m\alpha^2\beta}\left\{\left(b + b^{\dagger}\right)\left(b - b^{\dagger}\right) - \left(a + a^{\dagger}\right)\left(a - a^{\dagger}\right)\right\} \\ & - \left(K - \frac{\gamma^2}{4m}\right)\frac{\left(a - a^{\dagger}\right)\left(b - b^{\dagger}\right)}{4\alpha^2\beta^2}\end{aligned} \tag{3.144}$$

Let us define

$$A = \xi\left(a + b\right);\ B = \xi\left(a - b\right) \tag{3.145}$$

where ξ is a normalization constant. It can be verified that

$$\begin{aligned} A^{\dagger}A - B^{\dagger}B &= \xi^2\left\{\left(a^{\dagger} + b^{\dagger}\right)\left(a + b\right) - \left(a^{\dagger} - b^{\dagger}\right)\left(a - b\right)\right\} \\ &= \xi^2\left\{2b^{\dagger}a + 2a^{\dagger}b\right\} = \xi^2\left\{2ab^{\dagger} + 2a^{\dagger}b\right\}\end{aligned} \tag{3.146}$$

and

$$A^\dagger B^\dagger - AB = \xi^2\left\{\left(a^\dagger + b^\dagger\right)\left(a^\dagger - b^\dagger\right) - (a+b)(a-b)\right\}$$
$$= \xi^2\left\{a^\dagger a - aa - \left(b^\dagger b - bb\right)\right\} \tag{3.147}$$

These results are obtained, noting that

$$\left(a + a^\dagger\right)\left(b + b^\dagger\right) - \left(a - a^\dagger\right)\left(b - b^\dagger\right) = 2a^\dagger b + 2ab^\dagger \tag{3.148}$$

and

$$\left(b + b^\dagger\right)\left(b - b^\dagger\right) - \left(a + a^\dagger\right)\left(a - a^\dagger\right) = a^\dagger a^\dagger - aa - \left(b^\dagger b^\dagger - bb\right) \tag{3.149}$$

It can be verified that the Hamiltonian given by eq. 3.144 can be written as

$$\mathcal{H}_p = \hbar\Omega\left(A^\dagger A - B^\dagger B\right) + i\hbar\Gamma\left(A^\dagger B^\dagger - AB\right) \tag{3.150}$$

with

$$\xi = 1; \Gamma = \gamma / 2m \tag{3.151}$$

Equation 3.150 is a very important result in many ways. As pointed out in Ref. [21], if one retains only those states, $|\psi\rangle|$, which are annihilated by B, that is, $B|\psi\rangle| = 0$, then the eigenstates of $\mathcal{H}$ merge into those of the simple undamped harmonic oscillator when $\gamma \rightarrow 0$ ($\Omega \rightarrow \omega = \sqrt{K/m}$). Therefore the states generated by $B^\dagger$ represent the sink to which the energy dissipated by the quantum-damped oscillator flows. Hence, the B-oscillator thus represents the (single-mode) reservoir or heat bath coupled to the A-oscillator.

How do these results can be compared with the ideal quantum oscillator?

As shown in standards books of Quantum Mechanics (see e.g., Ref. [1]) an ideal quantum oscillator is expressed in terms of creation and annihilation operators, a and $a^\dagger$, respectively, the Hamiltonian is expressed as

$$\mathcal{H} = \hbar\Omega\left(a^\dagger a + 1/2\right) \tag{3.152}$$

3.8 ENTANGLEMENT AND CAUSALITY

Quantum mechanics equations imply an effect called by Einstein "spooky action at a distance" meaning that the properties of one system can instantly change the properties of the other system which seems to violate one of the main principles of the relativity theory which states that no signal can propagate with at a speed greater than the light velocity.

Albert Einstein disagreed with this result obtained from Quantum Mechanics claiming that it was not a complete theory and in a famous paper written in 1935 with B. Podolsky and N. Rosen, they designed a thought experiment (*gedanken experiment*) to show the incompleteness of the wave function in quantum mechanics to describe the reality.[22] This experiment is called the EPR paradox and can be summarized as follows: If two identical particles, identified as A and B, begin to move in opposite directions after a brief interaction, so that $\vec{\upsilon}_B = -\vec{\upsilon}_A$, and if at a given instant t_o, when A and B are arbitrarily far apart, the position of A and the momentum of B is measured, it is possible to know both the position and the momentum of A at t_o, since $\vec{p}_B = -\vec{p}_A$. Nevertheless, as known from the first courses of quantum mechanics, the Heisenberg uncertainty principle places a limitation on how precisely two complementary physical properties of a system can be measured simultaneously. Hence, according to the EPR paradox in order to comply with the uncertainty principle, this implies that somehow the second system would have information about the measuring event on the first system (so that the measurement on the second system would lose its precision in accordance to the uncertainty principle). But, if these two systems were sufficiently far apart from each other, then there could have been no way for the first system to have communicated any information to the second system in time. Therefore, from the EPR paradox, one could conclude that quantum mechanics is not a complete theory.

For relating the EPR paradox with the concept of time in quantum mechanics, we will first analyze how does the principle of causality work in this theory.

3.8.1 *CAUSALITY IN QUANTUM MECHANICS*

According to classical physics and our common view of reality, it is not possible for an effect to occur before its cause. This corresponds to the so-called principle of causality. In Einstein's theory of special relativity, this principle implies that an effect cannot occur from a cause that is not in the past light cone of that event. This fact is consistent with the accepted belief that causal influences cannot travel faster than light and with one of the basic postulates of the special relativity theory that there is not any object which can propagate faster than light.

Associated with the concept of causality is the concept of locality, meaning that it is impossible that a signal propagates faster than light. By the following example, it can be shown that NRQM causality is violated.

Let us first consider the case of a free particle of mass m and momentum $\vec{p}$ whose Hamiltonian is

$$\mathcal{H} = \frac{\vec{p}^{\,2}}{2m} \tag{3.153}$$

The amplitude of probability that a particle starting off at position state $|\vec{x}_o\rangle$ be detected at the position $|\vec{x}\rangle$ after time t is

$$AP(\vec{x},\vec{x}_o;t) = \langle\vec{x}|e^{-iHt}|\vec{x}_o\rangle = \langle\vec{x}|e^{-i\vec{p}^2/2m}|\vec{x}_o\rangle \tag{3.154}$$

By means of development as detailed in Ref. [23], it is obtained that:

$$AP(\vec{x},\vec{x}_o;t) = \left(\frac{m}{2\pi it}\right)^{3/2} e^{im(\vec{x}-\vec{x}_o)^2/2t} \tag{3.155}$$

This result for the amplitude of probability implies that the probability of finding a particle at two points separated by any distance and in an arbitrarily short time, is non-zero, which violates the causality principle from relativity. According to this result, it would be possible that a particle moves between two points with a velocity greater than the velocity of light violating this principle. In fact, if for simplicity, we consider the motion of a particle along the x-axis between two points with coordinates in space–time given by $P_1:(x_1 = 0, t_1 = 0)$ and $P_2:(x_2 = 2ct_2, t_2)$, clearly, the average velocity would be $2c$ implying that superluminal propagation would be possible.

In a space–time diagram, the position of these points is represented as shown in Figure 3.13.

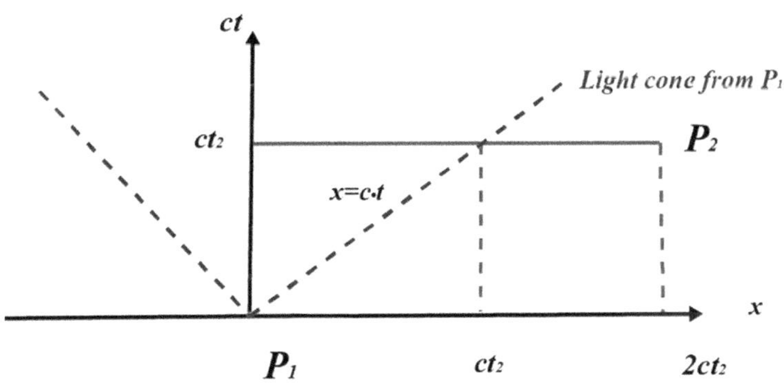

FIGURE 3.13 Space–time diagram showing two points P_1 and P_2 not causally connected. Dashed red lines correspond to the boundaries of the light cone for the particle at P_1.

As can be seen, the point P_2 is out of the light cone of P_1 meaning that it is not possible that these points correspond to two events causally connected. For example, the point P_1 could represent the absorption of a photon of an atom (atom 1) and P_2 the emission of a photon by another atom (atom 2) placed at a position such that these events are independent without any relation of causality between them.

The problem of non-causality presented by NRQM is solved by quantum field theory with the introduction of an antiparticle associated with each subatomic particle like an electron: the propagation of a particle across a space-like interval cannot be distinguished from the propagation of an anti-particle moving in the opposite direction. If we ask whether an observation made at a point P_1 placed at position $\vec{r}_1$ at time t_1 can affect an observation made at a point P_2 placed at position $\vec{r}_2$ at time t_2, the answer is that this is not possible since according to quantum field theory, the corresponding amplitudes for particle and antiparticle propagation exactly cancel each other and so causality is preserved.

3.8.2 ENTANGLEMENT

Entanglement is an effect resulting from the non-locality implicit in the NRQM which consists of the correlation between measurements made in two quantum systems, let us say particles such as electrons or photons, independent of the separation between these particles.

For instance, let us consider the states of polarization of a pair of photons moving away from each other after being produced both simultaneously as a pair at the source location. Each photon can be either in a polarization state $|H\rangle$ or $|V\rangle$ when measured by a polarizer. This is an entangled pair if, upon measurement, these two photons possess the same polarization states, for example, if the measurement on photon A results in the (+) state, then photon B must also be in the (+) state; if A is (-), then B is also (-). This is expressed by considering the following wave function for the entangled pair (also called an EPR pair):

$$|\psi\rangle = \frac{1}{\sqrt{2}}(|\leftrightarrow_1\rangle|\leftrightarrow_2\rangle + |\updownarrow_1\rangle|\updownarrow_2\rangle) \tag{3.156}$$

The kets inside the bracket of the above equation indicate vertical or horizontal polarization of the photons labeled with subscripts 1, 2. The states of polarization for a single photon are shown in Figure 3.14 and a combined state as that given by eq. 3.156 can be produced in the way shown in Figure 3.15 with two photons and two beam splitters.

FIGURE 3.14 Horizontal and vertical states of polarization of each photon of an entangled pair.

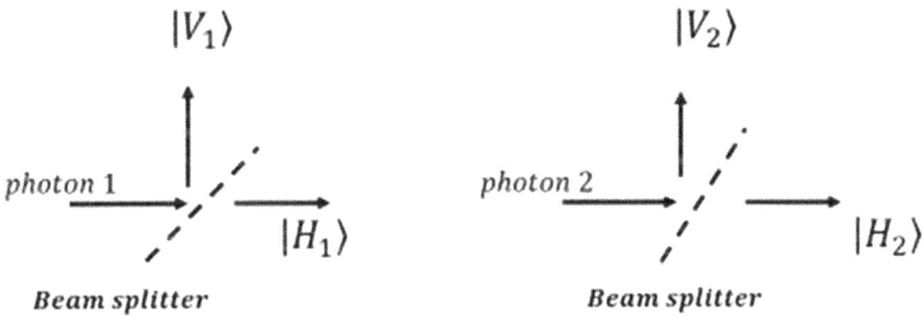

FIGURE 3.15 Each photon passing through a beam splitter can reach the horizontal or vertical polarization state, as shown.

The entangled state corresponding to eq. 3.156 can be obtained from the superposition as shown in Figure 3.16:

From eq. 3.156, we can see that if after a measurement of the state of polarization of photon 1 we obtain the state corresponding to ket $|\leftrightarrow\rangle$, that is horizontal polarization, the wave function collapses to the state $|\leftrightarrow_1\rangle|\leftrightarrow_2\rangle$ with probability ½. This means that instantaneously and independent of the separation between the photons when the measurement is made over photon 1, the photon 2 "knows" that the other must be in the state $|\leftrightarrow\rangle$. If the measurement of the state of polarization of photon 1 gives $|\updownarrow_1\rangle$, then the wave function collapses to the state $|\updownarrow_1\rangle|\updownarrow_2\rangle$ with probability ½. This also means that instantaneously and independent of the separation between the photons, due to the measurement made over photon 1 with the result $|\updownarrow_1\rangle$, the photon 2 acquires the state $|\updownarrow_2\rangle$. This looks like instantaneous transmission of information between photons violating the principle of the relativity theory which establishes that no signal can propagate with a velocity higher than light velocity in vacuum (c).

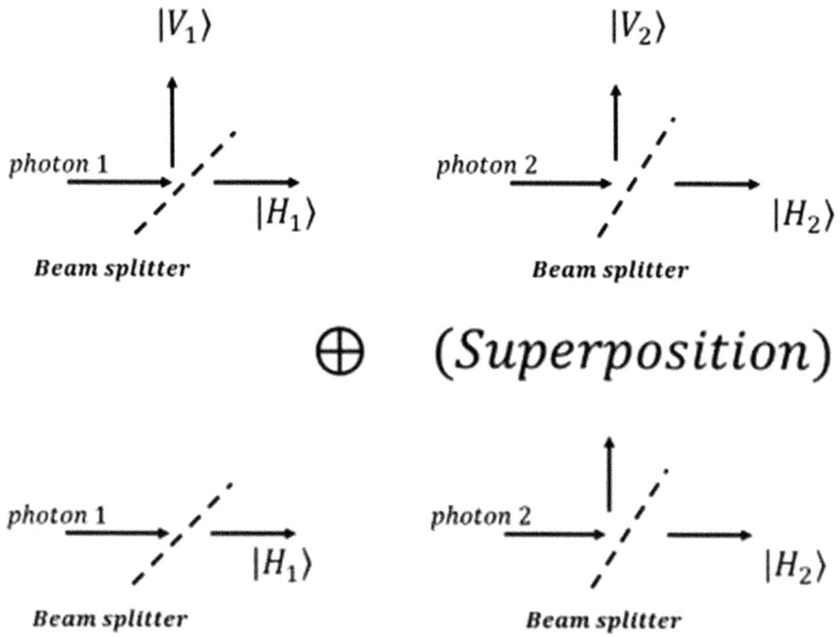

FIGURE 3.16 Superposition of polarization states for obtaining two entangled photons.

Another classical example of the EPR paradox based on the simplified version of David Bohm, considers the decay of the neutral pi meson resulting in a pair electron–positron:

$$\pi \rightarrow e^- + e^+ \tag{3.157}$$

If the pi meson is originally at rest, by conservation of momentum, the electron and positron produced in this process must fly in opposite senses with equal velocities by conservation of momentum. Since the pion has spin zero, by conservation of angular momentum, the electron and positron must be in the singlet configuration called an EPR pair,

$$\frac{1}{\sqrt{2}}(|\uparrow_e \downarrow_p\rangle - |\downarrow_e \uparrow_p\rangle) \tag{3.158}$$

In the above equation, the minus sign inside the brackets is necessary for the fulfillment of the exclusion principle considering that both electrons as positrons are fermions, and therefore they must satisfy the exclusion principle.

According to the principles of quantum mechanics when a measurement of the spin is made with the system being in the above combination of eigenstates, there is an instantaneous collapse to a state corresponding to the eigenvalue measured. In this case, if for example, a measurement of the spin of the electron is made and it is found to have the spin up, other simultaneous measurement of the spin of the positron will give its spin down and viceversa. This correlations called *entanglement* of quantum states is the same for any arbitrary distance between these two particles. This implies that there is again a contradiction with the relativity theory which says that no influence can propagate with a velocity greater than the velocity of the light. This is called the principle of *locality*. In a well-known paper written by Einstein, Podolsky and Rosen (Ref. [22]), they claimed that quantum mechanics gives an incomplete theory of reality and that in order to fully characterize the state of a system, besides the wave function, some other quantity called a hidden variable, is required. Nevertheless, in 1964 J.S. Bell proved that any local hidden variable theory must hold an inequality, called the Bell inequality.[24] After 1964 several experiments have been made proving that this inequality is violated while quantum-mechanical predictions are correct, demonstrating that nature is fundamentally nonlocal.

As shown in several publications and textbooks, in the above cases, there is no transmission of any useful information between two macro-scopically separated points. This can be proved by means of the following argumentation.

Suppose two persons A and B agree in advance to measure the component z of the spin (S_z). Then without asking A, by means of its measurement B can know with precision, the result of the measurement made by A. However, this must not be interpreted as that A and B are communicated. What B observes is a random sequence of positive and negative signs but not an information that could be considered as useful. When **A and B** become together, they only will be able to verify the correlations predicted by quantum mechanics comparing the respective notes.

Let us consider now that A agrees initially to measure the z component of the spin and B the x component (S_x). The results obtained by A will be completely uncorrelated with the results of the measurements made by B, so there is not any transferred information. *What would happen if A suddenly breaks his promise and without telling B starts measuring S_x?*, There will be now complete correlations between the results obtained both by A and B. However, B has not any possibility of inferring that A has changed the orientation of his analyzer. In this case, B will continue observing a random sequence of signs + and – when he reviews his notes, Therefore, again there will not be transferred information.

In addition to the above arguments, based on Ref. [25], we will give here a possible solution to the EPR paradox using the following arguments from time dilation of relativity theory:

As it will be discussed in Chapter 4, according to special relativity theory, a moving clock runs slower than a stationary clock. This corresponds to the following equation

$$\Delta t' = \Delta t\left(1-\frac{\upsilon^2}{c^2}\right)^{1/2} \tag{3.159}$$

where $\Delta t'$ is the time interval between two events measured by the clock moving with velocity υ and Δt is the corresponding time interval measured by a stationary clock.

A photon moves with light velocity, that is $\upsilon = c$ and therefore $\Delta t' = 0$, so that photon does not observe time passing at all; it is as the time stands still in its inertial frame. For an observer fixed to ground time passes by as photon is created in the past, and as it moves through in space in the present time. However, as seen by an imaginary observer fixed to the photon, the time at its past creation is exactly the same time as the present. Therefore, we can say that the photon at its past creation time is exactly the same, physically identical photon at the present time, and vice versa.

Relating the above argumentation to the EPR paradox, when we make a measurement of the polarization state of photon A at the present time, this measurement determines the state of the present photon A. But, the photon A at the present time is exactly the same photon that existed at the time of its creation. Thus, whatever the determination made to the present photon A, it also must have been made instantly to the photon A existing at the creation time.

Next, since the past photon A at the creation time is now known, consequently, this knowledge determines the state of the past photon B at the exact same creation time. Following the same argumentation line as above, the determination of the past photon B is equivalent with the instant determination of the current photon B.

To summarize, a measurement on the current photon A instantaneously determines the state of both photons A and B.

Therefore, the faster-than-light (FTL) interaction can occur between the two entangled systems, and this invalidates the assumptions made by EPR paradox and Bell's inequality.

Moreover, this FTL communication does not violate the causality principle implicit in special relativity theory because the present determination cannot go any further past than the creation time of the photon, and the present determination can never alter the past event. In order to influence the past event, it would require to make a measurement in the past to detect the effects of the present experiment. However, if such measurement had been made in the past, it would mean that the present measurement would have been not the first measurement. Therefore, the measurement made at present would not have been able to determine the state of the photon at its beginning. If it were possible that the present measurement could make the determination, this implies that in the past there was no measurement. This, in turns, implies that the past cannot be altered by the present measurement.

Also, even though the faster than light communication could have been done instantaneously toward a point placed at a great distance, it obeyed all rules set by the Special Relativity Theory, that is, no infinite energy was required because no physical entity has traveled faster than light.

We are led to distinguish two kinds of influences: the *causal* variety which produces real and detectable modifications in some of the physical properties of the receiver; the other type of influence corresponds to what in Ref. [2] are called ***"ethereal"*** for which there is not transmission of energy or information, and for which the only available evidence is a correlation in the data obtained from the two separate subsystems. Causal influences cannot propagate faster than light, but there are not proved reasons why ethereal ones should not. The influences related with the collapse of the quantum wave function after a measurement are of the latter type.

3.9 WHAT ABOUT TIME SYMMETRY IN QUANTUM OPTICS?

In introductory courses in quantum physics one learns the first principles about the interaction between light and matter, in particular the induced transitions between a ground and an excited energy level of the electron of an atom due to the absorption or emission of a photon. The processes of absorption and spontaneous emission of photons determined by the lifetime of excited states, suggest the possibility of time symmetry for an isolated field-atom system. In fact, Optics theory is governed by Maxwell's equations which obey time-reversal symmetry (Ref. [26]) and for perfect excitation of the atom the photon should have the same behavior as that of the time-reversed version of a spontaneously emitted photon. This holds true in free space, in a waveguide as well as for atoms in a resonator.[27] Considering that the interaction between an optical field and a two-level atom is one of the cases that best allow to illustrate the light–matter interaction, in this section, we analyze the time symmetry of this interaction considering the semiclassical Rabi model and the Jaynes–Cummings model.

We consider an atom at rest subject to an oscillating electric field which induces transitions between the ground and excited states. For simplicity, we consider only two quantum states for the atom (a two-level system) assuming that the frequency of the electric field is close to that corresponding to the difference of energy between these states.

A semiclassical model of the system, where the atom is represented as a quantum system but the field is classically represented, leads to oscillations in the probabilities of finding the atom in the ground or in the excited states. These oscillations, known as Rabi oscillations, show time translational invariance and time-reversal symmetry.

On the other hand, a full quantum model where the electric field is quantized, predicts oscillations which show a decay but also a revival phenomenon, breaking the two above time symmetries. However, the application of a controlled phase kick by means of an electromagnetic pulse, leads to a time reversal of the system's evolution.

3.9.1 THE SEMICLASSICAL MODEL OF RABI

The perturbation theory approach usually applied to describe the atom–field interactions, assumes that the initial atomic state population is kept essentially unchanged, that is, the probability amplitude to find the atom in any other state remains small. Nevertheless, a strong laser field of frequency near resonance with a pair of atomic levels will give rise to a large population transfer to the near resonant state but not to any other. In this case, only the two dominant states are considered and the problem can be solved by means of a semiclassical model known as Rabi model, so named because of the original setting for the study of the magnetic resonance as developed by Isidor Isaac Rabi several years ago.[28]

This model is briefly described here giving only the main results. Detailed analysis can be found in several books as, for example, in Ref. [29]. In Figure 3.17 based on that reference a two-level system is shown with a ground and excited states labeled as $|g\rangle$ and $|e\rangle$, respectively. The corresponding energies of these states are E_g and E_e, respectively, so that the resonance frequency for transition between these states is

$$\omega_o = \frac{E_e - E_g}{\hbar} \tag{3.160}$$

Let us consider that a monochromatic laser field with frequency $\omega < \omega_o$ is applied so that the detuning is $\Delta = \omega_o - \omega$

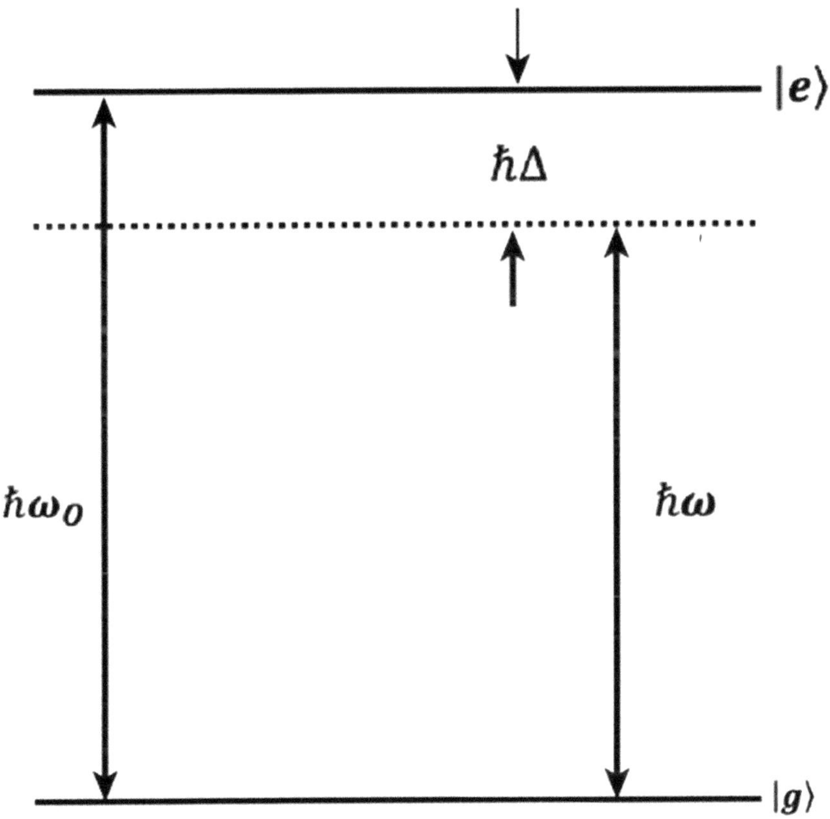

FIGURE 3.17 States and energy differences for a two-level quantum system.

The dynamics of this system is described by the Schrödinger equation as

$$\hbar\frac{d}{dt}|\psi\rangle = (\mathcal{H}_o + V_o cos\omega t)|\psi\rangle \tag{3.161}$$

In the above equation, $\mathcal{H}_o$ is the Hamiltonian operator for the isolated atom and the second term represents the interaction between the laser field and the atom where V_o is the corresponding potential which is given by

$$V_o = -\vec{d}\cdot\vec{E}_o \tag{3.162}$$

with $\vec{E}_o$ being the amplitude of the electric field vector corresponding to the laser field and $\vec{d}$ is the electric dipole operator. The transition matrix between the two states has nonzero components given by

$$\Im \equiv \langle e|V_o|g\rangle \tag{3.163}$$

The state of the system at any time is given by

$$|\psi(t)\rangle = C_g(t)e^{-iE_g t/\hbar}|g\rangle + C_e(t)e^{-iE_e t/\hbar}|e\rangle \tag{3.164}$$

The coefficients C_e and C_g are the amplitude probabilities for finding the system in the state $|g\rangle$ or $|e\rangle$, respectively. Combining the above equations, we obtain the following system of differential equation for these coefficients:

$$\dot{C}_g = -\frac{i}{2\hbar}\Im e^{i(\omega-\omega_o)t}C_e \tag{3.165}$$

$$\dot{C}_e = -\frac{i}{2\hbar}\Im e^{-i(\omega-\omega_o)t}C_g \tag{3.166}$$

$$\ddot{C}_e + i(\omega-\omega_o)\dot{C}_e + \frac{1}{4}\frac{\Im^2}{\hbar^2}C_e = 0 \tag{3.167}$$

Assuming a solution of eq. 3.167 of the form $e^{i\lambda t}$, we obtain the following values for the oscillation frequency between states $|g\rangle$ and $|e\rangle$:

$$\lambda_\pm = \frac{1}{2}\left\{\Delta \pm \left(\Delta^2 + \frac{\Im^2}{\hbar^2}\right)^{1/2}\right\} \tag{3.168}$$

If initially the atom is in its ground state $|g\rangle$, so that $C_g(0) = 1$ and $C_e(0) = 0$, we obtain:

$$C_e(t) = -i\frac{\Im}{\Omega_R\hbar}e^{-it\Delta/2}sin(\Omega_R t/2) \tag{3.169}$$

$$C_g(t) = e^{it\Delta/2}\left\{cos(\Omega_R t/2) - i\frac{\Delta}{\Omega_R}sin(\Omega_R t/2)\right\} \tag{3.170}$$

where $\Omega_R = [\Delta^2 + (\Im/\hbar)^2]^{1/2}$ is called the Rabi frequency. The probability to find the system in the state $|e\rangle$ is given by

$$P_e(t) = |C_e(t)|^2 = \frac{\Im^2}{\Omega_R^2 \hbar^2} sin^2\left(\frac{\Omega_R t}{2}\right) \tag{3.171}$$

Expressing the detuning frequency Δ in terms of $\Im/\hbar$ (which is the oscillation frequency of the atomic inversion) as $\Delta = \gamma/\Im\hbar$ with γ a real number, eq. 3.171 can be recast as

$$P_e(t) = \frac{1}{1+\gamma^2} sin^2\left(\frac{\Omega_R t}{2}\right) \tag{3.172}$$

Figure 3.18 shows the probability $P_e(t)$ versus $\Omega_R t$ for three values of the detuning frequency. It can be seen that according to the semiclassical Rabi model and without any dissipation, the atomic oscillations have both time reversal as translational symmetry.

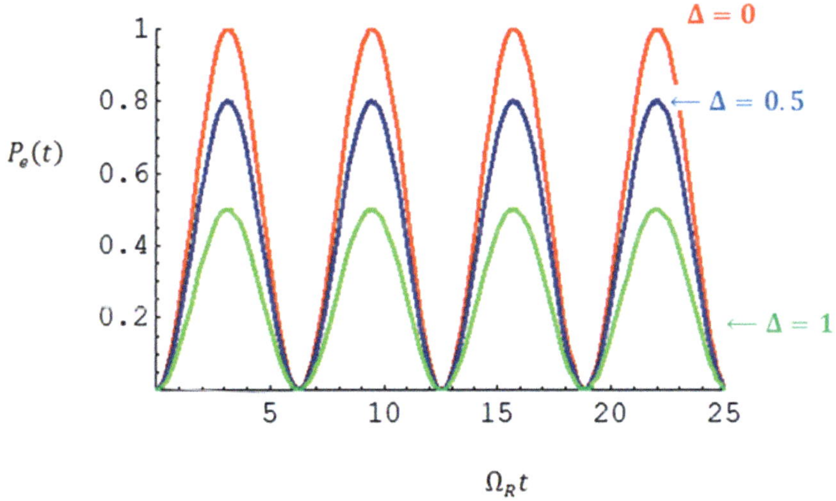

FIGURE 3.18 Atomic oscillations according to semiclassical Rabi model for the atom initially in the ground state.

3.9.2 COLLAPSE AND REVIVAL OF RABI OSCILLATIONS

The coupling between an atom and the quantized electromagnetic field in its environment leads to many fascinating phenomena important in the development of quantum mechanics, including first of all, spontaneous emission of photons. If an atom is in an excited state $|e\rangle$, it will emit a photon and irreversibly fall to a lower state $|g\rangle$ in a characteristic time (on the scale of nanoseconds for visible light, milliseconds for transitions in the microwave region). This process is irreversible because of the continuous density of states available to the photon, which once emitted, flies off at the speed of light for never returning.

Very interesting things happen if the atom is placed inside a cavity which prevents the photon from escaping. If one of the cavity modes has an energy with a value close to that corresponding to the atom transition frequency, then the atom and cavity can coherently exchange energy back and forth by oscillating between the excited and ground state emitting or absorbing a photon. This gives rise to the phenomenon of Rabi oscillations which is described in this section as follows.

In the semiclassical Rabi model, it has been assumed that the atom interacts with a single-mode free field (represented as a plane electromagnetic wave). Nevertheless, seldom a free atom interacts with a single mode field but rather with an infinite number of modes and an adequate description requires to consider a quantized model of the field where the electric and magnetic fields are represented by operators instead of continuum variables. In particular, a phenomenon of collapse and revival of Rabi oscillations can be observed if the atom interacts with light in a coherent state which is the state more similar to a classical light field and shows a Poisson photon number distribution with an average photon number $\bar{n}$. In Appendix 3.3, we give a brief introduction to the subject of quantization of the electromagnetic field and in Appendix 3.4, we describe in more details the phenomenon of collapse and revival of Rabi oscillations.

In quantum optics, a coherent state is defined as a state of the quantized electromagnetic field which presents maximal coherence and a behavior very similar to a classical field. Coherence is an ideal property of waves which means that there exists a constant interference, both temporally as spatially. It must be considered that in quantum mechanics, particles such as electrons are described by a wave function (which can be obtained from Schrödinger's equation). If there exists a definite phase relation between different states, the system is said to be coherent.

A coherent state of light can be represented the following equation

$$|\alpha\rangle = exp\left(-\frac{|\alpha|^2}{2}\right)\sum_{n=0}^{\infty}\frac{\alpha^n}{\sqrt{n!}}|n\rangle \tag{3.173}$$

In this case, if initially the atom is in the excited state, the probability that it will remain in this state is:

$$P_e(t) = \frac{1}{2}\left\{1 + e^{-\bar{n}} \sum_{n=0}^{\infty} \frac{\bar{n}^n}{n!} cos\left(2\lambda t\sqrt{n+1}\right)\right\} \tag{3.174}$$

where λ is a characteristic frequency proper of the Jaynes–Cumming model. As detailed in Appendix 3.4, this frequency is proportional to the amplitude of the applied electric field.

In the above equation, $\bar{n}$ is the expectation value of the number operator $\hat{n} = a^{\dagger}a$, or mean value of photons, and is given by

$$\bar{n} = \langle \alpha | a^{\dagger} a | \alpha \rangle = |\alpha|^2 \tag{3.175}$$

The probability given by eq. 3.174 for two values of $\bar{n}$ is shown in Figure 3.19.

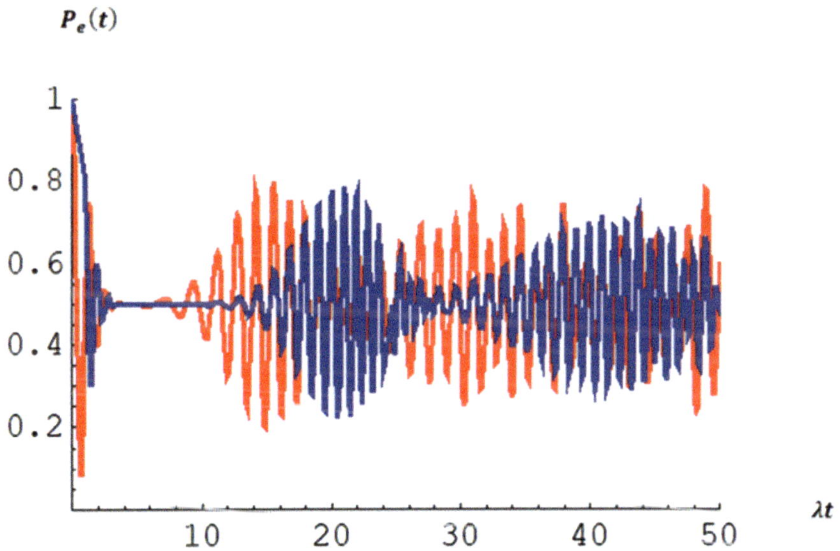

FIGURE 3.19 Probability that the atom remais in the excited state when it is subject to a laser field in a coherent state with average number of photons 5 (red curve) and 10 (blue curve).

From Figure 3.19, we observe collapses and revivals of the atomic populations and that as the mean value of photons increases the revivals occurs at a later time. This can be understood considering that in eq. 3.174, there is a sum of oscillating terms, each of them having a particular Rabi frequency $2\lambda\sqrt{n+1}$. If the neighboring terms are in phase with each other, it is expected that there will be a constructive interference. This should be so whenever neighboring phases differ by some multiple of 2π. Since only those modes with number of photons around $\bar{n}$ will contribute, revivals should occur for times t_R given by

$$\left[\Omega_{\bar{n}+1} - \Omega_{\bar{n}}\right] t_R = 2\pi j \tag{3.176}$$

where $j = 0, 1, 2, \ldots \ldots$

For $\bar{n} \gg 1$ the term inside brackets in the left-hand side of eq. 3.176 can be approximated as

$$\Omega_{\bar{n}+1} - \Omega_{\bar{n}} = 2\lambda\left\{\sqrt{\bar{n}+2} - \sqrt{\bar{n}+1}\right\} \approx 2\lambda\sqrt{\bar{n}}\left\{\left(1+\frac{1}{\bar{n}}\right) - \left(1+\frac{1}{2\bar{n}}\right)\right\} = \frac{\lambda}{\sqrt{\bar{n}}} \tag{3.177}$$

Combining eqs. 3.176 and 3.177 we deduce that revival of Rabi oscillations occur at times given by

$$t_R = \frac{2\pi j \sqrt{\bar{n}}}{\lambda} \tag{3.178}$$

where $j = 0, 1, 2, \ldots \ldots$

In Ref. [30], a scheme for inducing revival of Rabi oscillations by time reversal is proposed. For that purpose, the system is allowed to evolve for a given time T and then the atom is subject to a short electromagnetic pulse which allows to obtain a time reversal of the system's unitary evolution. At instant $2T$, the system is expected to "come back" exactly in the initial state, in the absence of decoherence. This scheme is analyzed in more detail in Appendix 3.4.

3.10 PT—SYMMETRIC QUANTUM MECHANICS

3.10.1 INTRODUCTION

Quantum mechanics is an area of physics that applies to the analysis of phenomena occurring at a microscopic scale. One of the main postulates of quantum mechanics is that energy is not actually continuous but instead needs to be transferred in packages of certain sizes, called quanta. The notion that the energy values of a physical system could be discrete was first suggested by Ludwig Boltzmann in 1877. In an attempt to solve the problem with the black body radiation, Max Planck proposed some years later that energy is emitted and absorbed in discrete amounts or quanta. This hypothesis was consistent with experimental observations. In 1905, Albert Einstein used a quantum-based model to describe the photoelectric effect. During the first 30 years of the 20th century, profuse theoretical and experimental research was made in order to verify the predictions given by the quantum models for light and atoms. A rigorous formulation of quantum mechanics was proposed later by Dirac, von Neumann, Hilbert and Weyl, among other notable physicists of the 20th century.

As discussed at the beginning of this chapter, a quantum system is described by an equation called the "Schrödinger equation" which is the quantum mechanical analog of Newton's second law of motion that describes the time evolution of a quantum mechanical system. The most general form the Schrödinger equation is as follows

$$i\hbar\frac{\partial\Psi}{\partial t} = \mathcal{H}\Psi \tag{3.179}$$

where Ψ is the wave function of the system and $\mathcal{H}$ is the Hamiltonian operator. The energy levels of the system can be determined by solving the eigenvalues equation for the stationary states

$$\mathcal{H}\Psi = E\Psi \tag{3.180}$$

A key feature of the formulations made by Dirac and other physicists of the first half of last century was that for obtaining consistent solutions for the energy and other observables the Hamiltonian needs to be Hermitian. This means that the Hamiltonian $\mathcal{H}$ is equal to its own adjoint, or in other words, to the matrix that results from transposing and conjugating its elements. The Hermiticity of $\mathcal{H}$ is one of the axioms of Quantum Mechanics which guarantees that all its eigenvalues are real, and also, that the corresponding time-evolution operator, here denoted as U_{ev}, is unitary, which means that its transpose is equal to its inverse, $U_{ev}^{T} = U_{ev}^{-1}$. The unitarity of U_{ev} is in turn necessary for the probability of the system to be conserved, which is fundamentally important for closed quantum mechanical systems.

In 1999, Bender, Boettcher and Meisinger published an article investigating the prospect that the requirement of Hermiticity of a Hamiltonian in a quantum mechanical system might be unnecessarily strict.[31] Bender and coworkers have shown that the assumption of Hermiticity can be relaxed under the circumstance that the Hamiltonian is invariant under the combined symmetry *PT* where *P* denotes the parity operation and *T* denotes time reversal. In Ref. [32], K. Jones-Smith and H. Mathur have formulated quantum mechanics for non-Hermitian Hamiltonians that are invariant under

PT for the case that time-reversal symmetry is odd ($T^2 = -1$), generalizing prior work for the even case ($T^2 = 1$).

In Ref. [33], Zhang and Wu analyze the theoretical framework of quantum systems with a non-Hermitian Hamiltonian emphasizing about its diagonalizability.

In this section, we present a brief analysis of those systems and explain how equivalent non-Hermitian Hamiltonian can be obtained in optical systems.

3.10.2 EIGENVALUES OF A PT-SYMMETRIC HAMILTONIAN

In general, in introductory courses on quantum mechanics, it is taught that the Hamiltonian operator must be Hermitian in order to obtain real values for the energy levels and that probability be conserved. If the Hamiltonian is Hermitian, then $\mathcal{H}^\dagger = \mathcal{H}$, where the symbol † denotes the usual Dirac Hermitian conjugation, meaning that the elements of its corresponding matrix are transposed and their complex conjugate is considered. Rather recently it has been recognized that the requirement of Hermiticity, which generally is considered as an axiom of quantum mechanics, may be replaced by the requirement of symmetry under space–time reflection symmetry (PT-symmetry) without losing any of the essential physical features of quantum mechanics.

Physical theories having a non-Hermitian PT-symmetric Hamiltonians show strange and unexpected properties at the classical as well as at the quantum level. A Hamiltonian is PT-symmetric if the operator PT commutes with this Hamiltonian ($[PT, \mathcal{H}] = 0$), where PT is defined as the successive application of the operators parity P and time reversal T defined by their action on the position and momentum operators as follows:

$$\left.\begin{aligned} P &: \hat{x} \to -\hat{x}; \hat{p} \to -\hat{p} \\ T &: \hat{x} \to \hat{x}; \hat{p} \to -\hat{p}; i \to -i \end{aligned}\right\} \tag{3.181}$$

where $\hat{p}$ and $\hat{x}$ are the momentum and position operators, respectively, and i denotes the imaginary unit ($\sqrt{-1}$).

While the Hermiticity property of a Hamiltonian is a sufficient condition for obtaining a real spectrum for the energy, this is not the case for the PT-symmetry of a Hamiltonian. In some cases, the spectrum of a Hamiltonian with PT-symmetry is not real, but it can be shown that if the eigenstate functions are the same for $\mathcal{H}$ as for PT, the spectrum of $\mathcal{H}$ will be entirely real.

EXAMPLE 3.9

Prove that a non-Hermitian Hamiltonian having PT-symmetry gives real values of energy.

SOLUTION

If the Hamiltonian $\mathcal{H}$ has PT-symmetry, $[\mathcal{H}, PT] = 0$ and if $\mathcal{H}$ and PT have a common eigenvector which we will designated as $|\phi\rangle$, then

$$\mathcal{H}|\phi\rangle = E|\phi\rangle \tag{3.182}$$

$$(PT)|\phi\rangle = \lambda|\phi\rangle \tag{3.183}$$

where E and λ are the eigenvalues of $\mathcal{H}$ and PT, respectively.

Applying the operator PT to both sides of eq. 3.182 we have

$$PT\mathcal{H}|\phi\rangle = PTE|\phi\rangle = E^* PT|\phi\rangle = E^*\lambda|\phi\rangle \tag{3.184}$$

Similarly, applying the operator $\mathcal{H}$ to both sides of eq. 3.183 leads to

$$\mathcal{H}PT|\phi\rangle = \mathcal{H}\lambda|\phi\rangle = \lambda\mathcal{H}|\phi\rangle = E\lambda|\phi\rangle \tag{3.185}$$

But from $[\mathcal{H}, PT] = 0$, we should have that

$$PT\mathcal{H}|\phi\rangle = \mathcal{H}PT|\phi\rangle \tag{3.186}$$

This implies that the right-hand sides of eqs. 3.184 and 3.185 must be equal, and therefore $E^* = E$ proving that if the Hamiltonian $\mathcal{H}$ has PT-symmetry, their eigenvalues E are real.

EXAMPLE 3.10

Consider the following Hamiltonian

$$\mathcal{H} = \begin{bmatrix} i\alpha & -\beta \\ -\beta & -i\alpha \end{bmatrix} \tag{3.187}$$

where α, β are real and positive values. In Examples 3.11 and 3.12, we will delve into the physical meaning of this Hamiltonian. For now, let us verify that although this Hamiltonian is non-Hermitian it has PT-symmetry.

SOLUTION

The Hermitian conjugate of this Hamiltonian is $\mathcal{H}^\dagger = \begin{bmatrix} -i\alpha & -\beta \\ -\beta & i\alpha \end{bmatrix}$ which is different from $\mathcal{H}$. Therefore, this Hamiltonian is not Hermitian.

Let us remember that transformation T performs complex conjugation and consider that the parity operator for this case is defined as[34]:

$$P = \begin{bmatrix} 0 & 1 \\ 1 & 0 \end{bmatrix} \tag{3.88}$$

Therefore, if $|\phi\rangle$ is an eigenket of the Hamiltonian $\mathcal{H}$

$$PT\mathcal{H}\,|\phi\rangle = \begin{bmatrix} 0 & 1 \\ 1 & 0 \end{bmatrix}\begin{bmatrix} -i\alpha & -\beta \\ -\beta & i\alpha \end{bmatrix} T\,|\phi\rangle = \begin{bmatrix} -\beta & i\alpha \\ -i\alpha & -\beta \end{bmatrix} T\,|\phi\rangle$$

and $\mathcal{H}PT|\phi\rangle = \begin{bmatrix} i\alpha & -\beta \\ -\beta & -i\alpha \end{bmatrix}\begin{bmatrix} 0 & 1 \\ 1 & 0 \end{bmatrix} T|\phi\rangle = \begin{bmatrix} -\beta & i\alpha \\ -i\alpha & -\beta \end{bmatrix} T|\phi\rangle$

Comparing these last two equations, we can see that the operator PT commutes with this Hamiltonian and therefore there is PT-symmetry.

The eigenvalues of the Hamiltonian given by eq. 3.187 are given by

$$\epsilon = \pm\sqrt{\beta^2 - \alpha^2} \tag{3.189}$$

These eigenvalues are real if $\beta > \alpha$ and this corresponds to the region of unbroken PT-symmetry.

EXAMPLE 3.11

Consider a quantum system described by the 1×1 matrix $\mathcal{H} = [a + ib]$ where a and b are real numbers.

a) Obtain the corresponding wave function $\Psi(t)$ for this system and the corresponding probability density $\wp(t) = \Psi^*(t)\Psi(t)$.
b) Obtain the time reversal versions of the Hamiltonian $\mathcal{H}$ and the corresponding time-reversed wave function.
c) A PT-symmetric Hamiltonian can be constructed by means of a combination of the Hamiltonian $\mathcal{H}$ and its time-reversed version $\mathcal{H}_{tr} = [a + ib]$.[35,36] Show that the combined Hamiltonian $\mathcal{H}_{combined}$ is not Hermitian but that it has PT-symmetry.

SOLUTION

a) From the Schrödinger equation (with $\hbar = 1$), we have
$i\frac{\partial\Psi}{\partial t} = [a+ib]\,\Psi \Rightarrow \Psi(t) = Ae^{(-ai+b)t}$, where A is a constant.

The corresponding probability density is given by $\wp(t) = \Psi^*(t)\Psi(t) = |A|^2 e^{2bt}$, which decreases in time if $b < 0$.

b) The time-reversal versions of the Hamiltonian $\mathcal{H}$ is given by $\mathcal{H}_{tr} = [a + ib]$ and from the Schrödinger equation $i\frac{\partial\Psi_{tr}}{\partial t} = [a-ib]\,\Psi_{tr}$, we obtain that the corresponding time-reversed wave function is $\Psi_{tr}(t) = Be^{(-ai-b)t}$, where B is a constant.

The corresponding probability density is given by
$\wp_{tr}(t) = \Psi_{tr}^*(t)\Psi_{tr} = |B|^2 e^{2bt}$, which grows in time if $b < 0$.

c) The combined Hamiltonian can be written as

$$\mathcal{H}_{combined} = \begin{bmatrix} a+ib & 0 \\ 0 & a-ib \end{bmatrix}$$

This Hamiltonian describes two subsystems that do no interact; one has gain and the other has loss. Clearly, this Hamiltonian is not Hermitian since

$$\mathcal{H}^{\dagger}_{combined} = \begin{bmatrix} a-ib & 0 \\ 0 & a+ib \end{bmatrix} \neq \mathcal{H}_{combined}$$

As made by Bender,[34] the parity operator $\wp$, which interchanges the two subsystems, can be defined as

$$\wp = \begin{bmatrix} 0 & 1 \\ 1 & 0 \end{bmatrix}, \text{ so that } \wp\mathcal{H}_{combined} = \begin{bmatrix} 0 & a-ib \\ a+ib & 0 \end{bmatrix}$$

Since the operators of parity and time reversal are independent of one another, the commute, that is $[\wp, T] = 0$. However, the combined Hamiltonian $\mathcal{H}_{combined}$ is symmetric under the combined operation $\wp, T$. In fact, it can be readily verified that

$$\wp T\mathcal{H}_{combined}(\wp T)^{-1} = \mathcal{H}_{combined} \Rightarrow [\wp T, \mathcal{H}_{combined}] = 0$$

EXAMPLE 3.12

The system described by the Hamiltonian $\mathcal{H}_{combined}$ is not in equilibrium because the wave function grows in time in one of the subsystems but in the other decays. Equilibrium can be achieved by coupling the two subsystems inserting a coupling parameter g, so that

$$\mathcal{H}_{coupled} = \begin{bmatrix} a+ib & g \\ g & a-ib \end{bmatrix} \tag{3.190}$$

a) Verify that this Hamiltonian is not Hermitian but that it has PT-symmetry.
b) Obtain the eigenvalues of $\mathcal{H}_{coupled}$ and determine the condition for transition from the region of unbroken symmetry to that of broken symmetry.

SOLUTION

a) It can be readily obtained that

$$\mathcal{H}^{\dagger}_{coupled} = \begin{bmatrix} a-ib & g \\ g & a+ib \end{bmatrix} \neq \mathcal{H}_{coupled}$$

Therefore, the Hamiltonian $\mathcal{H}_{coupled}$ is not Hermitian, but it has PT-symmetry since it can be verified that

$$\wp T\mathcal{H}_{coupled}\left(\wp T\right)^{-1} = \wp T\mathcal{H}_{coupled}\left(T\wp\right)^{-1} = \mathcal{H}_{coupled} \Rightarrow \left[\wp T, \mathcal{H}_{coupled}\right] = 0$$

b) The eigenvalues of $\mathcal{H}_{coupled}$ are obtained by making zero the determinant

$$\begin{vmatrix} a+ib-E & g \\ g & a-ib-E \end{vmatrix} = 0 \Rightarrow \left(a-E\right)^2 + b^2 - g^2 = 0$$

This results in the following eigenvalues: $E_{\pm} = a \pm \sqrt{g^2 - b^2}$

For weak coupling ($g^2 < b^2$) the eigenvalues are complex since the system is not in equilibrium: one eigenstate grows in time and the other decays. This corresponds to the region of broken PT-symmetry. As the coupling increases, g becomes greater than b and there is a transition to the region of unbroken PT-symmetry.

Note: The Hamiltonian given by eq. 3.187 is a particular case of the coupled Hamiltonian $\mathcal{H}_{coupled}$ of eq. 3.190) if we make $\alpha = -a, \beta = -g$.

3.10.3 COMPLEX POTENTIALS AND PARITY-TIME SYMMETRY IN OPTICS

In general, the non-relativistic Hamiltonian $\hat{\mathcal{H}} = \frac{\hat{p}^2}{2m} + V(\hat{x})$ for a particle of mass m (where $\hat{p}$ and $\hat{x}$ are momentum and position operators, respectively) is considered to have parity and time-reversal symmetry (PT-Symmetry) if the commutator $[PT, \hat{\mathcal{H}}] = 0$. This implies that the potential must satisfy the condition $V^*(-\hat{x}) = V(\hat{x})$. If, as we are used to, only real potentials are considered, this condition means that for PT-symmetry, it is sufficient that the potential be even under parity transformation, that is $V(-\hat{x}) = V(\hat{x})$.

EXAMPLE 3.13

Prove that a necessary condition for PT-symmetry of a non-relativistic Hamiltonian is that $PT\{V(\hat{x})\} = V^*(-\hat{x}) = V(\hat{x})$

SOLUTION

Considering that the action of the parity and time-reversal operators are

$$P\!: \hat{x} \rightarrow -\hat{x}, \hat{p} \rightarrow -\hat{p}$$

$$T\!: \hat{x} \rightarrow \hat{x}, \hat{p} \rightarrow -\hat{p}, i \rightarrow -i,$$

we obtain that

$$PT\left\{\hat{\mathcal{H}}\right\} = PT\left\{\frac{\hat{p}^2}{2m} + V(\hat{x})\right\} = \frac{(-\hat{p})^2}{2m} + V^*(-\hat{x}) = \frac{\hat{p}^2}{2m} + V^*(-\hat{x}). \quad (3.191)$$

This last expression shows that if $V^*(-\hat{x}) = V(\hat{x})$, this Hamiltonian is invariant under a PT transformation.

In optics, several physical processes are described by equations that have a form similar to those described by Schrödinger's equation of quantum mechanics. An example is analyzed in Ref. [37], where the propagation of an optical beam is described by means of the paraxial equation of diffraction which is formally equivalent to the propagation of a particle in a complex potential. Considering that the complex refractive index distribution $n(x) = n_R(x) + in_I(x)$ plays the role of an optical potential, a PT-symmetric system can be designed satisfying the conditions $n_R(x) = in_R(-x)$ and $n_I(x) = in_I(-x)$.

This potential can be obtained through the proper manufacturing to achieve the desired optical performance[35] and we will discuss how a Hamiltonian as that given by eq. 3.187 can be realized.

We consider propagation of an optical beam along the z direction and that the refractive index varies in the x direction so that through the spatial profile of its real part nR, its variation corresponds to an even function of position x whereas the gain/loss distribution associated with its imaginary component nI is an odd function. From Ref. [35], it can be deduced that nR can be approximated by a function proportional to sin^2kx, whereas nI can be approximated by a function proportional to – $sinkx$. This is qualitatively illustrated in Figure 3.20.

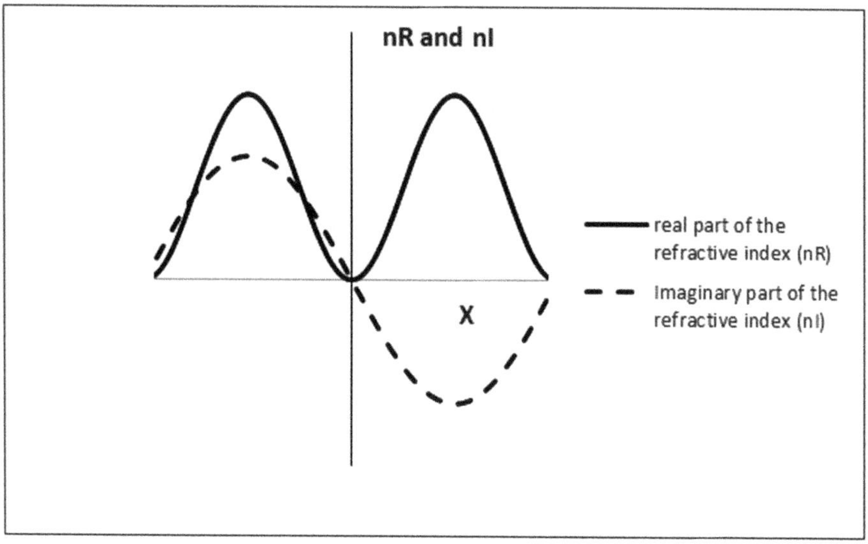

FIGURE 3.20 Real (n_R, red line) and imaginary (n_I, green line) parts of the refractive index distribution.

If the above conditions are realized, the electric field envelope E of the optical beam obeys the so-called paraxial equation of diffraction[37]

$$i\frac{\partial E}{\partial z}+\frac{1}{2k}\frac{\partial^2 E}{\partial x^2}+k_o\left[n_R(x)+in_I(x)\right]E=0 \tag{3.192}$$

Here E is the electric field envelope, $k = k_o n_o$ with n_o the substrate index and $k_o = 2\pi/\lambda$. eq. 3.101 corresponds to a case such as that considered in Ref. [36] where the propagation of a beam of light through an optical coupler is analyzed. This coupler is made of a chain of N waveguides lying in the (x, z) plane. It is assumed that the axes of the waveguides are straight in the first and last sections, but having an intermediate section in which the axes are bent along the propagation direction Z. We can appreciate that this last equation has a form similar to the Schrödinger's equation for the wave function corresponding to a particle of mass m subject to an electric potential, that is

$$i\frac{\partial\Psi}{\partial t} = -\frac{1}{2m}\frac{\partial^2\Psi}{\partial x^2} + V(x)\Psi \qquad (3.193)$$

where we have made $\hbar \equiv 1$ and we can observe the formal analogies $\leftrightarrow z$, $m \leftrightarrow k$ and $V(x) \leftrightarrow -k_o\,[n_R(x) + in_I(x)]$.

If two parallel optical waveguides interact with a coupling constant κ and experience equal and opposite loss and gain with strength γ, this gives rise to the following system of coupled equations for the electric field envelopes[37–39]

$$i\frac{dE_1}{dz} - i\frac{\gamma}{2}E_1 + \kappa E_2 = 0 \qquad (3.194)$$

$$i\frac{dE_2}{dz} + i\frac{\gamma}{2}E_2 + \kappa E_1 = 0 \qquad (3.195)$$

These last two equations describe an optical coupled system, with one of the two parallel channels being optically excited to provide gain for the guided light, whereas the neighbor arm experiences loss, replicating optically a process of coherent population transfer in quantum systems.

The system of eqs. 3.194 and 3.195 can be written in the form

$$i\frac{d|\psi\rangle}{dz} = \mathcal{H}|\psi\rangle \qquad (3.196)$$

In this case, the Hamiltonian $\mathcal{H}$ is

$$\mathcal{H} = \begin{bmatrix} i\gamma/2 & -\kappa \\ -\kappa & -i\gamma/2 \end{bmatrix} \qquad (3.197)$$

This last expression is of the same form of that given in eq. 3.187 with $\alpha = \gamma/2$ and $\beta = \kappa$.

In Ref. [40], a non-Hermitian version of the Rabi model is analyzed, where the dynamics of a two-level system is represented by a Hamiltonian which includes an imaginary part that oscillates in time. This leads to gain and loss which oscillate periodically. The model presents PT-symmetry, which can be broken if the drive is sufficiently strong and in that quoted reference the boundaries of the PT phase diagram are obtained considering the non-Hermitian Rabi model described by the following Hamiltonian

$$\mathcal{H} = \frac{\omega_o}{2}\sigma_z + 2i\lambda\sigma_x cos\omega t \tag{3.198}$$

3.11 CONCLUSIONS

The Theory of non-relativistic quantum mechanics is generally considered to be not ruled by a fundamental direction of time, since its fundamental dynamical equation, the Schrödinger's equation, is invariant under time reversal. Any time-reversal asymmetry is attributable to forces or interactions involving dissipation effects, or due to an external magnetic field.

If one accepts that after a measurement of an observable, the wave function composed by a set of superimposed states collapses to only one state which is the one corresponding to the measured eigenvalue (*Collapse of the Wave Function*), a quantum arrow could be associated with the process of measurement of a quantum observable. In fact, before a measurement of one observable be made, the state of the system is a superposition of wave functions. Each of these wave functions corresponds to one of the eigenvalues of this observable. Once the measurement is made, the wave function collapses irreversibly from a superposition of various states to just one of the possible states. According to the measurement theory formulated by Von Neumann in 1932, there are two kinds of evolutions that a quantum system can follow: one obeying the Schrödinger's time-dependent equation, and a stochastic one induced by the measurement.[41] This last kind of evolution gives a time-asymmetric process that could be associated with a direction of time. However, as Von Neumann postulates, it must be considered that the measurement that actually collapses the wave function is connected with the human mind: it only occurs when a conscious entity is observing the process. Therefore, the association of an arrow of time with the collapse of the wave function after making a measurement of an observable requires to be considered carefully and it is still an open theoretical and philosophical question.

Schrödinger's equation for a free particle or one subject to a time-reversal symmetric potential is symmetric under time reversal with an antiunitary transformation that must include conjugation in addition to flip the direction of time by doing $\rightarrow -t$. This means that if $\psi(\vec{r}, t)$ is a solution $\psi^*(\vec{r}, -t)$ also satisfies this equation. This symmetry is broken if an external magnetic field is applied.

It is worth to note that if time-reversal transformation is represented only as a change of the sign of time, that is making $\rightarrow -t$, Schrödinger's equation is non–time-reversal invariant and the time-reversed Hamiltonian would lead to negative energies which have no interpretation in NRQM. On the other hand, in relativistic quantum mechanics, negative energy solutions arise mathematically from Dirac's equation and one usually accepted interpretation of a negative energy solution is that in the case of an electron, it describes a particle which propagates backward in time, or equivalently a positive energy antiparticle propagating forward in time, called a positron as referred to in a paper of Richard Feynman published in 1949 (Ref. [42]). In that paper, Feynman analyzes the problem of the behavior of positrons and electrons in a given external potential, assuming that their mutual interaction can be neglected, and replacing the theory of holes by a new interpretation of the solutions of the Dirac equation. A "negative energy state" appears in a form which may be pictured in space-time as a wave traveling backwards in time, or equivalently, a positive energy antiparticle propagating forward in time.

For quantum systems whose evolution is determined by a set of external parameters varying cyclically, even though they return to the same initial values, the wave function of the system acquires a geometric phase which depends on the path followed by the external parameters in the parameter space. This geometric phase is due to a breakdown of time-reversal symmetry of some type. In case of a slow cyclic change of the external parameters, the acquired geometric phase is called the Berry Phase. The geometrical phase can be measured by observing points of constructive and destructive interference, as is the case if, for example, we take a beam of particles, let us say electrons, with that beam split in two. The propagation of one of the beams is through a region where there is an interaction with a magnetic field changing adiabatically, while the other beam does not undergo some interaction. When the two beams are recombined, an interference effect appears.

If a given Hamiltonian is invariant after being subject to a time-reversal operation and then to the parity reflection operator P (which changes the position vector $\vec{r}$ by $-\vec{r}$), new kinds of complex Hamiltonians are obtained having spectra that are still real and positive. This result allows to relax the requisite that the Hamiltonian operator must be Hermitian in order to obtain real values for the energy levels and that the theory be unitary (probability conserving). If the PT-symmetry of the Hamiltonian is not broken the Hamiltonian exhibits all of the features of a quantum theory described by a Hermitian Hamiltonian.

KEYWORDS

- **symmetry**
- **operator**
- **antiunitary**
- **Hamiltonian**
- **evolution**
- **geometric phase**
- **hermiticity**

REFERENCES

1. Susskind, S.; Friedman, A. *Quantum Mechanics—The Theoretical Minimum*; Basic Books: New York, 2014.
2. Griffiths, D. J. *Introduction to Quantum Mechanics*; Prentice Hall: Upper Saddel River, 1994.
3. Sakurai, J. J. *Modern Quantum Mechanics*, Rev. ed., Editor, San Fu Tuan; Wesley Publishing Company: Addison, 1994.
4. Fowler, R. H.; Nordheim, L. Electron Emission in Intense Electric Fields. *Proc. R. Soc. Lond. A* **1928,** *119* (781), 173–181.
5. Rojas, R.; Fuster, G. Gravitational Analogies for the Stopping Voltage and the Work Function in the Photoelectric Effect. *Phys. Teacher* **2021,** *59*, 196.
6. Pendry, J. B. Theory of Inverse Photoemission. *J. Phys. C: Solid State Phys*. **1981,** *14* 1381–1391.
7. Spicer, W. E. Photoemissive, Photoconductive, and Optical Absorption Studies of Alkali—Antimony Compounds. *Phys. Rev.* **1958,** *112*, 114.
8. Allen, P. M. G. Inverse Photoemission from Surfaces; Thesis Submitted for the Degree of PhD.; Imperial College of Science and Technology, University of London, 1986.
9. Wigner, E. Ueber die Operation der Zeitumkehr in der Quantenmechanik, Nachrichten von der Gesellschaft der Wissenschaften zu Göttingen. *Mathematisch-Physikalische Klasse* **1932,** 546–559. http://eudml.org/doc/59401
10. Guru, I. I. Time-Reversal Symmetry, Seven Time-Reversal Operators for Spin Containing Systems. Springer Tracts Modern Phys. *281*.
11. Wigner, E. *Group Theory and Its Applications to the Quantum Mechanics of Atomic Spectra*; New York, 1959.
12. Garrison, J. C.; Chiao, R. Y. Geometrical Phases from Global Gauge Invariance of Nonlinear Classical Field Theories. *Phys. Rev. Lett.* **1988,** *60* (1988), 165–168.
13. Berry, M. V. Quantal Phase Factors Accompanying Adiabatic Changes. *Proc. R. Soc. Lond. Ser. A Math. Phys. Sci.* **1984,** *392* (1802), 45–57.
14. Messiah, A.; Messiah, A. *Quantum Mechanics*, Vol. II, Chapter XVII, Section 12; John Wiley and Sons: New York, 1962.

15. Nenciu, G. On the Adiabatic Theorem of Quantum Mechanics. *J. Phys. A: Math. Gen.* **1980,** *13*.
16. Aharonov, Y.; Anandan, J. Phase Change During a Cyclic Quantum Evolution. *Phys. Rev. Lett.* **1987,** *58*, 1593–1596.
17. Chiao, R. Y.; and Wu, Y-S. Manifestations of Berry's Topological Phase for the Photon. *Phys. Rev. Lett.* **1986,** *57*, 933.
18. Aharonov, Y.; Bohm, D. Significance of Electromagnetic Potentials in the Quantum Theory. *Phys. Rev.* **1959,** *115*, 485.
19. Dalibard, J. Introduction to the Physics of Artificial Gauge Fields. Proceedings of the International School of Physics "Enrico Fermi", Course 191 "Quantum Matter and Ultralow Temperatures" (IOS, Amsterdam; SIF Bologna), 2016 (arXiv:1504.05520v1[cond-mat. quant-gas]).
20. Dalibard, J.; Gerbier, F.; Juzeliûnas, G.; Öhberg, P. Colloquium: Artificial Gauge Potentials for Neutral Atoms. *Rev. Modern Phys.* **2011,** *83*, 1523.
21. Celeghini, E.; Rasetti, M.; Vitiello, G. Quantum Dissipation. *Ann. Phys.* **1992,** *215*, 156–170.
22. Einstein, A.; Podolsky, B.; Rosen, N. Can Quantum-Mechanical Description of Physical Reality Be Considered Complete? *Phys. Rev.* **1935,** *47*, 777.
23. Peskin, M. A.; Schroeder, D. V. *An Introduction to Quantum Field Theory*; Perseus Books: Reading, 1995.
24. Bell, J. S. On the Einstein-Podolsky-Rosen Paradox. *Physics* **1964,** *1*, 195.
25. Lee, J. EPR Paradox Solved by Special Theory of Relativity. *Acta Physica Polonica A* **2014,** *125*, 1107.
26. Feynman, R. P.; Leighton, R. B.; Sands, M. *The Feynman Lectures on Physics*, Vol. 2; Addison-Wesley: London, 1966.
27. Leuchs, G.; Sondermann, M. Time Reversal Symmetry in Optics. *Phys. Scripta* **2012,** 85.
28. Rabi, I. I. Space Quantization in a Gyrating Magnetic Field. *Phys. Rev.* **1937,** *51*, 652.
29. Gerry, C.; Knight, P. *Introductory Quantum Optics*; Cambridge University Press: Cambridge, 2005.
30. Meunier, T.; Gleyzes, S.; Maioli, P.; Auffeves, A.; Nogues, G.; Brune, M.; Raimond, J. M.; Haroche, S. Rabi Oscillations Revival Induced by Time Reversal: A Test of Mesoscopic Quantum Coherence. *Phys. Rev. Lett.* **2005,** *94*, 010401-1–010401-4.
31. Bender, C.; Boettcher, S.; Meisinger, P. PT-Symmetric Quantum Mechanics. *J. Math. Phys.* **1999,** *40*, 2201–2229.
32. Jones-Smith, K.; Mathur, H. A New Class of Non-Hermitian Quantum Hamiltonians with PT Symmetry. *Phys.Rev.A* **2010,** *82*, 042101.
33. Zhang, Q.; Wu, B. Non-Hermitian Quantum Systems and Their Geometric Phases. *Phys. Rev. A* **2019,** *99*, 032121.
34. Bender, C. M. Introduction to PT -Symmetric Quantum Theory. *Contemp. Phys.* **2005,** *46*, 277–292.
35. Bender, C. M. Generalized PT Symmetry and Real Spectra, Letter to the Editor. *J. Phys. A: Math. Gen.* **2002,** 35, L467–L471
36. Berntson, B. K. PT -Symmetric Models in Classical and Quantum Mechanics. Dissertation, Imperial College London, Department of Physics, September 2012.
37. Rüter, C. E.; Makris, K. G.; El-Ganainy, R.; Christodoulides, D. N.; Segev, M.; Kip, D. Observation of Parity-Time Symmetry in Optics. *Nat. Phys.* **2010,** *6*, 192.

38. Longhi, S. Adiabatic Passage of Light in Coupled Optical Wave Guides. *Phys. Rev. E* **2006,** *73*, 026607.
39. El-Ganainy, R.; Makris, K. G.; Christodoulides, D. N.; Musslimani, Z. H. Theory of Coupled Optical PT-Symmetric Structures. *Opt. Lett.* **2007,** *32*, 2632.
40. Lee, T. E.; Joglekar, Y. N. PT -Symmetric Rabi Model: Perturbation Theory. *Phys. Rev. A* **2015,** *92*, 042103.
41. Von Neumann J. *Mathematische Grundlagen der Quantenmechanik*; Springer: Berlin, 1932.
42. Feynman, R. P. The Theory of Positrons. *Phys. Rev.* **1949,** *76*, 749.

APPENDIX 3.1: ADDITIONAL EXAMPLES

In this section, we present the analysis of some additional examples for illustrating the concepts about time reversal in Quantum Mechanics.

A quantum system consisting in a particle inside an infinite potential box

Consider a particle of mass m inside an infinite potential box of width a, so that for the region $0 < x \leq a$ the potential is zero and is infinite for $x = 0$ and $x = a$. Does this system have time-reversal symmetry?

ANSWER

The Hamiltonian is given by

$$\mathcal{H} = \frac{P^2}{2m} \tag{3.196}$$

where P denotes the momentum operator (for simplicity, we consider motion in a single dimension corresponding to the x-axis). This Hamiltonian is invariant under time reversal and the normalized solutions are

$$\Psi_n(x,t) = e^{\frac{-i}{\hbar}E_n t}\varphi_n(x) \tag{3.197}$$

where $\varphi_n(x) = \sin\left[\left(\frac{n\pi}{2a}\right)x\right]$ are real functions with n integer different of zero, which clearly are real and non-degenerate.

The time reversal of the solution $\Psi_n(x, t)$ is given by its complex conjugate and making the change t by $-t$ giving

$$\Psi_n^*(x,-t) = e^{\frac{+i}{\hbar}E_n(-t)}\varphi_n^*(x) = e^{\frac{-i}{\hbar}E_n t}\varphi_n(x) = \Psi_n(x,t) \tag{3.198}$$

Therefore, this system has time-reversal symmetry since $T\Psi_n(x, t) = \Psi_n(x, t)$ where T is the time-reversal operator which as has been explained in Section 3.2.1, acting over a complex wave function takes its complex conjugate and replaces t by $-t$.

Is the quantum one-dimensional harmonic oscillator a system with time-reversal symmetry?

ANSWER

As known [2–3], the Hamiltonian for this system is

$$\mathcal{H} = \frac{P^2}{2m} + \frac{1}{2} m\omega^2 X^2 \tag{3.199}$$

where X and P are the observables corresponding to position and momentum, respectively.

Since this Hamiltonian is time-independent, clearly is time-reversal symmetric. Furthermore, all eigenkets are nondegenerate and have time-reversal symmetry. This can be proved considering that the normalized eigenfunctions for this Hamiltonian are given by [2]

$$\Psi_n(x,t) = e^{\frac{-i}{\hbar}E_n t} \varphi_n(x) \tag{3.200}$$

where

$$\varphi_n(x) = \left(\frac{m\omega}{\pi\hbar}\right)^{1/4} \frac{1}{\sqrt{2^n n!}} H_n(\xi) e^{-\xi^2/2} \tag{3.201}$$

$\omega = \sqrt{k/m}$ is the frequency of oscillation, $\xi \equiv \sqrt{m\omega/\hbar}$, $E_n = \hbar\omega\left(n + \frac{1}{2}\right)$ is the energy corresponding to the level n and $H_n(\xi)$ is the Hermite polynomial of order n given by the Rodrigues formula

$$H_n(\xi) = (-1)^n e^{\xi^2} \left(\frac{d}{d\xi}\right)^n e^{-\xi^2} \tag{3.202}$$

In a similar way to example 3.12.1, the time reversal of the solution $\Psi_n(x, t)$ is given by its complex conjugate and making the change t by $-t$ giving

$$\Psi_n^*(x,-t) = e^{\frac{+i}{\hbar}E_n(-t)} \varphi_n(x) = e^{\frac{-i}{\hbar}E_n t} \varphi_n(x) = \Psi_n(x,t) \tag{3.203}$$

Therefore, this system has time-reversal symmetry.

If T is the time-reversal operator, determine conditions that the Hamiltonian must fulfill so that the time-reversed state $T|\psi\rangle$ obeys the same Schrödinger equation that the original state $|\psi\rangle$.

SOLUTION

If the original state satisfies the Schrödinger's equation, we have

$$i\hbar\frac{d\,|\,\psi\rangle}{dt} = \mathcal{H}\,|\,\psi\rangle \tag{3.204}$$

Does the time-reversed state $T|\psi\rangle$ satisfy the same equation?

For answering this question, in the above equation, we apply the time-reversal operator in both sides of the above equation obtaining that:

$$T(i\hbar\frac{d\,|\,\psi\rangle}{dt}) = T(\mathcal{H}\,|\,\psi\rangle) \tag{3.205}$$

Since time-reversal operation implies conjugation and the change of t by $-t$, we obtain

$$-i\hbar\frac{d(T\,|\,\psi\rangle)}{d(-t)} = i\hbar\frac{d(T\,|\,\psi\rangle)}{dt} = T\mathcal{H}\,|\,\psi\rangle \tag{3.206}$$

From this last equation, one can see that if $\mathcal{H}T = T\mathcal{H}$ or $[T, \mathcal{H}] = 0$, the time-reversed state satisfies the same equation than the original state since

$$i\hbar\frac{dT|\psi\rangle}{dt} = \mathcal{H}T|\psi\rangle \tag{3.207}$$

Therefore, if the Hamiltonian H has time-reversal symmetry, so that $[T, \mathcal{H}] = 0$, the time-reversed state $T|\psi\rangle$ satisfies the same equation than the original state $|\psi\rangle$.

If $\psi(\vec{r}, t)$ is a plane wave in three dimensions representing the wave function of a spinless particle, show that $\psi^*(\vec{r}, -t)$ is the wave function for the wave with the momentum vector reversed, and which satisfies the same three-dimensional Schrödinger's equation.

SOLUTION

Let us consider the wave function $\psi(\vec{r},t) = e^{\frac{i}{\hbar}(\vec{p}\cdot\vec{r}-Et)}$ which represents a free particle moving with momentum $\vec{p}$. Taking the complex conjugate and making $t \to -t$, we obtain $\psi^*(\vec{r},-t) = e^{-\frac{i}{\hbar}(\vec{p}\cdot\vec{r}-E(-t))} = e^{\frac{i}{\hbar}(-\vec{p}\cdot\vec{r}-Et)}$ which represents a plane wave with momentum reversed and it results from time reversing the wave function $(\vec{r}, t)$. In addition, it can be verified that $\psi^*(\vec{r}, -t)$ is a solution of the same Schrödinger's equation. In fact, starting from the three-dimensional Schrödinger's equation for a free particle, and applying the time-reversal operator to both sides we obtain:

$$T\left(i\hbar\frac{\partial\psi}{\partial t}\right)=T\left(-\frac{\hbar^2}{2m}\nabla^2\psi\right)\Rightarrow -i\hbar\frac{\partial\psi^*}{\partial(-t)}=-\frac{\hbar^2}{2m}\nabla^2\psi^*$$

implying that

$$i\hbar\frac{\partial\psi^*}{\partial t}=-\frac{\hbar^2}{2m}\nabla^2\psi^* \tag{3.208}$$

which is the same Schrödinger's equation satisfied by $\psi^*(\vec{r},t)$.

The wave function representing a state at $t = 0$ is given by a complex function $e^{i\vec{p}\cdot\vec{r}/\hbar}$. Explain why this function does not violate time-reversal invariance.

SOLUTION

The wave function of a plane wave $e^{i\vec{p}\cdot\vec{r}/\hbar}$ can be complex without violating time-reversal invariance, because it is degenerate with $e^{-i\vec{p}\cdot\vec{r}/\hbar}$. In fact, both states corresponds to the same Hamiltonian $\mathcal{H} = \vec{P}^2/2m$ where $\vec{P}$ is the momentum operator. Clearly, this Hamiltonian has time-reversal symmetry.

The time-reversal function of $e^{i\vec{p}\cdot\vec{r}/\hbar}$ is obtained by taking its complex conjugate and replacing $\vec{p}$ by $-\vec{p}$, that is

$$Te^{i\vec{p}\cdot\vec{r}/\hbar}=e^{-i(-\vec{p})\cdot\vec{r}/\hbar}=e^{i\vec{p}\cdot\vec{r}/\hbar} \tag{3.209}$$

Prove that the geometric phase is zero for a Hamiltonian $\mathcal{H}$ with real eigen-functions and depending on a set of parameters which vary cyclically with time. Is there time-reversal symmetry in this case?

SOLUTION

This can be proved as follows:

Let $|\varphi_n\rangle$ be an eigenket of the Hamiltonian $\mathcal{H}(\vec{R})$, where $\vec{R}$ denotes the set of time-dependent parameters. The corresponding geometric phase is determined by the Berry connection

$$\vec{\mathcal{A}}_n\left(\vec{R}\right)\equiv i\left|\varphi_n\left(\vec{R}\right)\right\rangle\left|\nabla_{\vec{R}}\varphi_n\left(\vec{R}\right)\right\rangle \tag{3.210}$$

so that the geometric phase is

$$\gamma_n=i\oint\langle\varphi_n\,|\,\nabla_{\vec{R}}\varphi_n\rangle\cdot d\vec{R} \tag{3.211}$$

Since $\langle\varphi_n|\varphi_n\rangle = 1$, we have that

$$\nabla_{\vec{R}}\langle\varphi_n|\varphi_n\rangle = \langle\nabla_{\vec{R}}\varphi_n(\vec{R})|\varphi_n(\vec{R})\rangle + \langle\varphi_n(\vec{R})|\nabla_{\vec{R}}\varphi_n(\vec{R})\rangle = 0 \Rightarrow$$

$$\langle\varphi_n|\nabla_{\vec{R}}\varphi_n\rangle^* + \langle\varphi_n|\nabla_{\vec{R}}\varphi_n\rangle = 0 \Rightarrow 2Re\{\langle\varphi_n|\nabla_{\vec{R}}\varphi_n\rangle\} = 0 \tag{3.212}$$

From eq. 3.212, we conclude that $\langle\varphi_n|\nabla_{\vec{R}}\varphi_n\rangle$ is a pure imaginary number. If $|\varphi_n\rangle$ is real, then $\langle\varphi_n|\nabla_{\vec{R}}\varphi_n\rangle$ must be zero. Therefore, for a Hamiltonian with real eigenfunctions and depending on a set of parameters denoted by $\vec{R}$, the geometric phase is zero. This implies that kets $|\psi_n\rangle$ and $T|\psi_n\rangle$ (where T is time-reversal operator) are equal (except a constant phase factor) and therefore $[T, \mathcal{H}] = 0$ so that there is time-reversal symmetry.

A particle of mass m is inside an infinite potential box of variable width w.

According to the Adiabatic Theorem of Quantum Mechanics [2, Chap 10], if the Hamiltonian of a system changes gradually from an initial form $\mathcal{H}^{initial}$ to another form $\mathcal{H}^{final}$ and the initial state corresponds to the nth eigenstate of $\mathcal{H}^{initial}$, under the Schrödinger equation, it will lead to the nth eigenstate of $\mathcal{H}^{final}$. This can be exemplified for the case of a particle confined to an infinite square well of width a and initially being in the state $\varphi_n^{initial}(x) = \sqrt{\frac{2}{a}} sin\left(\pi\frac{x}{a}\right)$. If the right wall is gradually moved so that the width of the well increases up to a new value $2a$, the adiabatic theorem states that after the expansion, the particle will be in the ground state of the expanded well which is given by $\varphi_n^{final}(x) = \sqrt{\frac{1}{a}} sin\left(\pi\frac{x}{2a}\right)$.

a) If an adiabatic expansion of the width from w_1 to w_2 occurs at a constant rate, so that $dw/dt = v$, determine the dynamic change of phase for this process?
b) Determine the geometric phase for this adiabatic expansion
c) If the box is contracted so that it comes back to its original size, determine the Berry phase for the cycle

SOLUTION

a) The dynamic phase change is given by

$$\theta_n(t)=\frac{1}{\hbar}\int_0^t E_n(t')\,dt'=\frac{1}{\hbar}\int_0^t \frac{n^2\pi^2\hbar^2}{2m\{w(t')\}^2}dt'=\frac{n^2\pi^2\hbar}{2m\upsilon}\int_{w_1}^{w_2}\frac{1}{w^2}d\omega=\frac{n^2\pi^2\hbar}{2m\upsilon}\left(\frac{1}{w_1}-\frac{1}{w_2}\right), \tag{3.213}$$

which is positive in this case since $w_1 < w_2$.
For obtaining this result, it has been considered that a particle of mass m confined into an infinite potential box, the energy is quantized with values given by $E_n = \frac{n^2\pi^2\hbar^2}{2mw^2}$ and that $\upsilon = \frac{dw}{dt} \Rightarrow dt = dw/\upsilon$.

b) In this case, the Hamiltonian depends on the parameter $R = w$ which is a time-dependent variable corresponding to the width of the potential box.
Since the expansion is adiabatic, at any instant, the system will be in the eigenstate of the instantaneous Hamiltonian. In this case, if this eigenstate corresponds to a width w of the box potential, the wave function is

$$\psi_n(x,t)= \varphi_n(x)e^{-iE_nt/\hbar} \tag{3.214}$$

with $\varphi_n(x)=\sqrt{\frac{2}{w}}sin\left(n\pi\frac{x}{w}\right)$ and $n=1,2,3,\ldots\ldots$

From eq. 3.91 and considering that in this case the parameter $R = w$ follows a non-closed path, we obtain that the geometric phase is given by

$$\gamma_n = i\int_{\vec{R}_{initial}}^{\vec{R}_{final}} \langle\psi_n|\nabla_{\vec{R}}\psi_n\rangle\cdot d\vec{R} = i\int_{w_1}^{w_2} dw \int_0^w dx\,\varphi_n\frac{\partial\varphi_n}{\partial R} \tag{3.215}$$

From eq. 3.214, we have that

$$\frac{\partial\varphi_n}{\partial R}=\sqrt{2}\left(-\frac{1}{2}\frac{1}{w^{3/2}}\right)sin\left(n\pi\frac{x}{w}\right)+\sqrt{\frac{2}{w}}\left(\frac{-n\pi x}{w^2}\right)cos\left(n\pi\frac{x}{w}\right) \tag{3.216}$$

Therefore, doing the integration of eq. 3.215 and after some algebra, we obtain that the geometric phase is given by

$$\gamma_n == -i\int_{w_1}^{w_2} dw\left(\frac{1}{2w}-\frac{1}{2w}\right)=0 \tag{3.217}$$

This result is in agreement with the fact that in this case, the eigenfunctions are real, so that the geometrical phase is zero. Therefore, as expected in this case, there is time-reversal symmetry.

c) By the same arguments given above, the Berry's phase is zero in this case.
For more details about the evolution of the wave function in an infinite potential well with time-dependent boundaries, the reader is invited to study the article of Kieran Cooney, *The infinite potential well with moving walls*, https://arxiv.org/abs/1703.05282v1[quant-ph], March 2017.

APPENDIX 3.2: BASIC CONCEPTS ABOUT QUANTUM MECHANICS

We summarize some basic concepts about Quantum Mechanics used in this chapter. For more details, the reader is advised to study the books specified in references [1–3].

1. States of a Quantum System and Hilbert Spaces.

The space of states of a quantum system is *not* a mathematical set; it is a *vector space*.

Before speaking about vector spaces, it is needed to clarify the term *vector*. As known, this term is used to indicate an object in ordinary space that has a magnitude and one direction. Such elements represent physical quantities which are called vectors, and they have three components, corresponding to the three dimensions of space. A physical quantity with magnitude and direction in ordinary space, is called a *3-vector*. A mathematical vector space is an abstract concept that is not necessarily related to the ordinary space we know. It may have any number of dimensions from 1 to ∞ and components corresponding to arbitrary numbers, or even more general things.

The vector space used to define quantum mechanical states is called a *Hilbert space*. The mathematical definition will not be given here, but let us only say that the term *Hilbert space* in Quantum Mechanics, refers to the space of states. In general, a Hilbert space may have an infinite number of dimensions.

2. The Principles of Quantum Mechanics

First Principle

In Quantum Mechanics, the observable or measurable quantities are represented by linear operators.

Second Principle

When making a measurement, the only possible results are the eigenvalues of the operator that represents the observable. The state for which the result is unambiguously λ_i is the corresponding eigenvector $|\lambda_i\rangle$.

Third Principle

States which are unambiguously distinguishable are represented by orthogonal vectors.

Fourth Principle

If $|A\rangle$ is the state-vector of a system and the observable L is measured, the probability of obtaining the value λ_i is:

$$P(\lambda_i) = \langle A|\lambda_i\rangle \langle\lambda_i|A\rangle \tag{3.218}$$

where $|\lambda_i\rangle$ is the eigenket corresponding to the eigenvalue λ_i of the observable L.

Minus First Law

If two identical isolated systems start out in different states, they will stay in different states. If two identical systems are in the same state at some point in time, then their stories and their future evolutions must also be identical.

3. *The Evolution Operator*

The basis dynamical assumption of quantum mechanics is that if we know the state of one given system at $t = 0$, then the state of this system for a later instant will be

$$|\Psi(t)\rangle = U_{ev}(t)|\Psi(0)\rangle \tag{3.219}$$

where U_{ev} is called the "time-development operator" or more usually the "quantum evolution operator."

Quantum Mechanics places two requirements on the quantum evolution operator:

1. $U_{ev}(t)$ must be a linear operator
2. Conservation of distinctions: If two identical systems are in the same state at some point in time, then their stories and their future evolutions must also be identical.

How can we distinguish states?

Any two states are differentiated if they are orthogonal. Suppose that $\Psi(0)\rangle$ and $\Phi(0)\rangle$ are two distinguishable states. This implies that they must be orthonormal, that is

$$\langle\Psi(0)|\Phi(0)\rangle = 0 \tag{3.220}$$

The conservation of distinctions imply that they will continue to be orthogonal for all time, meaning that at any instant t

$$\langle\Psi(t)|\Phi(t)\rangle = 0 \tag{3.221}$$

This last statement implies that $U_{ev}^{\dagger}U_{ev} = I$ and therefore the time-evolution operator is unitary.

4. The Hamiltonian

Based on reference [1], we present here a simplified derivation of the Schrödinger's equation. For that purpose, we replace the time interval t by an infinitesimal time interval η and consider the time evolution for this small interval.

There are two fundamental principles that must be considered in the analysis of incremental changes:

1. Unitarity: $U_{ev}^{\dagger}(\eta)U_{ev}(\eta) = I$ (3.222)
2. Continuity: This means that the state-vector does not change abruptly. To explain this more precisely, first consider the case in which η is zero. It should be obvious that in this case, the time-evolution operator is precisely the unit operator I. Continuity means that when η is very small, $U_{ev}(\eta)$ is approximately equal to the unit operator, with a difference of the order of η. Thus, we write

$$U_{ev}(\eta) = I - i\eta\mathcal{H} \tag{3.223}$$

One may wonder why does appear a minus sign and a factor i in front of the Hamiltonian operator $\mathcal{H}$. These factors are completely arbitrary for the moment.

From this last equation, we obtain that

$$U_{ev}^{\dagger}(\eta) = I + i\eta\mathcal{H}^{\dagger} \tag{3.224}$$

The requirement that the evolution operator be unitary implies that

$$\begin{gathered} U_{ev}^{\dagger}(\eta)U_{ev}(\eta) = I \rightarrow \left(I + i\eta\mathcal{H}^{\dagger}\right)\left(I - i\eta\mathcal{H}\right) \\ I + i\eta\mathcal{H}^{\dagger} - i\eta\mathcal{H} + \eta^{2}\mathcal{H}^{\dagger}\mathcal{H} = I \end{gathered} \tag{3.225}$$

Neglecting the term containing η^2, we deduce that $\mathcal{H}^\dagger = \mathcal{H}$, that is the hamiltonian operator $\mathcal{H}$ must be hermitian, meaning that it is an observable having a complete set of orthonormal eigenvectors and eigenvalues.

Which is the relation between $\mathcal{H}$ and U_{ev}?

If we consider the temporal evolution during a small interval η of a system initially at the state $|\Psi(0)\rangle$ we have

$$|\Psi(\eta)\rangle = (I - i\eta\mathcal{H})|\Psi(0)\rangle = |\Psi(0)\rangle - i\eta\mathcal{H}|\Psi(0)\rangle \tag{3.226}$$

This implies that

$$\frac{|\Psi(\eta)\rangle - |\Psi(0)\rangle}{\eta} = -i\mathcal{H}|\Psi(0)\rangle \tag{3.227}$$

Taking the limit as $\eta \rightarrow 0$ we get

$$\frac{\partial|\Psi(t)\rangle}{\partial t} = -i\mathcal{H}|\Psi(t)\rangle \tag{3.228}$$

Multiplying the left-hand side of this last equation by $\hbar = h/2\pi$ and identifying $\mathcal{H}$ with energy, the above equation becomes dimensionally consistent, since $\hbar$ has units Energy,·Time, and we obtain

$$\hbar\frac{\partial|\Psi(t)\rangle}{\partial t} = -i\mathcal{H}|\Psi(t)\rangle \tag{3.229}$$

This can be written as

$$i\hbar\frac{\partial|\Psi(t)\rangle}{\partial t} = \mathcal{H}|\Psi(t)\rangle \tag{3.230}$$

This is the Schrödinger's equation. Since $|\Psi(t)\rangle = U(t)\,|\Psi(t = 0)\rangle$, we obtain that the time-evolution operator also satisfies Schrödinger's equation

$$i\hbar\frac{\partial U_{ev}}{\partial t} = \mathcal{H}U_{ev} \tag{3.231}$$

The corresponding solution is

$$U_{ev}(t) = e^{-\frac{i}{\hbar}\mathcal{H}t} \tag{3.232}$$

4. *Expectation Values*

In Quantum Mechanics theory, the average values are referred to as expectation values.

Suppose we have a probability function for the outcome of an experiment measuring the value of an observable **L.** The outcome must be one of the eigenvalues λ_i of **L,** and the corresponding probability function is $P(\lambda_i)$.

The average of an observable $\boldsymbol{L}$ is expressed as

$$\langle \boldsymbol{L} \rangle = \Sigma_i \lambda_i \, P(\lambda_i) \tag{3.233}$$

Suppose that the normalized state of a quantum system is $|\mathrm{A}\rangle$. The expansion of $|\mathrm{A}\rangle$ in the orthonormal basis of eigenvectors of **L** leads to:

$$|\mathrm{A}\rangle = \Sigma_i \alpha_i |\lambda_i\rangle \tag{3.234}$$

Acting over $|\mathrm{A}\rangle$ with the linear operator **L** on both sides of eq. 3.269

$$\boldsymbol{L}|\mathrm{A}\rangle = \Sigma_i \alpha_i \boldsymbol{L}|\lambda_i\rangle \tag{3.235}$$

Now taking the inner product $\langle A|$ on both sides of eq. 3.235, we get

$$\langle A|\boldsymbol{L}|A\rangle = \Sigma_i (\alpha_i^* \alpha_i)\lambda_i \tag{3.236}$$

For obtaining the last equation, it has been considered that $|\lambda_i\rangle$ is the eigenfunction (or eigenket) of the operator $\boldsymbol{L}$ with the corresponding eigenvalue λ_i, so that $\boldsymbol{L}|\lambda_i\rangle = \lambda_i|\lambda_i\rangle$ and the fact that the vectors $|\lambda_i\rangle$ are orthonormals, that is $\langle \lambda_i|\lambda_j\rangle = \delta_{ij}$.

If we identify $(\alpha_i^* \alpha_i)$ with the probability $P(\lambda_i)$ of eq. 3.236, we obtain

$$\langle \boldsymbol{L} \rangle = \langle A|\boldsymbol{L}|A\rangle \tag{3.237}$$

The Uncertainty Principle

Let $|\psi\rangle$ be any ket representing one of the possible states in which a quantum system can be found and let us consider $\boldsymbol{A}$ and $\boldsymbol{B}$ as two operators representing observables associated with this state. In particular, these observables could be the position and momentum of a particle, or two of the three components of its spin.

If simultaneous measurements of observables $\boldsymbol{A}$ and $\boldsymbol{B}$ in the state $|\psi\rangle$ are made, it is not possible to get simultaneously absolute certainty in the results of the measurement of these observables and the product of the incertainties $\Delta\boldsymbol{A}$ and $\Delta\boldsymbol{B}$ cannot be smaller than a finite value. This is expressed as

$$\Delta\boldsymbol{A}\ \Delta\boldsymbol{B} \geq \frac{1}{2}\langle \psi|[\boldsymbol{A},\boldsymbol{B}]|\psi\rangle \tag{3.238}$$

This equation expresses the fact that the product of the uncertainties in the simultaneous measurement of the observables $\boldsymbol{A}$ and $\boldsymbol{B}$ in the state $|\psi\rangle$ cannot be smaller than half the magnitude of the expectation value of the commutator $[\boldsymbol{A}, \boldsymbol{B}]$ in this state.

In particular, if the observables $\boldsymbol{A}$ and $\boldsymbol{B}$ correspond to the momentum and position of a particle, respectively, since as known from introductory courses to Quantum Physics $[\boldsymbol{X}, \boldsymbol{P}] = i\hbar$, eq. 3.238 leads to

$$\Delta X \; \Delta \boldsymbol{P} \geq \frac{1}{2}\hbar$$

which is the usual form for expressing the Heisenberg Uncertainty Principle for the simultaneous determination of the position and the momentum of a particle, which is one of the fundamental and most famous principles of Quantum Mechanics. Even though we use the most advanced technology at our actual reach, and no matter how careful we are, the uncertainty in the position times and the uncertainty in the momentum will never be less than $\hbar$, a very small quantity of the order of 10^{-34} [Joule·seg], but different from zero.

APPENDIX 3.3: QUANTIZATION OF THE ELECTROMAGNETIC FIELD

For an accurate description of the interaction between particles and electromagnetic fields (light), it is relevant to use a quantum theory of electromagnetic fields. This is particularly necessary for low intensity of the light, where the number of light quanta (known as photons) is small and the fields cannot be considered continuous as is made in the classical description.

In this appendix, the traditional approach of quantization of light is briefly described, which begins with the identification of the coordinates and their conjugate momenta. Then the procedure continues considering that the fundamental coordinates and their momenta in the quantum mechanical system must have a commutator defined analogous to $[x, p_x] = i\hbar$ as in any simple QM system. Creation and annihilation operators are introduced having a unit commutator, such that $[a, a\dagger] = 1$.

It can be shown that a quantized field can be described as a system of decoupled harmonic oscillators with different wave vectors and wave polarizations, and that the basic quanta of the electromagnetic field, which are called photons, are created and annihilated in processes of emission and absorption of light by atoms or matter in general.

We begin with the idealized case of a radiation field confined to a cubic cavity with perfectly conducting walls.

The cavity is regarded merely as a region of space without any real boundaries, known as the "quantization cavity." We take running wave solutions of the field and subject them to periodic boundary conditions.

In free space, Maxwell's equation leads to the following wave equation for the magnetic field potential which clearly has time-reversal symmetry

$$\nabla^2 \vec{A} = \frac{1}{c^2}\frac{\partial^2 \vec{A}}{\partial t^2} \tag{3.239}$$

The solution of eq. 3.239 may be written as

$$\vec{A}(\vec{r},t) = \sum_{\vec{k},\lambda} \hat{e}_{\vec{k}\lambda}\alpha_{k,\lambda}(\vec{r},t) \tag{3.240}$$

where

$$\alpha_{k\lambda}(\vec{r},t) = A_{k\lambda}(t)exp(i\vec{k}\cdot\vec{r}) + A^*_{k\lambda}(t)exp(-i\vec{k}\cdot\vec{r}) \tag{3.241}$$

$$\vec{k} = \left(k_x = \frac{2\pi n_x}{L}, k_y = \frac{2\pi n_y}{L}, k_z = \frac{2\pi n_z}{L}\right) \tag{3.242}$$

is the wave vector with components k_x,k_y, and k_z which depend on integer numbers

$$n_x, n_y, n_z = \pm 1, \pm 2, \pm 2, \ldots \ldots \ldots \tag{3.243}$$

and with $\hat{e}_{\vec{k}\lambda}$ being unit polarization vectors for each mode $\vec{k}$ and a given transversal polarization state, specified by the index $\lambda = 1,2$.

The polarization vectors are perpendicular to each other and also to the corresponding direction of propagation given by the vector $\vec{k}$, so that

$$\hat{e}_{\vec{k}\lambda}\cdot\hat{e}_{\vec{k}\lambda'} = \delta_{\lambda\lambda'} \tag{3.244}$$

Each mode components of the vector potential are independent and they separately obey the equation:

$$k^2 A_{k\lambda}(t) + \frac{1}{c^2}\frac{\partial^2 A_{k\lambda}(t)}{\partial t^2} = 0 \tag{3.245}$$

Therefore, the mode coefficients and their complex conjugates satisfy an equation similar to that of the simple-harmonic oscillator:

$$\omega_k^{\,2} A_{k\lambda}(t) + \frac{\partial^2 A_{k\lambda}(t)}{\partial t^2} = 0 \tag{3.246}$$

where $\omega_k = ck$ is the dispersion relation for propagation in free space.

The electromagnetic field is quantized by conversion of the classical harmonic oscillator to its quantum mechanical counterpart. For that purpose, the terms $A_{k\lambda}(t)$ are expressed as

$$A_{k\lambda}(t) = A_{k\lambda}exp(-i\omega_k t) \tag{3.247}$$

In this way, the contribution of mode $\vec{k},\lambda$ to the vector potential becomes:

$$A_{k\lambda}(\vec{r},t) = A_{k\lambda} exp\left(-i\omega_k t + i\vec{k}\cdot\vec{r}\right) + A^*_{k\lambda} exp\left(i\omega_k t - i\vec{k}\cdot\vec{r}\right) \quad (3.248)$$

The electric and field become operators and they are obtained from:

$$\vec{E} = -\frac{\partial \vec{A}}{\partial t} \quad (3.249)$$

$$\vec{B} = \nabla \times \vec{A} \quad (3.250)$$

After a mathematical development, whose details can be found in well-known books about quantum optics (see, e.g., the book of Rodney Loudon, *The Quantum Theory of Light*, Third Edition, Oxford University Press), one can obtain that superposing all the electromagnetic modes, the electric and magnetic field inside the cavity are given by

$$\vec{E}(\vec{r},t) = \sum_{k,\lambda} \hat{e}_{\vec{k}\lambda} E_{k\lambda}(\vec{r},t) \quad (3.251)$$

$$E_{k\lambda}(\vec{r},t) = i\omega_k \left\{A_{k\lambda} exp\left(-i\omega_k t + i\vec{k}\cdot\vec{r}\right) + A^*_{k\lambda} exp\left(i\omega_k t - i\vec{k}\cdot\vec{r}\right)\right\} \quad (3.252)$$

$$\vec{B}(\vec{r},t) = \sum_{k\lambda} \frac{\vec{k}\times\hat{e}_{\vec{k}\lambda}}{\boldsymbol{k}} B_{k\lambda}(\vec{r},t) \quad (3.253)$$

$$B_{k\lambda}(\vec{r},t) = ik\left\{A_{k\lambda} exp\left(-i\omega_k t + i\vec{k}\cdot\vec{r}\right) + A^*_{k\lambda} exp\left(i\omega_k t - i\vec{k}\cdot\vec{r}\right)\right\} \quad (3.254)$$

The total energy of the electromagnetic radiation field inside the cavity is

$$W_R = \frac{1}{2}\int dV\left[\varepsilon_o\left(\vec{E}(\vec{r},t)\right)^2 + \mu_o^{-1}\left(\vec{B}(\vec{r},t)\right)^2\right] \quad (3.255)$$

where the integration is over the volume V of the cavity

After doing some algebra, we obtain that the total energy stored in the electromagnetic field reduces to a sum of time-independent contributions from the individual modes:

$$W_R = \varepsilon_o V \sum_{\vec{k}\lambda} \omega_k^2 \left(A_{\vec{k}\lambda} A^*_{\vec{k}\lambda} + A^*_{\vec{k}\lambda} A_{\vec{k}\lambda}\right) \quad (3.256)$$

In order to quantize the field, canonical variables $p_{\vec{k}\lambda}$ and $q_{\vec{k}\lambda}$ are introduced, so that

$$A_{\vec{k}\lambda} = \frac{1}{2\omega_k\left(\varepsilon_o V\right)^{1/2}}\left[\omega_k q_{\vec{k}\lambda} + ip_{\vec{k}\lambda}\right] \quad (3.257)$$

$$A^*_{\vec{k}\lambda} = \frac{1}{2\omega_k(\varepsilon_o V)^{1/2}}\left[\omega_k q_{\vec{k}\lambda} - ip_{\vec{k}\lambda}\right] \quad (3.258)$$

Upon substitution of eqs. 3.257 and 3.258 into eq. 3.256, we get the following Hamiltonian:

$$\mathcal{H} = \frac{1}{2}\sum_{\vec{k}\lambda}\left(p^2_{\vec{k}\lambda} + \omega_k^2 q^2_{\vec{k}\lambda}\right) \quad (3.259)$$

Each term of the sum in the Hamiltonian given by the above equation is the energy of a simple-harmonic oscillator with unitary mass. The quantization of the field is made by demanding that the canonical variables become operators satisfying the following commutation relations:

$$\left[q_{\vec{k}\lambda}, q_{\vec{k}'\lambda'}\right] = \left[p_{\vec{k}\lambda}, p_{\vec{k}'\lambda'}\right] = 0 \quad (3.260)$$

$$\left[q_{\vec{k}\lambda}, p_{\vec{k}'\lambda'}\right] = i\hbar\delta_{\vec{k}\vec{k}''}\delta_{\lambda\lambda'} \quad (3.261)$$

We introduce annihilation and creation operators defined as

$$\hat{a}_{\vec{k}\lambda} = \frac{1}{(2\hbar\omega_k)^{1/2}}\left\{\omega_k\hat{q}_{\vec{k}\lambda} + i\hat{p}_{\vec{k}\lambda}\right\} \quad (3.262)$$

$$\hat{a}_{\vec{k}\lambda}{}^{\dagger} = \frac{1}{(2\hbar\omega_k)^{1/2}}\left\{\omega_k\hat{q}_{\vec{k}\lambda} - i\hat{p}_{\vec{k}\lambda}\right\} \quad (3.263)$$

These operators satisfy the following commutation relations:

$$\left[\hat{a}_{\vec{k}\lambda}, \hat{a}_{\vec{k}'\lambda'}\right] = \left[\hat{a}_{\vec{k}\lambda}{}^{\dagger}, \hat{a}_{\vec{k}'\lambda'}{}^{\dagger}\right] = 0 \quad (3.264)$$

$$\left[\hat{a}_{\vec{k}\lambda}, \hat{a}_{\vec{k}'\lambda'}{}^{\dagger}\right] = \delta_{\vec{k}\vec{k}'}\delta_{\lambda\lambda'} \quad (3.265)$$

In terms of creation and annihilation operators, the Hamiltonian operator becomes

$$\hat{\mathcal{H}} = \sum_{\vec{k}\lambda}\hbar\omega_k\left(\hat{a}_{\vec{k}\lambda}{}^{\dagger}\hat{a}_{\vec{k}\lambda} + 1/2\right) = \sum_{\vec{k}\lambda}h\omega_k\left(\hat{n}_{\vec{k}\lambda} + 1/2\right) \quad (3.266)$$

where $\hat{n}_{\vec{k}\lambda} = \hat{a}_{\vec{k}\lambda}{}^{\dagger}\hat{a}_{\vec{k}\lambda}$ is the number operator for the mode $\vec{k}\lambda$. All of these modes are independent between them, and each one has a given number of photons associated with an eigenket denoted as $|n_{\vec{k}\lambda}\rangle$. For the jth mode, we define

$$\left.\begin{aligned} \hat{a}_{\vec{k}_j\lambda_j} &\equiv \hat{a}_j \\ \hat{a}^{\dagger}_{\vec{k}_j\lambda_j} &\equiv \hat{a}^{\dagger}_j \\ \hat{n}_{\vec{k}_j\lambda_j} &\equiv \hat{n}_j \end{aligned}\right\} \quad (3.267)$$

The Hamiltonian for the field is then

$$\hat{\mathcal{H}} = \sum_j \hbar\omega_j \left(\hat{a}_j^{\dagger}\hat{a}_j + 1/2\right) = \sum_j \hbar\omega_j \left(\hat{n}_j + 1/2\right) \quad (3.268)$$

and a multimode photon number state is defined as a product of the number states of all the modes, which is written as

$$|n_1\rangle|n_2\rangle|n_3\rangle\ldots\ldots \equiv |n_1, n_2, n_3, \ldots\ldots\rangle = |\{n_j\}\rangle \quad (3.269)$$

This is an eigenstate of $\hat{\mathcal{H}}$ with eigenvalue E, such that

$$\hat{\mathcal{H}}|\{n_j\}\rangle = E|\{n_j\}\rangle \quad (3.270)$$

The action of the annihilation operator of the jth mode on the multimode number state leads to

$$\hat{a}_j|n_1, n_2, n_3, \ldots n_j \ldots\rangle = \sqrt{n_j}\,|n_1, n_2, n_3, \ldots n_j - 1\ldots\rangle \quad (3.271)$$

Similarly, for the creation operator

$$\hat{a}_j^{\dagger}|n_1, n_2, n_3, \ldots n_j \ldots\rangle = \sqrt{n_{j+1}}\,|n_1, n_2, n_3, \ldots n_j + 1\ldots\rangle \quad (3.272)$$

For the vacuum state $\hat{a}_j|\{0\}\rangle = 0$ with = 1,2,3, … … …, with the multimode vacuum state denoted as $|\{0\}\rangle = |0,0,0 \ldots 0 \ldots\rangle$.

APPENDIX 3.4: REVIVAL OF RABI OSCILLATIONS APPLYING TIME REVERSAL

The coupling between an atom and the quantized electromagnetic field in its environment leads to many fascinating phenomena important in the history of quantum mechanics, including first of all, spontaneous emission of photons. If an atom is initially in an excited state $|e\rangle$, it will emit a photon and irreversibly fall to a lower state $|g\rangle$ in a characteristic time, which is in the scale of nanoseconds for visible light, and milliseconds for transitions in the microwave region. This process is irreversible because of the continuous density of states available to the photon, which once emitted, flies off at the speed of light never to return.

Very interesting things happen however if the atom is placed inside a cavity which prevents the photon from escaping [1–3]. If the discrete photon modes of the cavity do not match, the conditions for an atom transition, then spontaneous emission can be strongly inhibited. If however, one of the cavity modes has an energy close to the atom transition, then the atom and cavity can coherently exchange energy back and forth by oscillating between the excited and ground state emitting or absorbing a photon, that is making the process

$$|e, 0\rangle \Leftrightarrow |g, 1\rangle$$

One interesting case studied by Meunier et al. [4] is where the cavity contains, not a Fock state with definite number n of photons, but rather a coherent state with definite phase and an uncertain value for this number. Since the Rabi swapping rate varies with n, the oscillations of the different photon number components of the state get out of phase with each other and destructively interfere after about $\sqrt{n}$ oscillations. This apparent irreversibility occurs even though the photons do not escape. However, it is not truly irreversible because after about n oscillations, constructive interference is partially revived.

Taking as a reference what happens with nuclear magnetic resonance refocusing techniques [5], it is proposed to let the system evolve for a given time T and then submit the atom to a short electromagnetic pulse leading to a reversal of the system's unitary evolution. In the absence of decoherence, it is expected that at time $2T$ the system will come back exactly to the initial state (*remember that in quantum mechanics, particles such as electrons are described by a wave function; in case there exists a definite phase relation between different states, the system is said to be coherent*). Reference [5] shows the application of this scheme to the study of the phenomenon of atomic Rabi oscillation under the effect of a coherent field, with about 13 photons on the average. This oscillation undergoes a collapse originated by the dispersion of the Rabi frequencies associated with different photon numbers. This effect is reversible and after a time interval proportional to the field amplitude, the oscillation can "revive" spontaneously [6]. Applying a π-phase shift to the wave function of the system after the collapse time by means of a short electromagnetic pulse, an early induced Rabi revival is obtained.

In this appendix, we briefly present the theoretical basis for understanding these kind of experiments and go in depth in the Jayne-Cunnings model which predict the phenomenon of Rabi oscillations revival. For a detailed study of this subject the reader is advised to review Section 4.5 of the book of Gerry and Knight [7] and section 5.1 of the book of Rodney Loudon [8].

1. The Jaynes-Cumming model

We briefly discuss here the quantum electrodynamic version of the Rabi model, considering that a free atom really interacts with an infinite number of modes, so that its dynamical evolution is not adequately described considering only a single-mode field.

Let us begin considering a two-level atom with states $|e\rangle$ and $|g\rangle$ inside a cavity and interacting with a single-mode electromagnetic field of frequency ω and having an electric field operator of the form

$$\hat{\boldsymbol{E}}(z,t) = \boldsymbol{e}\left(\frac{\hbar\omega}{\varepsilon_o V}\right)^{1/2}\left(\hat{a}+\hat{a}^{\dagger}\right) sinkz \tag{3.273}$$

where $\boldsymbol{e}$ is an arbitrarily oriented polarization vector. Working in the Heisenberg picture, the operators $\hat{a}, \hat{a}^{\dagger}$ (creation and annihilation of photons, respectively) are time-dependent.

The interaction Hamiltonian is

$$\hat{\mathcal{H}}^{(1)} = -\hat{\boldsymbol{d}} \cdot \hat{\boldsymbol{E}} = dg(\hat{a} - \hat{a}^{\dagger}) \tag{3.274}$$

where

$$g = -\left(\frac{\hbar\omega}{\varepsilon_o V}\right)^{\frac{1}{2}} sinkz \tag{3.275}$$

and $\hat{\boldsymbol{d}} = q\langle g|\boldsymbol{R}|e\rangle$ is the dipole matrix element of the atomic transition between states $|e\rangle$ and $|g\rangle$, q is the electric charge and we define the scalar operator $\hat{d}$ as:

$$\hat{d} = \hat{\boldsymbol{d}} \cdot \boldsymbol{e} = d\,|\,g\rangle\langle e\,| + d^{*}\,|\,e\rangle\langle g\,| = d\hat{\sigma}_{-} + d^{*}\hat{\sigma}_{+} = d(\hat{\sigma}_{+} + \hat{\sigma}_{-}) \tag{3.276}$$

where $\hat{\sigma}_{+}$ and $\hat{\sigma}_{-}$ are the so-called transition operators defined as

$$\hat{\sigma}_{+} = |e\rangle\langle g| \text{ and } \hat{\sigma}_{-} = |g\rangle\langle e| = \hat{\sigma}_{+}^{\dagger} \tag{3.277}$$

In eqs. 3.276 and 3.277, we have set $d = \langle e|\hat{d}|g\rangle$ and for simplicity, we have assumed that d is real.

Therefore, the interaction Hamiltonian is

$$\hat{\mathcal{H}}^{(1)} = \hbar\lambda\left(\hat{\sigma}_{+} + \hat{\sigma}_{-}\right)\left(\hat{a} + \hat{a}^{\dagger}\right) \tag{3.278}$$

where we have defined $\lambda \equiv dg/\hbar$ which has units of frequency (in MKS d and g have units Coulomb·m and Joule/Coul·m, respectively).

If we define the level of the energy equal to zero halfway between the energies corresponding to the states $|g\rangle$ and $|e\rangle$ as shown in Figure 3.21, we may write the free atomic Hamiltonian as

$$\hat{\mathcal{H}}_{A} = \frac{1}{2}\left(E_{e} - E_{g}\right)\hat{\sigma}_{3} = \frac{1}{2}\hbar\omega_{o}\hat{\sigma}_{3} \tag{3.279}$$

where $\hat{\sigma}_3$ is the matrix

$$\hat{\sigma}_3 = |e\rangle\langle e| - |g\rangle\langle g| = \begin{pmatrix} 1 & 0 \\ 0 & -1 \end{pmatrix} \tag{3.280}$$

and ω_o is given by

$$E_e = -E_g = \frac{1}{2}\hbar\omega_o \tag{3.281}$$

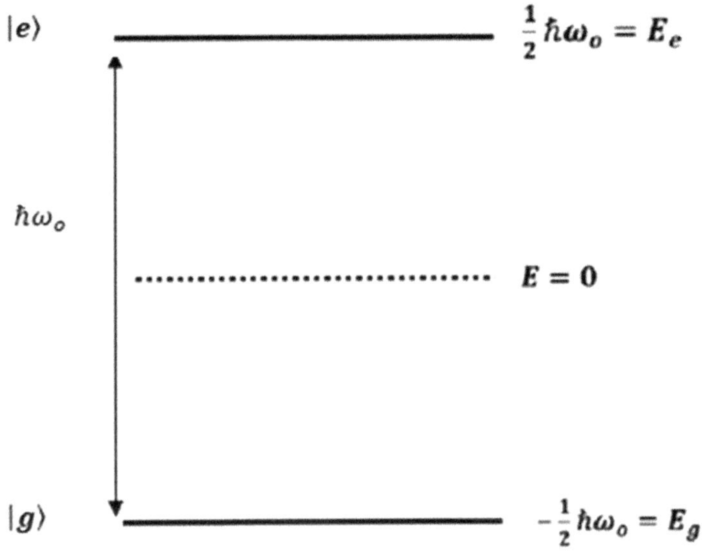

FIGURE 3.21 States and energies of a quantum two-level system.

Dropping the zero-point energy term, the free-field Hamiltonian is

$$\hat{\mathcal{H}}_F = \hbar\omega\hat{a}^\dagger\hat{a} \tag{3.282}$$

Thus, the total Hamiltonian is

$$\hat{\mathcal{H}} = \hat{\mathcal{H}}_A + \hat{\mathcal{H}}_F + \hat{\mathcal{H}}^{(1)} = \frac{1}{2}\hbar\omega_o\hat{\sigma}_3 + \hbar\omega\hat{a}^\dagger\hat{a} + \hbar\lambda\left(\hat{\sigma}_+ + \hat{\sigma}_-\right)\left(\hat{a} + \hat{a}^\dagger\right) \tag{3.283}$$

The evolution on time of the operators $\hat{a}$ and $\hat{a}^\dagger$ are proportional to $e^{-i\omega t}$ and $e^{+i\omega t}$, respectively, while for the operators $\hat{\sigma}_+$ and $\hat{\sigma}_-$, the corresponding factors are $e^{+i\omega_o t}$ and $e^{-i\omega_o t}$, respectively. This is obtained considering that the Heisenberg's dynamical equation for an arbitrary operator $\hat{O}$ is

$$\frac{d\hat{O}}{dt} = \frac{1}{i\hbar}\left[\mathcal{H}, \hat{O}\right] \tag{3.284}$$

and that the following commutation relations are valid:

$$[\hat{a}, \hat{a}^\dagger] = 1 \quad (3.285)$$

$$[\hat{\sigma}_3, \hat{\sigma}_+] = \pm 2\hat{\sigma}_\pm \quad (3.286)$$

Using the above equations, we obtain the time dependence of the term $\hbar\lambda(\hat{\sigma}_+ + \hat{\sigma}_-)(\hat{a} + \hat{a}^\dagger)$ in eq.(3.318), where the following factors appear:

$$\left.\begin{aligned} \hat{\sigma}_+\hat{a} &\sim e^{i(\omega_o-\omega)t} \\ \hat{\sigma}\_\hat{a}^\dagger &\sim e^{-i(\omega_o-\omega)t} \\ \hat{\sigma}_+\hat{a}^\dagger &\sim e^{i(\omega_o+\omega)t} \\ \hat{\sigma}\_\hat{a} &\sim e^{-i(\omega_o+\omega)t} \end{aligned}\right\} \quad (3.287)$$

For $\omega_o \approx \omega$, the last two terms of the above equation, change much more rapidly compared with the first two, and they can be neglected if the so-called rotating-wave approximation is used [9]. Furthermore, the analysis of these terms indicates that they do not have a physical sense since the principle of conservation of energy is not fulfilled, unlike the first two. In fact, the term $\hat{\sigma}_+\hat{a}^\dagger$ implies the emission of a photon as the atom makes a transition from the ground to the excited state. The term $\hat{\sigma}\_a$ implies the absorption of a photon as the atom makes a transition from the excited to the ground state. Due to these reasons, these terms are discarded and the total Hamiltonian can be written as

$$\hat{\mathcal{H}} = \frac{1}{2}\hbar\omega_o\hat{\sigma}_3 + \hbar\omega\hat{a}^\dagger\hat{a} - \hbar\lambda\left(\hat{\sigma}_+\hat{a} + \hat{\sigma}\_\hat{a}^\dagger\right) \quad (3.288)$$

The interaction described by this last Hamiltonian is referred to as the Jaynes-Cummings model [10]. This Hamiltonian can be written as $\hat{\mathcal{H}} = \hat{\mathcal{H}}_I + \hat{\mathcal{H}}_{II}$ with

$$\hat{\mathcal{H}}_I = \hbar\omega\hat{N}_e - \frac{\hbar\omega}{2}\hat{P}_E \quad (3.289)$$

$$\hat{\mathcal{H}}_{II} = \frac{1}{2}\hbar\hat{\sigma}_3\Delta + \hbar\lambda\left(\hat{\sigma}_+\hat{a} + \hat{\sigma}\_\hat{a}^\dagger\right) \quad (3.290)$$

In the above two equations:

$$\hat{P}_E = |e\rangle\langle e| + |g\rangle\langle g| = 1 \quad (3.291)$$

$$\hat{N}_e = \hat{a}^\dagger a + |e\rangle\langle e| \quad (3.292)$$

$\hat{P}_E$ is called the *electron number* and when none of the other atomic states can be populated is a conserved quantity since $[\hat{\mathcal{H}}, \hat{P}_E] = 0$.

$\hat{N}_e$ is called the *excitation number* and also is a conserved quantity since $[\mathcal{H},\hat{N}_e]=0$.

$\Delta=\omega_o-\omega$ is called the *detuning* and is the difference between the frequency corresponding to transitions between the two states and the frequency of the incident light.

Clearly, all the essential dynamics comes from $\hat{\mathcal{H}}_{II}$ since it contains time-dependent operators so that for obtaining the evolution of this two-level quantum system, the corresponding Schrödinger's equation is:

$$i\hbar\frac{d|\psi(t)\rangle}{dt}=\hat{\mathcal{H}}_{II}|\psi(t)\rangle \tag{3.293}$$

Writing the state vector as $|\psi(t)\rangle=C_i(t)|i\rangle+C_f(t)|f\rangle$ and considering that initially the atom is in the excited state $|e\rangle$ and the field in the number state $|n\rangle$, according to Ref. [2], the following is the solution of eq. 3.328 in the resonant case with detuning $\Delta=0$:

$$|\psi(t)\rangle=cos\left(\lambda t\sqrt{n+1}\right)|e\rangle|n\rangle+sin\left(\lambda t\sqrt{n+1}\right)|g\rangle|n+1\rangle \tag{3.294}$$

The probability that the system does not leave the initial state $|i\rangle=|e\rangle|n\rangle$ is given by

$$P_i(t)=|\langle i|\psi(t)\rangle|^2=cos^2\left(\lambda t\sqrt{n+1}\right) \tag{3.295}$$

and the probability that it makes a transition to the state $|f\rangle=|g\rangle|n+1\rangle$ is

$$P_f(t)=|\langle f|\psi(t)\rangle|^2=sin^2\left(\lambda t\sqrt{n+1}\right) \tag{3.296}$$

Let us now consider again that initially, the atom is in the excited state but the field is in a coherent state given by

$$|\psi(0)\rangle_{field}=\sum_{n=0}^{\infty}C_n|n\rangle \tag{3.297}$$

where for a coherent state

$$C_n=e^{-\bar{n}/2}\frac{\alpha^n}{\sqrt{n!}} \tag{3.298}$$

with $\bar{n}$ being the average number of photons and α a complex number with magnitude $\sqrt{\bar{n}}$ and arbitrary phase.

The state of the system at any instant t can be obtained from eq. 3.294 considering all the states $|n\rangle$ of the field, so that

$$|\psi(t)\rangle=\sum_{n=0}^{\infty}C_n cos\left(\lambda t\sqrt{n+1}\right)|e\rangle|n\rangle+\sum_{n=0}^{\infty}C_n sin\left(\lambda t\sqrt{n+1}\right)|g\rangle|n+1\rangle \tag{3.299}$$

The probability that the system remains in the initial state $|i\rangle = |e\rangle|n\rangle$ is given by

$$P_i(t) = |\langle i|\psi(t)\rangle|^2 = \sum_{n=0}^{\infty} |C_n|^2 cos^2\left(\lambda t\sqrt{n+1}\right) \qquad (3.300)$$

and the probability that it makes a transition to the state $|f\rangle = |g\rangle|n+1\rangle$ is

$$P_f(t) = |\langle f|\psi(t)\rangle|^2 = \sum_{n=0}^{\infty} |C_n|^2 sin^2\left(\lambda t\sqrt{n+1}\right) \qquad (3.301)$$

Figure 3.22 shows the time revival effect computing by means of eq. 3.301, the probability to find the analyzed two-level system in the ground state $|f\rangle = |g\rangle|n+1\rangle$ for 3 values of the average number of photons $\bar{n}$: 13.4 (red curve), 19 (green curve) and 25 (blue curve). In this figure, time is expressed units of the Rabi oscillation period $T_R = \pi\sqrt{\bar{n}}/\lambda$ [4]. It can be seen that as this number increases, the same occurs with the revival time in agreement with the expression $t_r = \frac{2\pi}{\lambda}\sqrt{\bar{n}}$ obtained in Section 3.9.2.

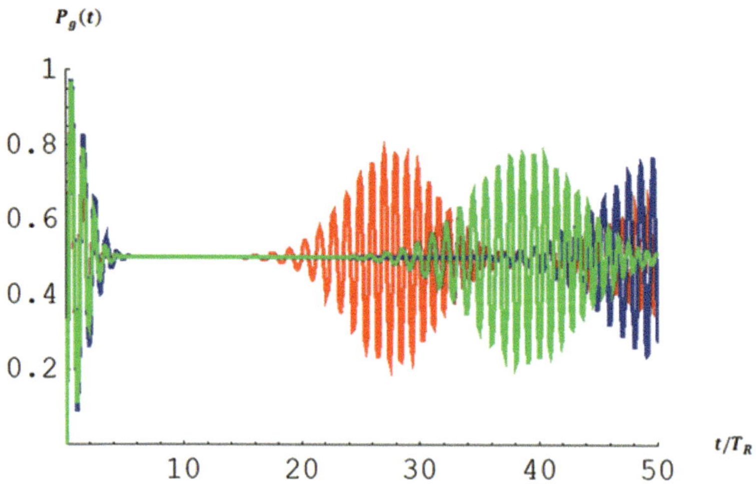

FIGURE 3.22 Probability of finding the two-level system in the ground state $|f\rangle = |g\rangle|n+1\rangle$, for 3 values of the average number of photons: 13.4 (red curve), 19 (green curve) and 25 (blue curve).

2. *Time Reversing the Collapse of Rabi Oscillations*

In the so-called interaction picture (where the operators are time-dependent while the basis of the space remain fixed), the system's evolution is ruled by the Jaynes–Cummings (JC) Hamiltonian given by eq. 3.290 which for zero detuning is

$$\hat{\mathcal{H}}_{II} = \hbar\lambda\left(\hat{\sigma}_+ \hat{a} + \hat{\sigma}_- \hat{a}^\dagger\right) \tag{3.302}$$

According to the echo sequence presented in Ref. [4], the atom first evolves from time 0 to T when it is subjected to the action of the unitary evolution operator $U_{ev1} = exp(-i\hat{\mathcal{H}}_{II}\ T/\hbar)$. At time T, the atom undergoes a controlled phase kick corresponding to the unitary operation $U_\pi = |e\rangle\langle e| - |g\rangle\langle g|$ which corresponds to the Pauli matrix $\hat{\sigma}_3$ given by eq. 3.280. The Jaynes-Cummings evolution then resumes for the remaining time $\tau - T$ with the evolution operator $U_{ev2} = exp(-i\hat{\mathcal{H}}_{II}(\tau - T)/\hbar)$. The overall evolution operator U_{ev} is then

$$U_{ev} = U_{ev2}\hat{\sigma}_3\ U_{ev1} = \hat{\sigma}_3\ exp\left[-i\hat{\mathcal{H}}_{II}\left(2T - \tau\right)/\hbar\right] \tag{3.303}$$

Equation 3.303 shows that at time $\tau = 2T$, the unitary evolution returns the system to its initial state, except that due to the $\hat{\sigma}_3$ term, a global π-phase shift arises between the amplitudes associated with states $|e\rangle =$ and $|g\rangle$. The proof of eq. 3.303 is as follows:

We first prove that the evolution after the kick given at time T is the time-mirror image of the evolution between 0 and T. For that purpose, we write the term $U_{ev2}\hat{\sigma}_3 U_{ev1}$ of eq. 3.303 as

$$U_{ev2}\hat{\sigma}_3\ U_{ev1} = \hat{\sigma}_3{}^2 U_{ev2}\hat{\sigma}_3\ U_{ev1} = \hat{\sigma}_3\left(\hat{\sigma}_3 U_{ev2}\hat{\sigma}_3\right)U_{ev1} \tag{3.304}$$

In eq. 3.304, it has been considered that $\hat{\sigma}_3^2$ es equal to the identity matrix 2×2.

Using the definition of $\hat{\sigma}_+$ and $\hat{\sigma}_-$ given by eq. 3.277, we can write the Hamiltonian of eq. 3.302 as:

$$\hat{\mathcal{H}}_{II} = \hbar\lambda\left\{\begin{pmatrix} 0 & 0 \\ \hat{a} & 0 \end{pmatrix} - \begin{pmatrix} 0 & \hat{a}^\dagger \\ 0 & 0 \end{pmatrix}\right\} = \hbar\lambda\begin{pmatrix} 0 & -\hat{a}^\dagger \\ \hat{a} & 0 \end{pmatrix} \tag{3.305}$$

In this last equation, we have considered that in the basis $\{|g\rangle, |e\rangle\}$, the operators $\hat{\sigma}_+$ and $\hat{\sigma}$ have the following associated matrix:

$$\hat{\sigma}_+ = \begin{bmatrix} 0 & 0 \\ 1 & 0 \end{bmatrix}, \hat{\sigma}_- = \begin{bmatrix} 0 & 1 \\ 0 & 0 \end{bmatrix} \tag{3.306}$$

The transformation $\hat{\sigma}_3\hat{\mathcal{H}}_{II}\hat{\sigma}_3$ gives

$$\hat{\sigma}_3\hat{\mathcal{H}}_{II}\hat{\sigma}_3 = \hbar\lambda\begin{pmatrix} 1 & 0 \\ 0 & -1 \end{pmatrix}\begin{pmatrix} 0 & -\hat{a}^\dagger \\ \hat{a} & 0 \end{pmatrix}\begin{pmatrix} 1 & 0 \\ 0 & -1 \end{pmatrix} = \hbar\lambda\begin{pmatrix} 0 & \hat{a}^\dagger \\ -\hat{a} & 0 \end{pmatrix} = -\hat{\mathcal{H}}_{II} \tag{3.307}$$

Therefore, the term $\hat{\sigma}_3 U_{ev2}\hat{\sigma}_3$ in eq. 3.304 becomes

$$\hat{\sigma}_3 U_{ev2}\hat{\sigma}_3 = exp\left[-i\hat{\mathcal{H}}_{II}\left(T - \tau\right)/\hbar\right] \tag{3.308}$$

Combining the above results, we obtain that

$$U_{ev} = \hat{\sigma}_3\left(\hat{\sigma}_3 U_{ev2}\hat{\sigma}_3\right)U_{ev1} = \hat{\sigma}_3 exp\left[-i\hat{\mathcal{H}}_{II}\left(T-\tau\right)/\hbar\right]exp\left(-i\hat{\mathcal{H}}_{II}T/\hbar\right)$$
$$= \hat{\sigma}_3 exp\left[-i\hat{\mathcal{H}}_{II}\left(2T-\tau\right)/\hbar\right]$$

which corresponds to eq. 3.303.

REFERENCES FOR APPENDIX 3.4

1. Mabuchi, H.; Doherty, A. Cavity Quantum Electrodynamics: Coherence in Context. *Science* **2002,** *298*, 1372.
2. Raimond, J.; Brune, M.; Haroche, S. Manipulating Quantum Entanglement with Atoms and Photons in a Cavity. *Rev. Mod. Phys.* **2001,** *73*, 565.
3. Wallraff, A.; Schuster, D. I.; Blais, A.; Frunzio, L.; Huang, R.-S.; Majer, J.; Kumar, S.; Girvin, S. M.; Schoelkopf, R. J. Strong Coupling of a Single Photon to a Superconducting Qubit Using Circuit Quantum Electrodynamics. *Nature* **2004,** *431*, 162–167.
4. Meunier, T.; Gleyzes, S.; Maioli, P.; Auffeves, A.; Nogues, G.; Brune, M.; Raimond, J. M.; Haroche, S. Rabi Oscillations Revival Induced by Time Reversal: A Test of Mesoscopic Quantum Coherence. *Phys. Rev. Lett.* **2005,** *94*, 010401.
5. Levitt, M. *Spin Dynamics: Basics of Nuclear Magnetic Resonance*; Wiley: Chichester, 2001.
6. Eberly, J. H.; Narozhny, N. B.; Sanchez-Mondragon, J. J. Periodic Spontaneous Collapse and Revival in a Simple Quantum Model. *Phys. Rev. Lett.* **1980,** *44*, 1323.
7. Gerry, C. C.; Knight, P. L. *Introductory Quantum Optics*; Cambridge University Press: Cambridge, 2004.
8. Loudon, R. *The Quantum Theory of Light*, 3rd ed.; Oxford University Press: Oxford, 2000.
9. Slichter, C. P. *Principles of Magnetic Resonance*, 3rd ed.; Springer: Berlin, 1992.
10. Jaynes, E. T.; Cummings, F. W. Comparison of Quantum and Semiclassical Radiation Theories with Application to the Beam Maser. *Proc. IEEE* **1963,** *51*, 89–109.

CHAPTER 4

Time in Special Relativity Theory

ABSTRACT

Special relativity theory is one of the great discoveries of the Physics during the first years of the twentieth century and originated a deep impact on the way that the physical phenomena must be described for objects moving with a speed comparable with the light velocity, as is the case of particles emitted in nuclear reactions. One of the great changes introduced by this theory was that time must be considered on equal footing with respect to the spatial coordinates and combined with them, to form a frame called space time where events occur differently for observers moving with a relative velocity having a magnitude comparable with light velocity. In particular, the results of measurements of the time interval between two events are not the same for these observers and, even more, the sequence as they occur can be different if they are not causally connected.

In this chapter, we analyze the repercussions that this theory has on the predictions about the duration of events and its corresponding sequence in space time according to different inertial observers. The existence of superluminal particles, such as tachyons, gives rise to paradoxes such as those discussed at the end of this chapter.

4.1 INTRODUCTION

In Chapter 1 we have studied what happens with the macroscopic description of processes under time-reversal using the laws of classical mechanics and classical electrodynamics. We have shown that in the absence of dissipation there is a symmetry under time reversal, that is the description of the dynamics does not change if t is changed by $-t$ provided the velocity changes its sign.

Looking at Time from a Physics Perspective. Patricio Robles (Author)

In this chapter, we study how do the results of measurements of time intervals between events and its sequence change when some process is observed from two different frames in motion with relative velocity comparable with light velocity. The analysis here presented is at a macroscopic scale of dimensions so that quantum effects are not considered.

When the velocity of an object increases and its value becomes comparable with light velocity, according to special relativity theory (SRT) effects such as time dilatation and length contraction arise, which are not observable at the usual velocities we find in our lives.

Special relativity theory is one of the great discoveries of the Physics during the first years of the twentieth century and originated a deep impact in the form the description and conceptualization of the physical phenomena were done up to this moment. Its impact on the scientific development was similar to that originated by quantum mechanics. As all new theory meaning a substantial change of concepts that up to these years were widely accepted, it originated a great reluctance in its acceptance by the scientific community from the beginning of the above century.

This theory was developed mainly through the research work of Albert Einstein (German physicist, 1879–1955), Hendrik Lorentz (Dutch physicist, 1853–1928), and Hermann Minkowski (German mathematician 1864–1909). In 1905, Einstein formulated the so called Special Relativity Theory (SRT) which deals with the physical description of events occurring at different positions and instants of time according to observers in motion with a constant relative velocity (reference frames fixed to these observers are called inertial frames).

It is usually called special or restricted relativity theory since it does not include the effects of gravity. As we shall see in Chapter 5, these effects produce a deformation of space time with important consequences over the concept of time.

One of the main postulates of the theory of special relativity is that the laws of physics are the same in all inertial reference systems from which the same effect is observed, and that there is not an absolute reference frame. In connection with time, that is the issue analyzed in this book, that means that there is not an absolute time, as is the case with Newtonian's mechanics, and time intervals are different for inertial observers moving with a relative velocity of magnitude comparable to that of light.

We begin this chapter presenting the fundamental principles of the SRT and their consequences over measurements of time and length, mainly time dilation and Lorentz contraction of lengths. In particular, we discuss situations where events that occur simultaneously for an observer, cannot be simultaneous for other observer and even more, the sequence of events not causally connected can be different for two inertial observers. We review the mathematical formulation of SRT expressed by Lorentz transformations, analyzing in particular how time intervals, energy, and an electromagnetic field are seen by different inertial observers, searching the similarities between the corresponding transformations.

At the end of this chapter we include a discussion about the paradoxes that would be originated if the existence of superluminal particles, like tachyons, was possible. According to several publications about this subject and quoted here, what is really not possible according to the theory of relativity, is not the movement of particles with a speed greater than that of light, but rather the acceleration to reach speeds of this magnitude, passing through the value c. It must be considered that according to the theory of relativity known until now, for a speed equal to c, the Lorentz transformations are not applicable and that the energy of a particle with a small but not zero mass, such as an electron for example, becomes infinite. The possibility of existence of such particles gives rise to the same kind of unsolved paradoxes as those associated with backward time travel, mainly the communication with the past if tachyon existed. Another paradox is concerned with the fact that a tachyon traveling backward in time would have negative energy for some observers.

4.2 FUNDAMENTAL POSTULATES OF RELATIVITY TEORY

This theory is based on the following fundamental postulates[1–4]:

First Postulate

The laws of physics hold equally for all inertial observers.

This means that there are an infinite number of reference systems moving with constant relative velocity between them for which any physical phenomenon occurs in an identical way. These equivalent coordinate systems are called ***inertial frames.***

Second Postulate

The speed of light in vacuum has a finite value that is the same for all inertial reference frames and is independent of the relative motion between the source and the observer.

4.3 TIME AND LENGTH MEASUREMENTS

4.3.1 *TIME DILATION*

The measurement of the time interval between two events does not give the same result for two inertial observers in relative motion. This can be shown by means of a thought experiment based on the so called "Light Chronometer."[5]

Let us consider a chronometer whose operation is described by the time interval between the emission of a luminous pulse (TIC) and the reception of this pulse after being reflected in a mirror (TAC) as shown in Figure 4.1.

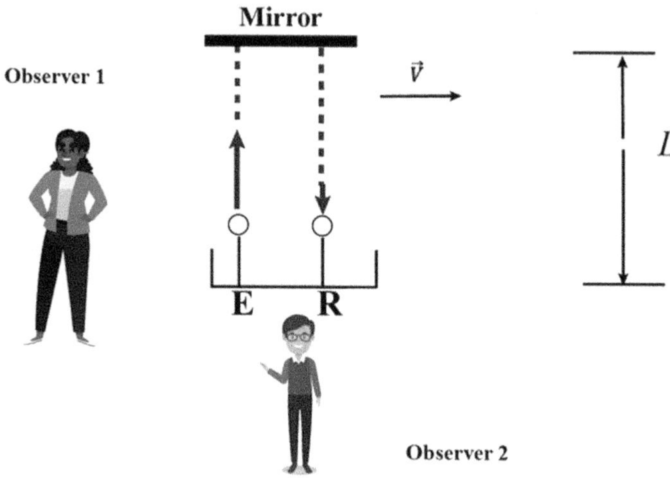

FIGURE 4.1 Luminous clock moving with respect to observer 1. A light pulse is emitted from a source placed at E (Tic). This pulse is reflected in a mirror and the reflected beam is received by a detector placed at R (Tac). Observer 2 is fixed to the luminous clock. Both observers measure the time interval between a Tic and a Tac.

For observer 2 the chronometer is at rest, so that the time interval of the TIC-TAC is

$$\tau_{CR} = 2D/c \tag{4.1}$$

This is a proper time since it is an interval of time measured between two consecutive events occurring at the same spatial position.

For observer 1 the chronometer moves with velocity $\vec{V}$ which we will asume parallel to the X axis, and for him this implies that the interval for TIC-TAC is τ_{CM} given by (see Figure 4.2)

$$\left(\frac{c\tau_{CM}}{2}\right)^2 = D^2 + \left(\frac{V\tau_{CM}}{2}\right)^2 \Rightarrow \tau_{CM} = \gamma\frac{2D}{c} = \gamma\tau_{CR} \tag{4.2}$$

where $\gamma = 1/\sqrt{1 - V^2/c^2}$ is called the Lorentz factor.

Since $|\vec{V}| = V$ cannot be higher than c this implies $\gamma > 1$and therefore the time interval of the Tic-Tac τ_{CM} is greater than τ_{CR}. That means that for observer 1 the elapsed time interval between a "Tic" and a "Tac" is greater than the detected by observer 2. Therefore, time goes slower than for observer 2 in comparison with observer 1. This effect predicted by Special Relativity Theory is called time dilation.

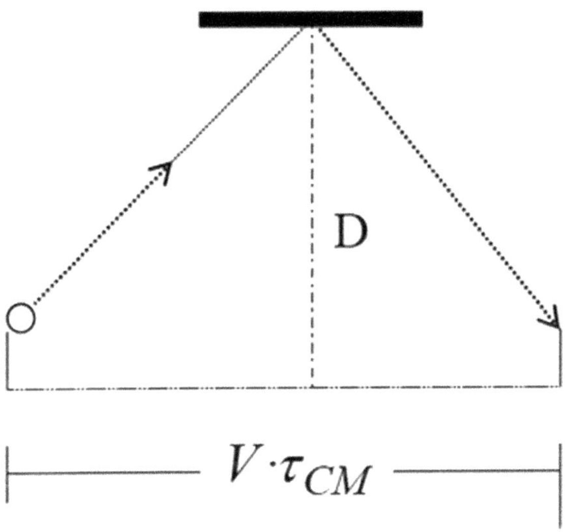

FIGURE 4.2 Light rays as seen by observer 1 for whom the time interval between a Tic and a Tac is τ_{CM}.

4.3.2 *LORENTZ CONTRACTION OF LENGTHS*

Another consequence of SRT is that the values of lengths are not the same for two inertial observers, in particular for spatial dimensions parallel to the relative velocity vector between these observers.

This can be shown by using again the luminous clock, but considering now that this device is moving in a direction perpendicular to the plane of its mirrors relative to the laboratory observer, as shown in Figure 4.3. This figure shows a system composed by a light source a mirror and a detector.

A pulse of light emitted by the source travels a distance D_m toward a mirror where there is total reflection so that the reflected light is received by a detector placed together with the source.

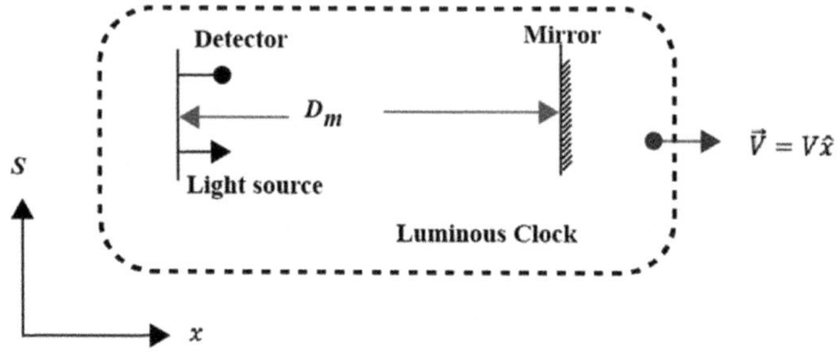

FIGURE 4.3 Luminous clock moving with respect to a reference frame S with velocity V along the x axis.

Figure 4.4 shows the position of the source and mirror at three instants. At the instant t_o the source emits a pulse toward the mirror in the x direction. This pulse is received by the mirror at instant t_r and is reflected back to the source. The reflected pulse is detected at instant t_d.

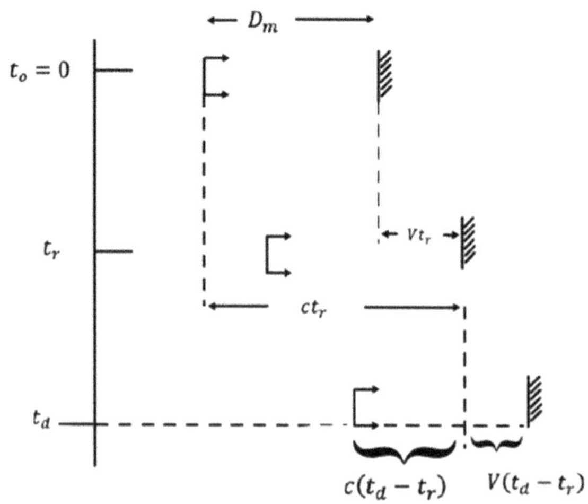

FIGURE 4.4 Sequence of events according to an observer fixed to the laboratory.

In Figure 4.4 it has been considered that at times t_r and t_d, the pulse of light has traveled distances ct_r and ct_d, respectively, measured with respect to the point of emission. From

Figure 4.4 we can obtain the following relations between time and distances:

$$ct_r = D_m + V_{tr} \tag{4.3}$$

which implies

$$t_r = \frac{D_m}{c - V} \tag{4.4}$$

Furthermore from Figure 4.4 we can obtain the following equations:

$$D_m = c(t_d - t_r) + V(t_d - t_r) = (c+V)(t_d - t_r) \tag{4.5}$$

$$\Rightarrow t_d = \frac{D_m}{c+V} + t_r = \frac{D_m}{c+V} + \frac{D_m}{c-V} \tag{4.6}$$

Since t_d is a period of time measured by an observer whose luminous clock is moving (reference frame S), it is related with the corresponding time t_{cr} measured by other observer at rest with respect to the clock (reference frame S') by means of the relation $t_{cr} = t_d/\gamma$ where $\gamma = 1/\sqrt{1-(V/c)^2}$. Furthermore if D_r is the distance between the source and the mirror according to reference frame S' we have $t_{cr} = 2D_r/c$ corresponding to the interval between the emission of the light signal, reflection at the mirror, and detection by the detector D.

Combining the above equations we obtain

$$\frac{2D_r}{c}\gamma = \frac{D_m}{c+V} + \frac{D_m}{c-V} = D_m\left(\frac{2c}{c^2 - V^2}\right) = \frac{2D_m}{c}\gamma^2$$

$$\Rightarrow D_m = \frac{D_r}{\gamma} = D_r\sqrt{1 - \frac{V^2}{c^2}} \tag{4.7}$$

Therefore, according to an observer at the laboratory (S), the distance D_m between the source and the mirror is lower than the distance D_r measured by an observer S' moving with the clock. This result is the expression of the relativistic effect corresponding to length contraction: For an observer for whom an object is in motion, its length looks contracted in a direction parallel to the vector corresponding to the relative velocity.

4.4 THE SPACE TIME AND LORENTZ TRANSFORMATIONS

For describing any physical event (i.e., the disintegration of a particle or a ball thrown by a boy), it is necessary to specify the location where this event occurs and the corresponding instant. Mathematically, this is expressed by means of three spatial coordinates (x,y,z) and a fourth variable which is the time represented as the variable t. The combination (x,y,z,t) corresponds to a point in a space of four dimensions, called the ***space time.***

Therefore, for an inertial observer A a given event occurs at a point in space time with coordinates (x,y,z,t).

For an observer B moving with respect to observer A along a rectilinear path and with constant speed V the same event occurs in a point with coordinates (x',y',z',t'). Note that in this theory the time t is not an absolute quantity and independent of the reference frame as has been considered in the physics of Newton.

Let us consider that observer B fixed to reference frame S' moves with respect to observer A (fixed to reference frame S) with constant speed V along the axis X as shown in Figure 4.5.

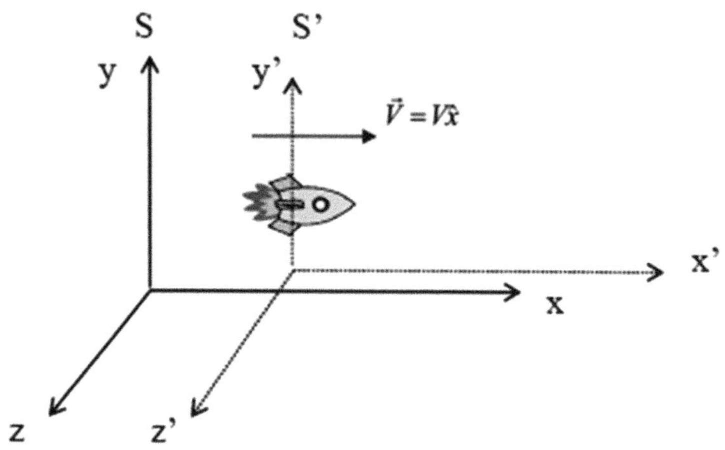

FIGURE 4.5 Reference frame S' moving with respect to S as shown in this figure.

For the corresponding measurements of time let us assume that at instants $t=t'=0$ the origins of the reference systems S and S' coincide and consider two simultaneous events E1 and E2 at S, so that in space time their coordinates are:

$$E_1 : \{x_1 = 0, y_1 = 0, z_1 = 0; t_1 = 0\}$$

$$E_2 : \{x_2 = x, y_2 = 0, z_2 = 0; t_2 = 0\}$$

For an observer fixed to system S' the system S is moving toward the left (negative x axis) with velocity –V and any length is contracted. Therefore, a distance x according to observer S is seen at S' as x/γ so that

$$x' = \frac{x}{\gamma} - Vt' \tag{4.8}$$

Similarly for an observer fixed to S there is also a length contraction, so that

$$x = \frac{x'}{\gamma} + Vt \tag{4.9}$$

From eqs 4.8 to 4.9 we have

$$x' = \gamma(x - Vt) \tag{4.10}$$

$$t' = \frac{1}{V}\left(-x' + \frac{x}{\gamma}\right) = \frac{1}{V}\left(-\gamma(x - Vt) + \frac{x}{\gamma}\right) = \gamma\left(t - \frac{x}{V}\left(1 - \frac{1}{\gamma^2}\right)\right) = \gamma\left(t - \frac{xV}{c^2}\right) \tag{4.11}$$

In summary we have the following transformation of coordinates in space time for this one-dimensional case:

$$x' = \gamma(x - Vt) \tag{4.12}$$

$$t' = \gamma\left(t - \frac{xV}{c^2}\right) \tag{4.13}$$

These transformations are called Lorentz transformations and can be generalized for relative motion in three dimensions (see for instance Chapter 11 of Ref. [4]).

Equations 4.12 and 4.13 can be expressed in matrix form for the transformation of coordinates from S to S' as follows:

$$\begin{bmatrix} x' \\ y' \\ z' \\ t' \end{bmatrix} = \begin{bmatrix} \gamma & 0 & 0 & -\gamma V \\ 0 & 1 & 0 & 0 \\ 0 & 0 & 1 & 0 \\ \frac{-\gamma V}{c^2} & 0 & 0 & \gamma \end{bmatrix} \begin{bmatrix} x \\ y \\ z \\ t \end{bmatrix} \tag{4.14}$$

For the transformation of coordinates from S' to S:

$$\begin{bmatrix} x \\ y \\ z \\ t \end{bmatrix} = \begin{bmatrix} \gamma & 0 & 0 & +V\gamma \\ 0 & 1 & 0 & 0 \\ 0 & 0 & 1 & 0 \\ \frac{+\gamma V}{c^2} & 0 & 0 & \gamma \end{bmatrix} \begin{bmatrix} x' \\ y' \\ z' \\ t' \end{bmatrix} \tag{4.15}$$

EXAMPLE 4.1

Derive the effect of time dilation from Lorentz transformations.

SOLUTION

The time dilation effect result can be obtained directly from Lorentz transformations. For example, let us consider two observers fixed to reference frames S and S' as shown in Figure 4.5 and two events E_1 and E_2 occurring in system S' at points $(x_1', y_1', z_1'; t_1')$ and $(x_2' = x_1', y_2' = y_1', z_2' = z_1'; t_2' = t_1' + \Delta t')$, respectively. For an observer fixed to systems S according to eq 4.15 the instants of occurrence of the same events are

$$t_1 = \gamma\left(t_1' + \frac{\beta}{c}x_1'\right) \tag{4.16}$$

$$t_2 = \gamma\left(t_2' + \frac{\beta}{c}x_2'\right) \tag{4.17}$$

From the two above equations and considering that $x_2' = x_1'$ we get

$$\Delta t = \gamma \Delta t' \tag{4.18}$$

This last equation implies that the time interval between two events is greater for observer S than the corresponding value measured by observer S', result in concordance with the experiment described in Section 4.3.1 by means of the "light chronometer."

EXAMPLE 4.2

Derive the effect of length contraction from Lorentz transformations.

SOLUTION

Let us imagine that in system S' there is a rectilinear rod of length L_o at rest whose extremes are placed at x_1' and x_2' as shown in Figure 4.6, so that $x_2' - x_1' = L_o$.

Let us imagine that this length is known for the observer at S' so that he can choose where to put the extremes of the rod and knows completely the values of x_1' and x_2'. The length L_o is a ***proper length*** since it is the corresponding value for an observer for which the rod is at rest.

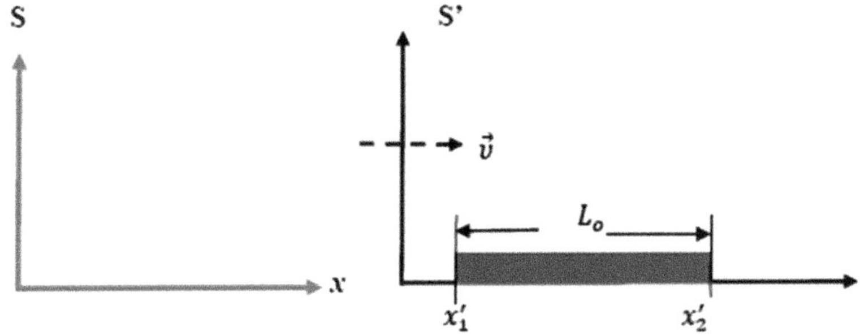

FIGURE 4.6 Lengths of a rod according to two inertial observers S and S'.

For an observer fixed to S the rod moves toward the right along a direction parallel to the positive X axis with a velocity vector of magnitude υ.

Using Lorentz transformations we have

$$x'_2 - x'_1 = \gamma(x_2 - \upsilon t_2) - \gamma(x_1 - \upsilon t_1) = \gamma(x_2 - x_1) - \gamma\upsilon(t_2 - t_1) \tag{4.19}$$

According to S system the length L of the rod can be determined by means of the distance between its extremes where two simultaneous events take place, as for example the emission of light pulses. Therefore we have $t_2 = t_1$ and the above expression becomes

$$x'_2 - x'_1 = L_o = \gamma(x_2 - x_1) = \gamma L \tag{4.20}$$

Therefore for the observer at S which sees the rod moving its length is $L = L_o/\gamma$ resulting in a value smaller than the value L_o corresponding to the observer at S'. This effect is called Lorentz contraction.

4.5 ABOUT THE SIMULTANEITY OF EVENTS

According to SRT the simultaneity of two events is not Lorentz invariant. That means that if for observer fixed to reference frame S two events E_1 and E_2 occur simultaneously, in general this will not occur for an observer fixed to reference frame S' and in motion with respect to S.

Let us assume that for observer fixed to system S' two events occur simultaneously at different positions, so that in space time:

Event E_1 has coordinates (x'_1, y'_1, z'_1, t'_1)

Event E_2 has coordinates $(x'_2, y'_2, z'_2, t'_2 = t'_1)$

For a second observer fixed to inertial system S' moving with velocity $\vec{\upsilon} = \upsilon\hat{x}$ ith respect to S:

Event E_1 has coordinates (x_1, y_1, z_1, t_1)

Event E_2 has coordinates (x_2, y_2, z_2, t_2)

From Lorentz transformation for time and considering that in this case $t'_1 = t'_2$ one obtains

$$t_2 - t_1 = \gamma\left(t'_2 + \frac{\upsilon x'_2}{c^2}\right) - \gamma\left(t'_2 + \frac{\upsilon x'_1}{c^2}\right) = \frac{\gamma\upsilon}{c^2}\left(x'_2 - x'_1\right) \tag{4.21}$$

If the events are not causally connected, so that event E_2 does not require the previous occurrence of event E_1, we have that event E_2 occurs after E_1 if $x'_2 > x'_1$ and $\vec{\upsilon}$ is parallel to the positive X axis. If the relative velocity between S' and S changes sign we would have

$$t_2 - t_1 = \gamma\left(t'_2 + \frac{\upsilon x'_2}{c^2}\right) - \gamma\left(t'_1 + \frac{\upsilon x'_1}{c^2}\right) = -\frac{\gamma\upsilon}{c^2}\left(x'_2 - x'_1\right) \tag{4.22}$$

In that case we have that event E_2 occurs before event E_1 if $x'_2 > x'_1$

What happens if now the two events are causally connected so that the condition for the occurrence of event E_2 is that previously event E_1 has occurred?

In this case for observer S' the two events cannot occur simultaneously if the corresponding positions are different. In fact, according to the second principle of special relativity it is impossible that an event produced at position x'_1 be detected instantaneously at position x'_2 and determine the occurrence of an event at this place, since no signal can travel faster than light velocity. That means that $t'_2 - t'_1$ cannot be lower that $\frac{(x'_2 - x'_1)}{c}$. According to the observer fixed at S the time interval between the events is

$$t_2 - t_1 = \gamma\left\{\left(t'_2 - t'_1\right) + \frac{\beta}{c}\left(x'_2 - x'_1\right)\right\} \tag{4.23}$$

Considering that for causally connected events $t'_2 - t'_1 \le \frac{(x'_2 - x'_1)}{c}$, eq 4.23 implies that

$$t_2 - t_1 \le \gamma\frac{x'_2 - x'_1}{c}(1 + \beta) \tag{4.24}$$

Therefore, the sequence of events does not change even though the relative velocity between S and S' changes sign (that is β changes to $-\beta$) since $|\beta| \leq 1$.

4.6 SPACE TIME AND THE LIGHT CONE

In the special relativity theory any event is described by its location in a specific reference frame given by four coordinates in a four-dimensional space, called space time, with one of them being the temporal coordinate corresponding to its time of occurrence and the other three are spatial coordinates corresponding to its location.

The Lorentz transformations show how the description of a given event is seen by different inertial observers. If for simplicity, we consider relative motion only along one direction, say the x axis, the position of each event can be considered as a point in a two-dimensional reference system where one axis corresponds to time and the other to the position along the x direction.

Let us consider two reference frames S (laboratory) and S' (Arnold) with this last moving in a car along the x axis with velocity V with respect to the first. For observer S the events are described by their position in a coordinate system with horizontal and vertical axes x and ct, respectively, as shown in Figure 4.7 while for observer A (Arnold) the axes are x' and ct' as shown by dotted black lines in this figure. This is explained considering that for S' the axis ct' corresponds to $x'=0$ and according to the Lorentz transformation $x' = \gamma(x - Vt)$ this implies that $x = Vt$. Similarly, the axis $x'=0$ corresponds to $t'=0$ and from the Lorentz transformation $t' = \gamma\left(t - \frac{Vx}{c^2}\right)$ this implies $t = Vx/c^2$.

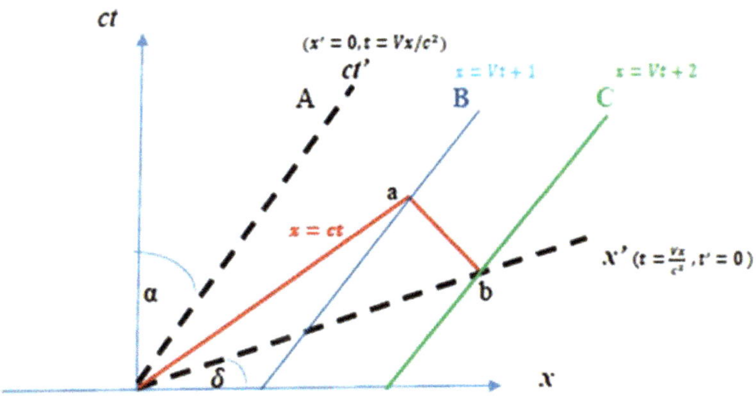

FIGURE 4.7 Orientation of axes x' and ct' (frame reference S') with respect to axes x and ct (frame reference S).

Source: Adapted from Ref. [6]

EXAMPLE 4.3

With respect to Figure 4.7 show that $\alpha = \delta = tan^{-1}\,(V/c)$

SOLUTION

The axis ct' corresponds to $x' = 0$. From Lorentz transformation $x' = \gamma(x - Vt)$ we have $x = Vt$. Therefore

$$\frac{x}{ct} = \frac{Vt}{ct} \Rightarrow \frac{x}{ct} = \frac{V}{c} < 1 \tag{4.25}$$

This implies that there is an angle α between the direction of the axis ct' and the axis ct as shown in Figures 4.7 and 4.8 with α given by $tan\alpha = \frac{x}{ct} = V/c$.

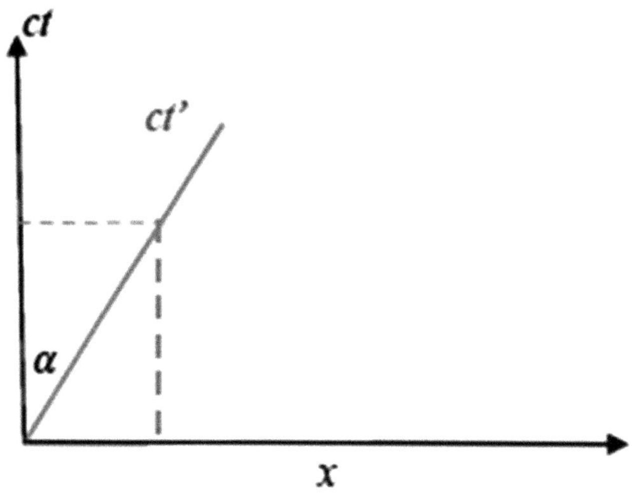

FIGURE 4.8 Orientation of axis ***ct'*** with respect to axis ***ct***.

The axis x' corresponds to t' = 0. From Lorentz transformation $t' = \gamma\left(t - \frac{Vx}{c^2}\right)$ this implies $t = \frac{Vx}{c^2} \Rightarrow \frac{ct}{x} = \frac{V}{c}$. Therefore, the axis x' is at an angle δ with respect to axis x as can be seen in Figures 4.7 and 4.9 with δ given by $\delta = \frac{ct}{x} = V/c$.

Therefore $\alpha = \delta = tan^{-1}\,(V/c)$.

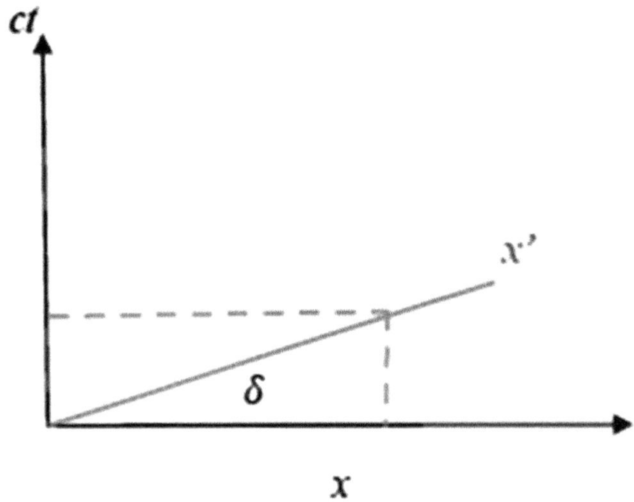

FIGURE 4.9 Orientation of axis x' with respect to axis x.

EXAMPLE 4.4

With respect to Fig.4.7 suppose that in addition to observer A (Arnold), there are two more observers, B (Bruno) and C (Conrad), all of them in the same moving frame and which are placed in an adjacent way, so that they are separated from each other along the x axis by a distance d as measured in the rest frame S (see Figure 4.10).

Suppose Arnold sends out a light signal to Bruno from the origin. At some unknown point Conrad will *also* send out a light signal toward Bruno. Let us suppose that Conrad somehow arranges to do this in a way that both signals reach Bruno at the same instant. If Arnold's light signal starts at the origin, Bruno will receive it at the point a in Figure 4.7. ***From what point must Conrad send his light signal, if it is to reach at Bruno simultaneously with the signal sent by Arnold ?***

This is an example based on Ref. [6] and the reader is invited to solve it by two methods, one in a graphic form using Figure 4.7 and the other formulating the appropriate set of equations.

SOLUTION

Method 1 Using Figure 4.7 as done in Ref. [6].

The coordinates position x_a and time t_a of point a according to frame reference S are found from

$$x_a = ct_a = Vt_a + d \Rightarrow t_a = \frac{d}{c-V}, x_a = \frac{cd}{c-V} \tag{4.26}$$

Line *ab* of Figure 4.7 corresponds to points in the space time of S' for which events occur simultaneously and this is the reference frame to which our three observers are fixed.

The equation of line *ab* is $x + ct = constant$, where

$$constant = x_a + ct_a = \frac{2cd}{c-V} \tag{4.27}$$

The intersection of line *ab* with line C (world line of Conrad) is obtained from the system of equations

$$\left.\begin{aligned} x_b - Vt_b &= 2d \\ x_b + ct_b &= \frac{2cd}{c-V} \end{aligned}\right\} \tag{4.28}$$

$$\Rightarrow x_b = \frac{2c^2 d}{c^2 - V^2}; \; t_b = \frac{2Vd}{c^2 - V^2} \tag{4.29}$$

Equation 4.29 gives the position x_b and instant t_b at which Conrad must emit the light pulse toward Bruno, so that this pulse arrives simultaneously with the pulse emitted by Arnold.

METHOD 2 USING EQUATIONS OF KINEMATICS

Figure 4.10 shows the initial condition, where the cars of Arnold, Bruno, and Conrad begin to move with the same velocity V along the x axis and maintain a distance $\boldsymbol{d}$ between adjacent observers.

At $t = 0$ Arnold sends out a light pulse to Bruno which according to a fixed oberserver at laboratory (frame reference S) is received by him at t_a. This instant is found from

$$x_B(t_a) = d + Vt_a = ct_a \Rightarrow t_a = \frac{d}{c-V}, \; x_B(t_a) = \frac{c}{c-V}d \tag{4.30}$$

At $t = t_b$ Conrad sends out another light pulse to Bruno received by him at instant t_c given by

$$t_c = t_b + \frac{L}{c} \tag{4.31}$$

where L is the distance traveled by the pulse emitted by Conrad between the instant of emission (t_b) up to the reception by Bruno at t_c. This distance is given by

$$L = c(t_c - t_b) = d - V(t_c - t_b) \tag{4.32}$$

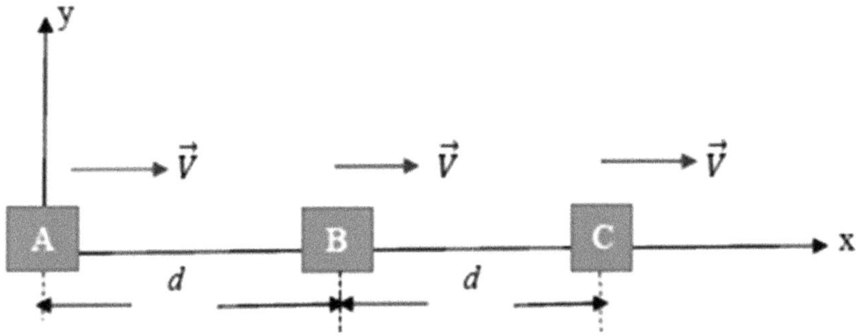

FIGURE 4.10 Arnold, Bruno, and Conrad begin to move at t=0 with speed V as shown.

Considering that $t_c = t_a$ as observed from a reference frame S where Arnold, Bruno, and Conrad move with speed V as shown in Figure 4.10, we obtain

$$t_b = t_c - \frac{d}{c+V} = t_a - \frac{d}{c+V} = \frac{d}{c-V} - \frac{d}{c+V} = \frac{2Vd}{c^2 - V^2} \tag{4.33}$$

which is the same result obtained by the graphic method illustrated in Figure 4.7. This is another example where events that must occur simultaneously in one reference frame (emission of light pulses to be received simultaneously by Bruno according to S'), must occur in a sequential form for other observer fixed to S with a time difference given by eq 4.33.

A concept used in special relativity theory is the representation of past and future and causality using the idea of a light cone as shown in Figure 4.11.

The trajectory of any material object in space time is as the path AB in Figure 4.11, called its world line. For times $t > 0$ this path lies inside the upper half-cone and that region of the space time is called the future. Similarly, the lower half-cone corresponds to the past. The regions in blue are called elsewhere. A particle at O can never reach or come from a point in space time in these regions.

The division of space time into the past-future regions and elsewhere can be emphasized if we consider the invariant separation or interval s_{12} between two events located at points ***E1*(x_1, t_1)** and ***E2* (x_2, t_2)**. The square of the invariant interval is defined as

$$s_{12}{}^{2} = c^2 \left(t_2 - t_1\right)^2 - \left|x_2 - x_1\right|^2 \tag{4.34}$$

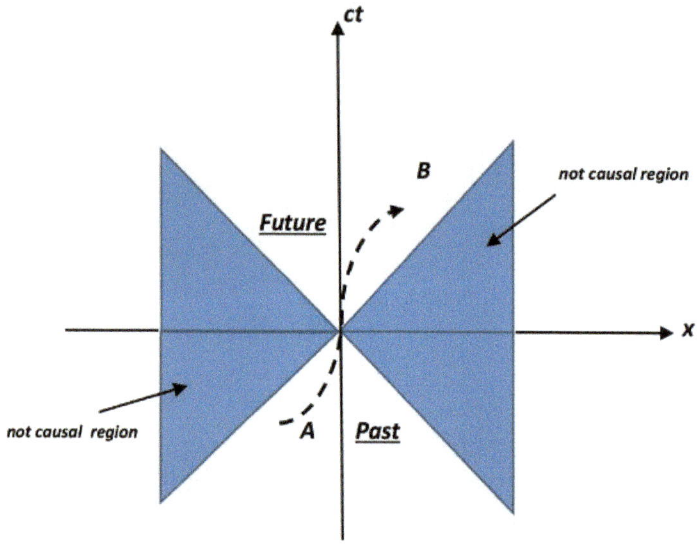

FIGURE 4.11 World line and the light cone. The curve between points A and B is the path followed in space time by any person or object according to the principle of Causality.

If we consider any two events denoted as E1 and E2, we have the following possibilities:

1. If $s_{12}^2 > 0$ we say that these events have a time-like separation. We can always find a Lorentz transformation to a new reference frame S' such that $x_1' = x_2'$.
2. If $s_{12}^2 < 0$ the events are said to have a spacelike separation. Analogously to the previous case, we can always find a Lorentz transformation to a new reference frame S" such that $t_1' = t_2'$.
3. If If $s_{12}^2 = 0$ the events have a light like separation and the only possibility to connect them is by means of light signals.

EXAMPLE 4.5

Prove that for two events with time like separation, we can find a Lorentz transformation to a new reference frame such that $x_1' = x_2'$.

SOLUTION

Let us consider two events ***E1*** and ***E2*** that according to frame reference system S occur at points (x_1, t_1) and (x_2, t_2) in space time, such that $t_2 > t_1$ and $x_2 > x_1$ For a frame reference S' moving in a direction parallel to the x-axis with velocity V with respect to S these events occur at points $x_1' = t_1'$ and (x_2', t_2'). Using Lorentz transformation for the position we have

$$x_2' = \gamma(x_2 - Vt_2) \tag{4.35}$$

$$x_1' = \gamma(x_1 - Vt_1) \tag{4.36}$$

The condition $x_2' = x_1'$ implies

$$x_2 - x_1 = V(t_2 - t_1) \Rightarrow V = \frac{x_2 - x_1}{t_2 - t_1} = c\frac{x_2 - x_1}{c(t_2 - t_1)} \tag{4.37}$$

For $s_{12}^2 > 0$ it is necessary that $c(t_2 - t_1) > x_2 - x_1$. Therefore, the last term at the right of eq 4.37 is lower than 1 implying that V < c, so that according to the special relativity theory it is possible to find a frame reference system S' moving with V < c, and in this case for an observer fixed to this frame the events occur at the same point.

EXAMPLE 4.6

Prove that for two events with space-like separation, we can find a Lorentz transformation to a new reference frame such that $t_1' = t_2'$.

SOLUTION

Let us consider two events ***E1*** and ***E2*** that according to frame reference system S occurs at points (x_1, t_1) and (x_2, t_2) in space time, such that $t_2 > t_1$ and $x_2 > x_1$ For a frame reference S' moving along the x axis with velocity V with respect to S, these events occur at points (x_1', t_1') and (x_2', t_2'). Using Lorentz transformation for time we have

$$t_2' = \gamma\left(t_2 - \frac{Vx_2}{c^2}\right) \tag{4.38}$$

$$t_1' = \gamma\left(t_1 - \frac{Vx_1}{c^2}\right) \tag{4.39}$$

The condition $t_2' = t_1'$ implies

$$t_2 - t_1 = \frac{V}{c^2}(x_2 - x_1) \Rightarrow V = c\left\{\frac{c(t_2 - t_1)}{x_2 - x_1}\right\} \tag{4.40}$$

For $s_{12}^2 < 0$ it is necessary that $c(t_2 - t_1) > x_2 - x_1$. Therefore, the term inside {} brackets in eq 4.40 is lower than 1 implying that V < c, so that according to the special relativity theory it is possible to find a frame reference system S' moving with V < c, showing that in this case for an observer fixed to this frame the events occur simultaneously.

4.7 SOME REFLECTIONS ABOUT CAUSALITY AND TIME

A principle that is part of the way we conceive reality is related to causality and nobody questions the fact that an effect never occurs before its corresponding cause, assuming that the cause-effect relationship is fully established and undisputed. For instance, it is not possible for a child to be born before the birth of its mother, or that a detector receives a light signal before it is emitted by some source.

For phenomena observed on a macroscopic scale and adequately described by classical mechanics and electrodynamics, the principle of causality is compatible with the theory of relativity which implies that it is not possible the transmission of any signal faster than light speed in vacuum. This is consistent with the Lorentz transformations that mathematically express the postulates of the special theory of relativity (SRT), and is applicable for all observers that move relative to each other with constant velocity (inertial observers).

Albert Einstein is considered as the main author of this theory, which was formulated at the beginning of the 20th century. So far all the predictions based on this theory have been experimentally confirmed and it is one of the pillars of physics of our times, such as quantum field theory.

As stated, this theory implicitly implies that the principle of causality is valid for all inertial reference systems and that there are no objects that can move with a speed greater than c (speed of light in a vacuum with a value of approximately 300,000 km/s[5]). For example, if for an observer A two events E1 and E2 are concatenated such that the occurrence of *E2* is only possible if previously event *E1* has occurred, the events will occur in the same chronological sequence for another observer B in relative motion with respect to A. This is shown in Figure 4.11 by a diagram in space time where the horizontal and vertical axes correspond to the spatial coordinate along the x-axis and to time, respectively. For simplicity only motion parallel to the x axis is considered. The diagonal line that passes through the origin of this coordinate system has a slope of value equal to c. Events inside the blue area are causally connected and the order as they occur in time is the same for all inertial observers, independently of the relative velocities between them. However, an event inside this area is not causally connected with an event external to this region, and for two inertial observers the sequence with which events occur that are not causally connected can be different.[2]

The curve that joins points A and B passing through the origin in the space time represents the succession of events that occur in the life of a person like us, in different places and instants of time.

The way in which the principle of causality is implicit in the theory of relativity, and how this relates to the way we perceive time in our lives, can be illustrated through the following example: Is it possible to find a reference frame S' for which a mother (named Anne) is younger than her son (named John), and if this is possible, what should be its speed with respect to a fixed reference frame at land?

Suppose also that Anne was born in 1950 in city A and that in 1978 she moved to city J, the year her son John was born. For purposes of the example, let us consider that the distance between both cities is 3000 km. According to reference frame S the events are:

E1: Anne was born in November 28, 1950 (*t1*) at city ***A*** (*x1=0*); ***E2:*** John was born in November 29, 1978 (*t2*) at city ***J*** (*x2*=3000 km).

Therefore according to system S: $\Delta t = t_2 - t_1 = 28$ *years;* $\Delta x = x_2 - x_1 =$ 3000 *Km*

Using the Lorentz transformations for time, it follows that for an observer in the system S' the time difference is given by the following expression:

$$\Delta t' = t_2' - t_1' = \frac{1}{\sqrt{1 - \frac{V^2}{c^2}}}\left(\Delta t - \frac{V}{c^2}\Delta x\right) \qquad (4.41)$$

From the above equation, we deduce that the speed V with which S' must move with respect to S to change the temporal order of these events (that is $t_2' < t_1'$ meaning that event *E2* occurs before event *E1*) must be much greater than c by a factor of the order of 9.10^{10}. Nevertheless, this calculation has not any sense since the SRT considers that velocities greater than c are not possible and using its equations we would get imaginary values. For instance, we would obtain imaginary values for the mass of an object, a result which according to our conception of the reality does not have any physical sense.

For dealing with theoretical situations where some particles could move with velocity greater than c (as is the case of particles called tachyons whose existence up to now has not been verified experimentally), extended relativity theories have been formulated as analyzed in Refs. [7–9].

We can conclude from this brief analysis that it seems that light rules our existence and is closely related with the pass of our lives through the time. Perhaps, we could say that the origin of our universe coincides with the simultaneous origin of light and time. But if that were the case, the question that arises and whose answer, at least I do not know, is what existed before.

4.8 CAUSALITY AND SUPERLUMINAL SIGNALS

From a conceptual point of view, causality is not intrinsically implied by the equations of relativity theory. However, it is postulated as a principle that needs to be satisfied, stating that a cause always precedes its effect.

It is usually believed that SRT forbids the propagation of a signal with a speed higher than the light velocity in vacuum (superluminal signal). However, SRT does not explicitly contain this constraint. The original papers and notes written by Einstein only state that the light speed in vacuum is constant in all inertial reference frames. The concept of a speed limit, rather arises from the fact that according to the equation for the relativistic energy of an object, this energy takes an infinite value when this object moves with the light speed and becomes imaginary if that speed is exceeded.

In 1907, Einstein discussed with Wilhelm Wien the possibility of superluminal velocities in dispersive and absorptive media, arguing that according to the electromagnetic theory of dispersion, it is not possible the propagation of optical signals with a speed greater than c. Using the addition theorem of velocities (that will be discussed in Section 4.9 of this chapter) he showed that "no action can exist which can be utilized for arbitrary signaling and which has a propagation velocity greater than light speed un vacuum" (phrase taken verbatim from Ref. [10]). In example 4.8 of this chapter we will show, as an example, the demonstration made by Einstein to discard superluminal propagation of any optical signal.

In this section, we relate the above discussion with the causality principle, considering events with *space like* separations. As has been shown in Example 4.6 of this chapter, for two events *E1* and *E2* with space like separation and not causally connected, it is possible to find a Lorentz transformation to a new coordinate frame where these events occur simultaneously. From Lorentz transformations, it can be shown that it is possible to find a coordinate frame where the sequence of occurrence of these events change, that is, if in a reference frame *S* the event *E1* occurs before the event *E2*, then there exists another reference frame S' where that chronology is changed (*E2* occurs before *E1*). For more illustration, let us consider that event *1* corresponds to the emission of an optical signal which propagates along a rectilinear path and through an ideal medium (with absorption and dispersion negligible), toward a receptor placed at a distance Δx which immediately upon the reception of that signal emits an order to open an electric switch. According to our common sense the only expected sequence under which these events can occur is *E2* after *E1*.

In the space time associated with reference frame S, let us consider that these events have the following coordinates:

$$E1: (x_c, t_c) \tag{4.42}$$

$$E2: (x_e, t_e) \tag{4.43}$$

We assume that $t_e > t_c$ and that the two events are "space like," that is

$$x_e - x_c > c(t_e - t_c) \tag{4.44}$$

Let us describe these two events in another reference frame S' which moves with respect to S with speed υ along the X axis with $\lfloor \upsilon \rfloor < c$. According to an observer fixed to system S' the time interval between these events is

$$t'_e - t'_c = \gamma\left\{(t_e - t_c) - \frac{\upsilon}{c^2}(x_e - x_c)\right\} \tag{4.45}$$

From eq 4.44 to 4.45 it can be seen that it is possible to find a relative speed υ with magnitude lower than *c*, so that in S' the event *E1* (the cause) occurs after the effect (event *E2*). For instance, if $\Delta x = x_e - x_c = \lambda \cdot c\Delta t$ with λ a factor greater than 1, from eqs. 4.44 to 4.45 we have that for $\Delta t' = t'_e - t'_c < 0$, it is necessary that $\upsilon > c/\lambda$ which is a speed lower than c. The origin of this conflict with our experience is on the assumption that the two events can be causally connected and also that they are space-like events. A situation like this was considered as impossible by Einstein, and according to the arguments here presented, it can be inferred that space-like events cannot be linked causally.

4.9 RELATIVISTIC TRANSFORMATION OF VELOCITIES

We have seen that according to the Special Relativity Theory, the speed of a particle cannot reach a value higher than c, which corresponds to the light velocity in vacuum, a limit which only can be reached by the light or its associated quantum, the photon.

But let us imagine a reference frame fixed to a particle moving with a velocity equal to that of the light in vacuum. What would be the relative velocity of this particle with respect to an electromagnetic wave or a photon? A question of this type was made by Einstein in the years when relativity theory was being elaborated.

According to Galileo's classical composition of velocities, the answer would be zero if the particle is moving in the same direction and sense corresponding to the motion of the electromagnetic signal. If the motion of the particle is nearly **–c** we would have that the electromagnetic signal is moving with ***2c*** with respect to the particle. Alternatively, we would say that with respect to the electromagnetic wave the particle is moving with ***2c*** !!!. This would led us to very strange conclusions. For instance, according to the effect of time dilation discussed in Section 4.3.1 with the help of the light chronometer, if observer 1 is moving away with respect to this device with speed greater than ***c*** he would get imaginary values for the intervals between a TIC and a TAC, or equally strange, he would get an imaginary distance between the detector and the mirror (see eqs 4.2 and 4.7). Mathematically, this occurs because according to SRT the Lorentz factor $\gamma = 1/\sqrt{1 - V^2/c^2}$ is imaginary for $V > c$ and it is generally accepted that this theory precludes the possibility of transmitting energy at velocities greater than c. Nevertheless, as shown by Feinberg,[3] a description of superluminal particles, called tachyons, is possible using the formalism of relativistic quantum field theory. This description is not within the scope of this book and in this chapter, we will continue with the study of special relativity theory and its implications on time measurements, analyzing in particular the application of Lorentz transformations for obtaining the relativistic transformation of velocities. For doing that, let us consider two observers fixed to the reference frames S and S′ moving relative to each other with velocity $\vec{V}$. These observers look a particle moving with velocities $\vec{\upsilon}$ and $\vec{\upsilon}'$, respectively, as shown in Figure 4.12.

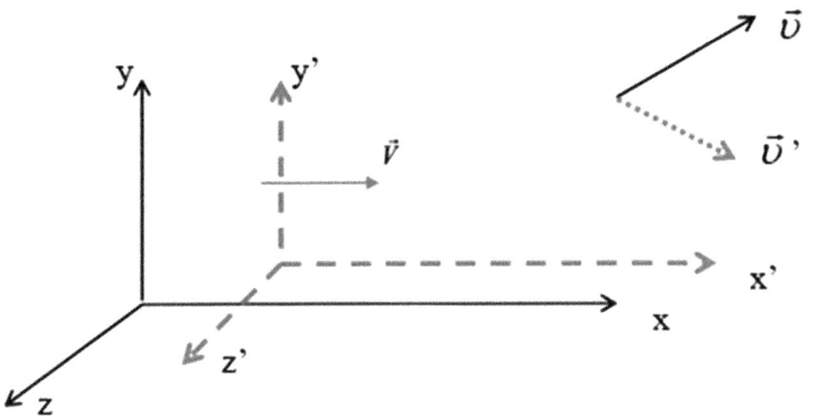

FIGURE 4.12 The motion of a particle seen by observers S and S' with a relative velocity $\vec{V}$ as shown.

The respective components of the velocities for the observers fixed at S and S' are:

$$with\,respect\,to\,S:\begin{cases}\upsilon_x=\dfrac{dx}{dt}\\ \upsilon_y=\dfrac{dy}{dt}\\ \upsilon_z=\dfrac{dz}{dt}\end{cases}\quad with\,respect\,to\,S':\begin{cases}\upsilon'_x=\dfrac{dx'}{dt}\\ \upsilon'_y=\dfrac{dy'}{dt}\\ \upsilon'_z=\dfrac{dz'}{dt}\end{cases}\tag{4.46}$$

Using the Lorentz transformations according to eq 4.14, we obtain that the transformation of the component of the velocities is

$$\begin{cases}dx'=\gamma\left(dx-Vdt\right)\\ dy'=dy,\,dz'=dz\\ dt'=\gamma\left(dt-\dfrac{V}{c^2}dt\right)\end{cases}\Rightarrow\begin{cases}\upsilon'_x=\dfrac{\upsilon_x-V}{1-\dfrac{\upsilon_xV}{c^2}}\\ \upsilon'_y=\dfrac{\upsilon_y/\gamma}{1-\dfrac{\upsilon_xV}{c^2}}\\ \upsilon'_z=\dfrac{\upsilon_z/\gamma}{1-\dfrac{\upsilon_xV}{c^2}}\end{cases}\text{or}\begin{cases}\upsilon_x=\dfrac{\upsilon'_x+V}{1+\dfrac{\upsilon'_xV}{c^2}}\\ \upsilon_y=\dfrac{\upsilon'_y/\gamma}{1+\dfrac{\upsilon'_xV}{c^2}}\\ \upsilon_z=\dfrac{\upsilon'_z/\gamma}{1+\dfrac{\upsilon'_xV}{c^2}}\end{cases}\tag{4.47}$$

From eq 4.47 it can be verified that if the relative velocity between the observers is parallel to the X direction, and with magnitude negligible with respect to the light velocity, these equations lead to the classical transformations of Galileo for velocities.

If the observed particle is a photon which according to observer S is moving along the +X direction, it can be verified that with respect to both inertial observers, the velocity of the photon is c and not $c - V$ as would be obtained from Galileo's transformations.

EXAMPLE 4.7

Using the relativistic transformation of velocities try to give an answer to a question that, as it is usually said, Einstein formulated in the years when Relativity Theory was being elaborated: *If one could travel with a velocity equal to that of the light in vacuum chasing a photon, with what velocity the motion of the photon would be seen?*

SOLUTION

Let us imagine that according to an observer fixed to a frame of reference S a rocket moves with speed V along the X axis and inside the rocket is an astronaut. A light beam emitted by some external source is moving in the same direction, which would be the velocity of the photons of that light beam with respect to the astronaut?

Using the first equation of system 4.47 we obtain

$$\upsilon'_x = \frac{\upsilon_x - V}{1 - \frac{\upsilon_x V}{c^2}} = \frac{c - V}{1 - \frac{cV}{c^2}} = \frac{c - V}{1 - \frac{V}{c}} = c \tag{4.48}$$

Here, we have made $\upsilon_x = c$ since this is the velocity of each photon of the light beam. Note that this last result does not depend on the velocity of the rocket, and is not zero as would be deduced from Galileo's transformations for the velocities.

EXAMPLE 4.8

In this example, we discuss a case analogous to that considered by Albert Einstein, to prove that it is not possible that an optical signal has a speed of propagation greater than the light velocity in vacuum.[10]

Let us consider a coach of a train of length λ, where from one of its ends a passenger throws a ball to another passenger located at the other end. The ball moves with speed W with respect to the coach, which in turn moves with speed υ, with respect to the ground. Consider that the ball moves in the opposite sense with respect to the velocity of the coach, so that the magnitude of W is greater than that corresponding to υ.

Which is the velocity of the ball with respect to ground?

Answer: If for simplicity, we consider all the motions along the x-axis, using the first of the equations given by 4.47 we obtain that this velocity is $V = \dfrac{W - \upsilon}{1 - \dfrac{W\upsilon}{c^2}}$.

If it was possible a superluminal propagation of the ball, so that $W > c$, is it possible to find a speed υ of the coach lower than c, so that the time of flight of the ball would be negative?

Answer: The time of the flight of the ball is obtained from $T = \lambda / V$ so that we get

$$T = \frac{\lambda}{W - \upsilon}\left(1 - \frac{W\upsilon}{c^2}\right)$$

This time is negative for $\upsilon > c\left(\dfrac{c}{W}\right)$ and if we assume that it is possible a propagation of the ball with $W > c$, we infer that it is possible to find a value lower than c for the speed υ of the coach.

An analogous result was obtained by Einstein, where he supposed "a material strip extended along the x-axis of a reference frame S, relative to which a certain action can be propagated with the velocity W (as judged from the material strip)." He concluded that, assuming $W > c$, it is possible to choose υ so that $T < 0$, and that *"this result signifies that we must consider as possible a transmission mechanism that allows the intended action to precede the cause"* (phrase taken verbatim from Ref. [10]). The opinion of Einstein was that from a logical point of view, this result does not contain any contradiction, even though this is out of our experience.

Anyway, the question of causality and relativity theory is still a matter of philosophical discussion.

4.10 TIME DILATION OF THE LIFETIME OF A PARTICLE

The effects of time dilation and Lorentz contraction predicted by the SRT are illustrated in several books by means of the example of a particle like a muon that is emitted in some processes. The muon is a particle created due to the interaction of cosmic rays with the upper atmosphere having a lifetime τ with a value of about $2\mu s$ and moving with a velocity close to that corresponding to light. Considering that muons are created at a height H of about 100 km with respect to the surface of the earth, one wonders if it is possible that they reach the ground.

Using classical dynamics we would say that this is impossible since even though we consider that a muon moves with velocity c, the distance it can go down before disintegration is:

$$h = c\tau = 300000\left[km/s\right]\cdot 2\cdot 10^{-6}\left[s\right] = 0.6\,km \ll 100\,km$$

Nevertheless, it is known that muons are able to reach the ground, and this can be explained by SRT considering the effect of time dilation as follows:

Let us consider an observer fixed to the earth's surface for which there are two events (see Figure 4.13):

Event 1: muon at instant $t_1 = 0$ is at a height $y_1 = H$ with respect to the surface of the ground. That is, the coordinates of this event in space time are ($y_1 = H$, $t_1 = 0$).

Event 2: At time $t_2 = T$ the muon has falled over the ground. That is, the coordinates of this event in space time are ($y_2 = 0$, $t_2 = T$).

Using Lorentz transformation to a reference frame fixed to the muon we have

$$t_2' - t_1' = \Delta t = \gamma\left\{\left(t_2 - t_1\right) - \frac{\beta}{c}\left(y_2 - y_1\right)\right\} \tag{4.49}$$

From eq 4.49 and considering that $t_2 - t_1 = T$ and $y_2 - y_1 = H$ we obtain

$$\tau = \gamma\left\{T - \frac{\beta}{c}H\right\} \tag{4.50}$$

Since $H = \upsilon T$ and $= \upsilon/c$, where υ is the falling speed of the muon, from the above equations and with $\gamma = 1/\sqrt{1-\beta^2}$ we get

$$\tau = T\sqrt{1-\beta^2} = T/\gamma \tag{4.51}$$

eq 4.51 corresponds to the relativistic effect of time dilation, and implies that according to an observer fixed to the muon, the elapsed time up to its impact with the ground, that is τ, is lower than the time T measured by another observer fixed to the ground and for whom the muon moves downward with speed υ. From this same equation, it can be deduced the following expression for the velocity required for the muon to reach the earth from a site at height H where it is created, with an elapsed time equal to its time life τ:

$$\upsilon = c\frac{H}{\sqrt{H^2 + c^2\tau^2}} \tag{4.52}$$

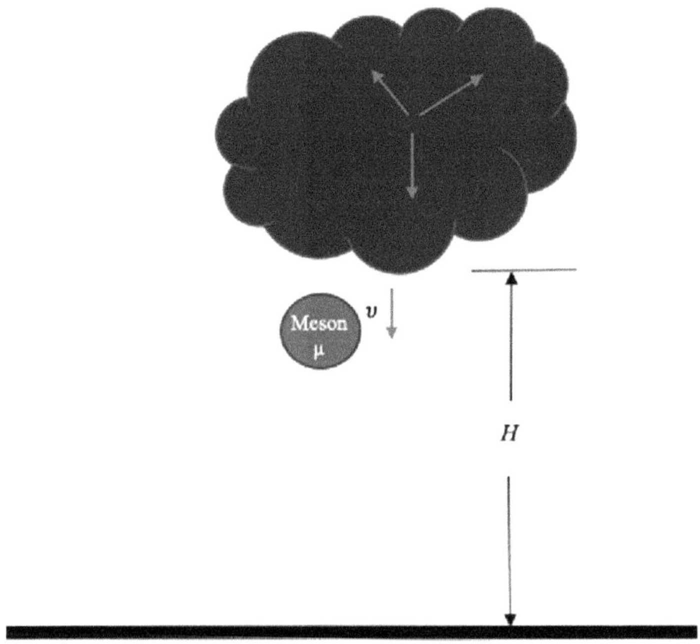

FIGURE 4.13 Muon μ produced at a height H over the surface of the ground and falling with speed υ.

Figure 4.14 shows the required falling speed of the muon versus the height H of the point where it is created, so that it could reach the ground before its disintegration with a time life of 2μ.

An alternative description of this process can be made considering that according to the relativistic length contraction, for an observer fixed to the muon the surface of the ground moves upward toward him with a speed υ

traveling through a distance $H' = \upsilon\tau = \frac{H}{\gamma} = \frac{\upsilon T}{\gamma} \Rightarrow \tau = T/\gamma$, which coincides with eq 4.51.

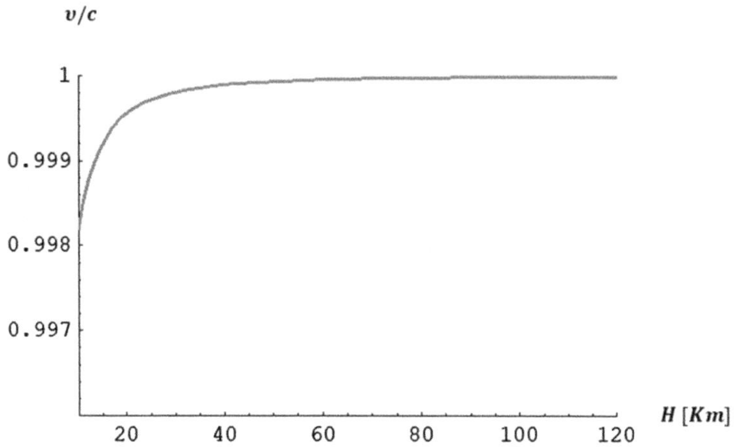

FIGURE 4.14 Speed required for a muon (in units of c) versus traveled distance before its disintegration at $\tau = 2\mu s$.

4.11 TIME AND RELATIVISTIC DOPPLER EFFECT

As we have seen, in relativity theory time becomes a fourth coordinate and it is transformed from one inertial reference frame to another, as occurs with the spatial coordinates.

Since time intervals between the same two events are different in different inertial frames, it is expected that a parameter like the period associated with an electromagnetic wave or a periodic train of pulses, also changes from one inertial reference frame to others. This is the relativistic doppler effect, which we will show in the nonrelativistic limit tends to the classical doppler effect.[11,12]

Let us consider a situation where an electromagnetic signal is emitted by a source located in a spaceship moving away from the earth with velocity $\vec{V}$. This signal consists in a succession of pulses of duration Δt with period T_e such that $\Delta t \gg T_e$ is shown in Figure 4.15. For simplicity we assume that with respect to a frame reference fixed to earth the rocket moves along the +X direction (Figure 4.16).

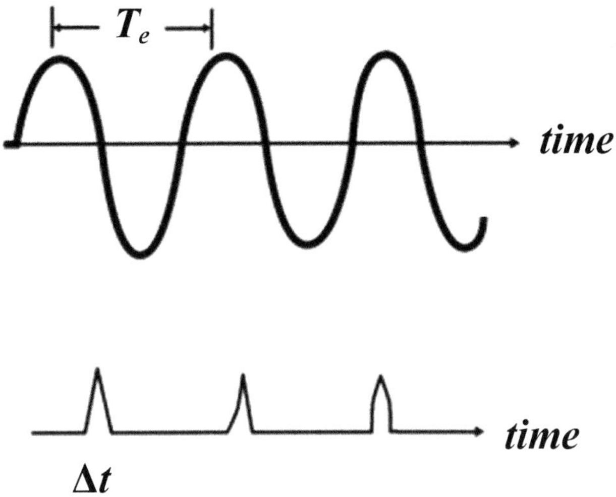

FIGURE 4.15 Signal emitted from a rocket moving away from the Earth with velocity $\vec{V}$ (see text).

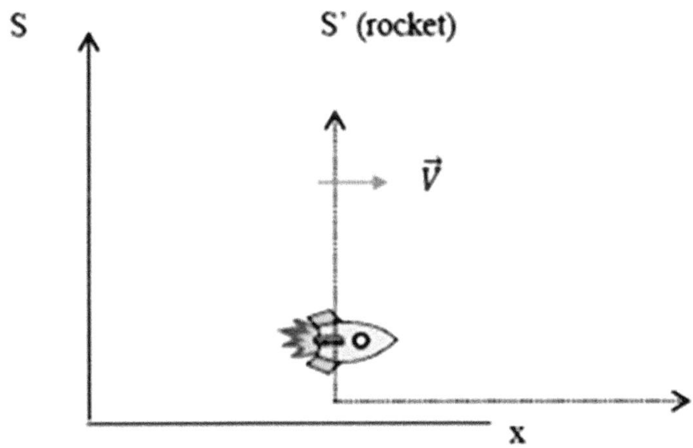

FIGURE 4.16 Rocket moving away from earth with velocity vector of magnitude V.

Let us consider that at $t = t' = 0$ the origins of systems S and S' coincide, and that at this instant the first pulse is emitted from the rear of the rocket toward the earth. Calling this event as *Event 1,* in space time its corresponding coordinates are: ($x_1 = 0$, $t_1 = 0$) in S and ($x'_1 = 0$, $t'_1 = 0$) in S'.

The second pulse is emitted from the same point of the rocket toward the earth at time T_e according to the system S' fixed to the rocket (*Event 2*), so that the corresponding coordinates for this event are ($x_2' = 0$, $t_2' = T_e$). Using Lorentz transformations the coordinates of this event for S are

$$x_2 = \gamma\left(x_2' + Vt_2'\right) = \gamma VT_e \tag{4.53}$$

$$t_2 = \gamma\left(t_2' + \frac{V}{c^2}x'_{\overline{ABC}}\right) = \gamma T_e \tag{4.54}$$

Considering that this second pulse travels with velocity c toward the earth during a time interval $\Delta t = x_2/c$, it will be detected at S at the instant T_d given by

$$T_d = t_2 + \frac{x_2}{c} = \gamma\left(1 + \frac{V}{c}\right)T_e = \left\{\frac{1+\beta_e}{1-\beta_e}\right\}^{1/2} T_e \tag{4.55}$$

where we have defined $\beta_e = V/c$.

From this last equation, we can see that the period of the train of pulses measured on earth is greater than the period measured at the rocket. Since $\lambda = c\,T$ we conclude that if the wavelength of the spectrum of the radiation emitted from atoms at the rocket is measured on earth, the value obtained will be greater. This effect is called ***Red Shift***.

In terms of frequency, since $v_d = 1/T_d$, the nonrelativistic limit of eq 4.55 can be obtained as follows:

$$v_d = \left(1-\beta_e\right)^{\frac{1}{2}}\left(1+\beta_e\right)^{-\frac{1}{2}} = \left(1 - \frac{\beta_e}{2} + \ldots\right)\left(1 - \frac{\beta_e}{2} + \ldots\right)$$

$$= \left(1 - \frac{\beta_e}{2} - \frac{\beta_e}{2} + \frac{\beta_e^{\,2}}{4} + \ldots\right) \tag{4.56}$$

If $V \ll c$ the terms with exponents greater than 2 in the above expansion can be neglected, and we obtain the expression for the classical Doppler effect:

$$v_d = (1 - \beta_e v_e \tag{4.57}$$

4.12 LORENTZ TRANSFORMATIONS OF OTHER PHYSICAL QUANTITIES

4.12.1 CUADRIVECTORS

In SRT any event occurs at a given point in the space specified by the position vector $\vec{r} = (x, y, z)$, and at a given instant specified by time t. The three spatial coordinates can be combined with t to define the array $\mathbb{R} = (ct; \vec{r})$ which has four quantities: one is proportional to the time and the other three are the components of the position vector $\vec{r}$. An array of this type is called a ***quadrivector***. The Lorentz transformations we have considered up to now describe the transformation from one inertial frame to another of the three spatial coordinates plus time, considered as a fourth coordinate.

The quadrivector $\mathbb{R}$ is usually written as $x_\mu = (x_0, x_1, x_2, x_3)$ where the index μ varies between 1 and 4, in which $x_0 = ct$, $x_1 = x$, $x_2 = y$ and $x_3 = z$.

Lorentz transformations can be expressed by means of the following linear transformation:

$$x' = \sum_{\nu=0}^{3} a_{\mu\nu} x_\nu \tag{4.58}$$

Any set of four quantities denoted as A_μ which is transformed into A'_μ by a transformation like that given by eq 4.58 is called a 4-vector or quadrivector.

There are numerous physical quantities that when put together form a 4-vector. This is the case of the energy and momentum of a particle which form the quadrivector

$$\wp = \left(\frac{E}{c}; \vec{p}\right) \tag{4.59}$$

Other examples are the quadrivector charge density-current density and the electromagnetic quadrivector potential given, respectively, by

$$\mathbb{j} = (c\rho, \vec{J}) \tag{4.60}$$

$$\mathbb{A} = \left(\frac{\Phi}{c}, \vec{A}\right) \tag{4.61}$$

In eq 4.60 ρ is the volumetric charge density and $\vec{J}$ is the current density vector. In eq 4.61 Φ is the scalar potential and $\vec{A}$ is the vector potential.

An arbitrary 4-vector can be written as $(G_o; \vec{G})$ where G_o is a scalar quantity and $\vec{G}$ is a vector with components G_1, G_2, G_3. The generalized Lorentz transformation for this arbitrary 4-vector is[4]:

$$G'_0 = \gamma\left(G_0 - \vec{\beta}\cdot\vec{G}\right) \tag{4.62}$$

$$G'_{//} = \gamma\left(G_{//} - \beta G_0\right) \tag{4.63}$$

$$\vec{G}'_\perp = \vec{G}_\perp \tag{4.64}$$

where the subscripts // and $\perp$ indicate components of the vector $\vec{G}$ parallel and perpendicular, respectively, to the direction of the relative velocity vector $\vec{V} = c\vec{\beta}$ between the two inertial frames S and S'.

EXAMPLE 4.9

As a more general result than that obtained in example 4.6, based on the above equations, obtain the Lorentz transformation for the time if the relative velocity between two inertial frames S and S' has an arbitrary direction. Considering two events having a space-like separation, and two events which according to an observer fixed to S occur at times t_1 and t_2 with $t_2 > t_1$. *Is it possible to find an inertial frame where these events occur simultaneously?*

SOLUTION

From eq 4.62 we have that

$$t' = \gamma\left(t - \frac{\vec{V}}{c^2}\cdot\vec{r}\right)$$

Let $(t_1, \vec{r}_1)$ and $(t_2, \vec{r}_2)$ be the points in space time corresponding to the two events as observed from reference frame S. For the observer at reference frame S' these events occur at points $(t'_1, \vec{r}'_1)$ and $(t'_2, \vec{r}'_2)$ with $t'_2 = t'_1$, as required. Therefore, from the last equation we have that S' must move with respect to S with a relative velocity $\vec{V} = c\vec{\beta}$ so that

$$\Delta t = \frac{\vec{V}}{c^2}\cdot\Delta\vec{r}$$

where $\Delta t = t_2 - t_1$ and $\Delta\vec{r} = \vec{r}_2 - \vec{r}_1$. Solving for $\vec{V}$ we get that its magnitude must be

$\left|\vec{V}\right| = c\left(\frac{c\cdot t}{\left|\Delta\vec{r}\right|cos\theta}\right)$, where θ is the angle between $\vec{V}$ and $\Delta\vec{r}$.

For $\theta = 0$ it is possible to obtain a magnitude of $\vec{V}$ lower than c, so that it is possible to find an inertial frame where these events occur simultaneously. For other values of this angle, the possible directions of relative motion are plotted in Figure 4.17 for different values of the relation $|\Delta\vec{r}/c\Delta t|$. Below the dashed line one can find the combinations of values of $|\vec{V}|$ and θ allowing that these events occur simultaneously for an observer fixed to S' so that $|\vec{V}| < c$ is satisfied.

With these generalized transformations the "scalar product" of two 4-vectors $(G_o; \vec{G})$ and $(H_o; \vec{H})$ defined as $G_0 H_o - \vec{G}\cdot\vec{H}$ is invariant so that

$$G'_0 H'_0 - \vec{G}'\cdot\vec{H}' = G_0 H_o - \vec{G}\cdot\vec{H} \tag{4.65}$$

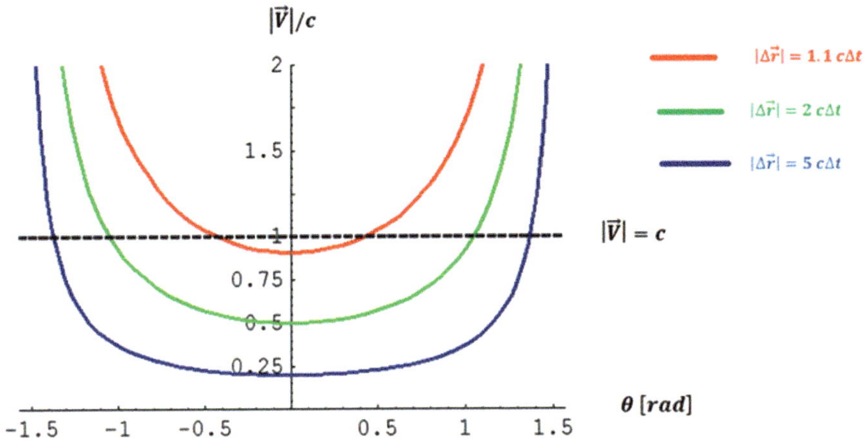

FIGURE 4.17 Magnitude of the relative velocity between S and S' versus the angle θ for simultaneity of events at S'. Results are given for three values of the relation $|\Delta\vec{r}/c\Delta t|$.

This leads to the invariance of the so called "module" of a 4-vector defined so that:

$$G_0'^2 - \left|\vec{G}'\right|^2 = G_0^2 - \left|\vec{G}\right|^2 \tag{4.66}$$

EXAMPLE 4.10

Consider two inertial frames S and S' with relative velocity $\vec{V}$ as shown in Figure 4.18. Based on the above equations, obtain the Lorentz transformation for energy and momentum of a particle of mass m which with respect to frame S is moving with constant momentum $\vec{p} = \hat{x}\, p cos\theta + \hat{y}\, p sin\theta$. Discuss if there is some analogy with Lorentz transformations for position and time.

SOLUTION

Using eqs 4.62 to 4.64 with $G_0 = E/c$, $G_{//} = p_x = p cos\theta$ and $\vec{G}_\perp = p_y = p sin\theta$, we get

$$E' = \gamma(E - Vp_x) \tag{4.67}$$

$$p_y' = \gamma\left(p_x - \frac{V}{c^2}E\right) \tag{4.68}$$

$$p_y' = p_y \tag{4.69}$$

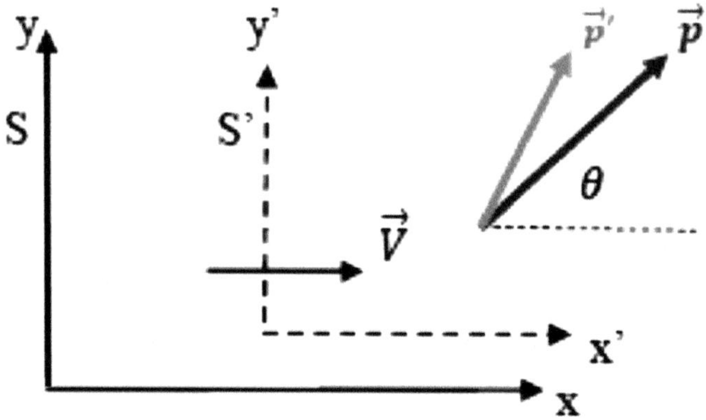

FIGURE 4.18 Particle with momenta p and p' with respect to reference frames S and S', respectively.

Dividing each side of eq 4.67 by c and multiplying each side of eq 4.68 by c, we can obtain as a possible analogy $E/c \leftrightarrow ct$ and $cp_x \leftrightarrow x$, since they obey equations of the same form.

4.12.2 TRANSFORMATION OF THE ELECTROMAGNETIC FIELD

When speaking about an electromagnetic field, we refer to a measurable effect produced by distributions of charges and currents. This effect is described by the corresponding electric field and magnetic field vectors denoted as $\vec{E}$ and $\vec{B}$, respectively. These fields are ruled by Maxwell's equations summarizing the theoretical content of classical electrodynamics.

Under Lorentz transformations Maxwell's equations are invariant in form what is called covariance. This implies that quantities included in these equations, such as ρ, $\vec{J}$, $\vec{E}$, and $\vec{B}$ must transform appropiately under Lorentz transformations.

As shown in several books about relativity theory (see e.g., Ref. [13]), $\vec{E}$ and $\vec{B}$ do not transform as indicated by eqs 4.63 and 4.64 which give the Lorentz transformations applied to the component of a 4-vector. The correct procedure must consider the transformation of a special matrix with elements composed by components of the vectors $\vec{E}$ and $\vec{B}$. This matrix is a second-rank tensor known as the electromagnetic field-strength tensor which is obtained as follows:

From Maxwell's equations

$$\vec{E} = -\nabla\Phi - \frac{\partial \vec{A}}{\partial t} \tag{4.70}$$

$$\vec{B} = \nabla \times \vec{A} \tag{4.71}$$

and considering the 4-vector electromagnetic potential

$$\mathbb{A} = \left(\frac{\Phi}{c}, \vec{A}\right) = A^{\mu} = \left(\frac{\Phi}{c}, A_x, A_y, A_z\right) \tag{4.72}$$

we obtain the antisymmetric second-rank tensor $F^{\alpha\beta} = \partial^{\alpha} A^{\beta} - \partial^{\alpha} A^{\beta}$ corresponding to the following matrix:

$$F = \begin{bmatrix} 0 & E_x/c & E_y/c & E_z/c \\ -E_x/c & 0 & B_z & -B_y \\ -E_y/c & -B_z & 0 & B_x \\ -E_z/c & B_y & -B_x & 0 \end{bmatrix} \tag{4.73}$$

Considering an inertial reference frame S' moving with respect to S with velocity V in a direction parallel to the x-axis, the Lorentz transformations for the time and the spatial coordinates can be expressed by means of the following matrix:

$$A = \begin{bmatrix} \gamma & -\gamma\beta & 0 & 0 \\ -\gamma\beta & \gamma & 0 & 0 \\ 0 & 0 & 1 & 0 \\ 0 & 0 & 0 & 1 \end{bmatrix} \tag{4.74}$$

Using this last matrix we obtain that the Lorentz transformation for the electromagnetic field for this case is obtained from the components of the following matrix:

$$F' = AFA^T = \begin{bmatrix} 0 & E_x/c & \gamma\left(E_y/c - \beta B_z\right) & \gamma\left(E_z/c + \beta B_y\right) \\ -E_x/c & 0 & \gamma\left(-\beta E_y/c + B_z\right) & \gamma\left(-\beta E_z/c - B_y\right) \\ \gamma\left(-E_y/c + \beta B_z\right) & \gamma\left(\beta E_y/c - B_z\right) & 0 & B_x \\ \gamma\left(-E_z/c - \beta B_y\right) & \gamma\left(\beta E_z/c + B_y\right) & -B_x & 0 \end{bmatrix} \tag{4.75}$$

The comparison between eqs 4.73 and 4.75 leads to the following Lorentz transformations for the components of the electromagnetic field for this particular case:

$$E'_x = E_x, B'_x = B_x \tag{4.76}$$

$$E'_y = \gamma\left(E_y - VB_z\right), E'_z = \gamma\left(E_z + VB_z\right) \tag{4.77}$$

$$B'_y = \gamma\left(B_y + \frac{V}{c^2}E_z\right), B'_z = \gamma\left(B_z - \frac{V}{c^2}E_y\right) \tag{4.78}$$

EXAMPLE 4.11

Prove that Lorentz transformations applied to the electromagnetic field lead to an equation like eq 4.66 relating the electric and magnetic fields for two reference frames S and S':

$$\vec{E}'^2 - c^2\vec{B}'^2 = \vec{E}^2 - c^2\vec{B}^2 \tag{4.79}$$

SOLUTION

Let us consider that reference system S' moves with speed V with respect to the *x* axis as shown in Figure 4.19 where the corresponding electric and magnetic field vectors are arbitrarily drawn.

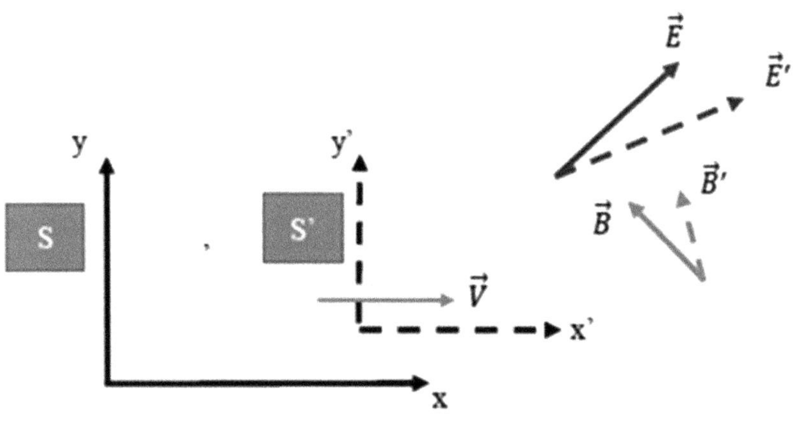

FIGURE 4.19 An electromagnetic field observed from reference frames S and S'.

In system S the components of the electromagnetic field perpendicular to the motion can be written as the vectors $\vec{E}_\perp$ and $\vec{B}_\perp$ expressed for this case as

$$\vec{E}_\perp = \hat{y}E_y + \hat{z}E_z \tag{4.80}$$

$$\vec{B}_{\perp} = \hat{y}B_y + \hat{z}B_z \tag{4.81}$$

Combining the above equations with eqs 4.76–4.78 and according to Chapter 11 of Ref. [4], we can write the following expressions for the Lorentz transformations of the components of the electric and magnetic field vectors perpendicular to the relative velocity.

$$\vec{E}'_{\perp} = \gamma\left\{\vec{E}_{\perp} + c\vec{\beta}\times\vec{B}_{\perp}\right\} = \gamma\left\{\vec{E}_{\perp} - \vec{B}_{\perp}\times c\vec{\beta}\right\} \tag{4.82}$$

$$\vec{B}'_{\perp} = \gamma\left\{\vec{B}_{\perp} - \frac{\vec{\beta}}{c}\times\vec{E}_{\perp}\right\} \tag{4.83}$$

The corresponding components parallel to the relative velocity are the same for both reference frames, that is $\vec{E}'_{//} = \vec{E}_{//}$ and $\vec{B}'_{//} = \vec{B}_{//}$.

From eqs 4.82to 4.83:

$$\vec{E}'_{\perp}\cdot\vec{E}'_{\perp} = \gamma^2\left(\vec{E}_{\perp} + c\vec{\beta}\times\vec{B}_{\perp}\right)\cdot\left(\vec{E}_{\perp} + c\vec{\beta}\times\vec{B}_{\perp}\right)$$

$$= \gamma^2\left\{\vec{E}_{\perp}^{\,2} + c\vec{\beta}\times\vec{B}_{\perp}\cdot\vec{E}_{\perp} + c\vec{E}_{\perp}\cdot\vec{\beta}\times\vec{B}_{\perp} + c^2\left(\vec{\beta}\times\vec{B}_{\perp}\right)^2\right\}$$

$$= \gamma^2\left(\vec{E}_{\perp}^{\,2} + V^2\vec{B}_{\perp}^{\,2}\right)$$

$$\vec{B}'_{\perp}\cdot\vec{B}'_{\perp} = \gamma^2\left(\vec{B}_{\perp} - \frac{\vec{V}}{c^2}\times\vec{E}_{\perp}\right)\cdot\left(\vec{B}_{\perp} - \frac{\vec{V}}{c^2}\times\vec{E}_{\perp}\right)$$

$$= \gamma^2\left\{\vec{B}_{\perp}^{\,2} - \frac{\vec{V}}{c^2}\times\vec{E}_{\perp}\cdot\vec{B}_{\perp} - \vec{B}_{\perp}\cdot\frac{\vec{V}}{c^2}\times\vec{E}_{\perp} + \frac{1}{c^4}\left(\vec{V}\times\vec{E}_{\perp}\right)^2\right\}$$

$$= \gamma^2\left(\vec{B}_{\perp}^{\,2} + \frac{\beta^2}{c^2}\vec{E}_{\perp}^{\,2}\right)$$

Therefore, we have

$$\vec{E}'_{\perp}\cdot\vec{E}'_{\perp} - c^2\,\vec{B}'_{\perp}\cdot\vec{B}'_{\perp} = \gamma^2\left(\vec{E}_{\perp}^{\,2} + V^2\vec{B}_{\perp}^{\,2} - c^2\vec{B}_{\perp}^{\,2} - \beta^2\vec{E}_{\perp}^{\,2}\right)$$

$$= \gamma^2\left(\left(1-\beta^2\right)\vec{E}_{\perp}^{\,2} + \left(V^2 - c^2\right)\vec{B}_{\perp}^{\,2}\right)$$

$$= \gamma^2\left(\frac{1}{\gamma^2}\vec{E}_{\perp}^{\,2} - c^2\left(1-\beta^2\right)\vec{B}_{\perp}^{\,2}\right)$$

$$= \vec{E}_{\perp}^{\,2} - c^2\vec{B}_{\perp}^{\,2}$$

Since $\vec{E}' = \vec{E}'_{\perp} + \vec{E}'_{//}$ and $\vec{B}' = \vec{B}'_{\perp} + \vec{B}'_{//}$, after some algebra we obtain the result given by eq 4.79.

We could be tempted to set a certain analogy of the result given by eq 4.79 with the invariance of the quantity $\vec{r}^2 - c^2t^2$ expressed by Lorentz transformations in space time. Furthermore, we can appreciate some similarity in the form of eq 4.83 with the Lorentz transformation for time t which for arbitrary direction of the relative velocity of system S' with respect to S can be written as[8]:

$$t' = \gamma\left(t - \frac{\vec{\beta}\cdot\vec{r}}{c}\right) \tag{4.84}$$

In fact, if just for a while we do not consider the important difference that eq 4.83 expresses a transformation between vector quantities with cross product involved, while eq 4.84 corresponds to a transformation between scalar quantities with a dot product included, from these two equations we could set the analogy $\leftrightarrow \vec{B}_{\perp}$.

EXAMPLE 4.12

Consider an infinitely long wire charged with linear density λ at rest with respect to the laboratory (frame S) and an observer moving with speed υ in a direction parallel to the wire (frame S'). Let us choose the axis x along this last direction. Show that for this observer there is a magnetic field coincident with that obtained applying Ampere's law.

SOLUTION

For the observer fixed to frame S there is only an electric field whose components are perpendicular to the x axis, whereas according to the observer fixed to S' the charges of the wire are in motion, so that it is expected that this latter will detect both a magnetic and an electric field.

From eq 4.79 we can write

$$\vec{E}'^2 - c^2\vec{B}'^2 = \vec{E}^2$$

Combining with eqs 4.76 and 4.77 we have

$$\vec{E}'^2 = E_y'^2 + E_z'^2 = \gamma^2\left(E_y^2 + E_z^2\right) = \gamma^2\vec{E}^2.$$

Therefore, after some algebra

$$\vec{B}'^2 = \frac{\vec{E}^2}{c^4}\gamma^2 V^2$$

As known, the magnitude of the electric field produced by an infinite long wire charged with linear density λ at a point placed at distance r from the axis of this wire is given by

$$\left|\vec{E}\right| = \frac{\lambda}{2\pi \in_o r}$$

Therefore we obtain that the magnitude of the magnetic field detected by observer fixed to S' is

$$\left|\vec{B}'\right| = \frac{\gamma(\lambda V)}{2\pi \in_o c^2 r} = \frac{\mu_o \gamma(\lambda V)}{2\pi \in_o r}$$

Considering the relativistic effect of length contraction, for the observer fixed at S' the charge density is $\lambda' = \dfrac{Q}{L/\gamma} = \gamma\lambda$ for a charge Q uniformly distributed along a length. Therefore, the term $\gamma(\lambda V)$ in this last expression corresponds to the current I' asociated with the motion of the charges according to this last observer, so that he detects a magnetic field of magnitude $\left|\vec{B}'\right| = \dfrac{\mu_o I'}{2\pi \in_o r}$ which corresponds to the form obtained from Ampere's law for a rectilinear wire of infinite length through which a current I' circulates.

EXAMPLE 4.13

Prove that $\vec{E} \cdot \vec{B}$ is an invariant quantity under a Lorentz transformation. Can we say that the energy density in the electromagnetic field is a Lorentz invariant?

SOLUTION

Using eqs 4.82 and 4.83, we have that for the observer fixed to the system S':

$$\vec{E}' \cdot \vec{B}' = \left(\vec{E}'_\perp + \vec{E}'_{//}\right) \cdot \left(\vec{B}'_\perp + \vec{B}'_{//}\right) = \vec{E}'_{//} \cdot \vec{B}'_{//} + \vec{E}'_\perp \cdot \vec{B}'_\perp$$

$$= \vec{E}_{//} \cdot \vec{B}_{//} + \gamma^2 \left\{ \left(\vec{E}_\perp + c\vec{\beta} \times \vec{B}_\perp\right) \cdot \left(\vec{B}_\perp - \frac{\vec{V}}{c^2} \times \vec{E}_\perp\right) \right\}$$

$$= \vec{E}_{//} \cdot \vec{B}_{//} + \gamma^2 \left\{ \vec{E}_\perp \cdot \vec{B}_\perp - \frac{1}{c^2}\left(\vec{V} \times \vec{B}_\perp\right) \cdot \left(\vec{V} \times \vec{E}_\perp\right) \right\}$$

From the identity vector algebra:

$$\left(\vec{a}\times\vec{b}\right)\cdot\left(\vec{c}\times\vec{d}\right)=\left(\vec{a}\cdot\vec{c}\right)\left(\vec{b}\cdot\vec{d}\right)-\left(b\cdot\vec{c}\right)\left(\vec{a}\cdot\vec{d}\right)$$

we have that

$$\left(\vec{V}\times\vec{B}_{\perp}\right)\cdot\left(\vec{V}\times\vec{E}_{\perp}\right)=\left(\vec{V}\cdot\vec{V}\right)\left(\vec{B}_{\perp}\cdot\vec{E}_{\perp}\right)-\underbrace{\left(\vec{B}_{\perp}\cdot\vec{V}\right)\left(\vec{V}\cdot\vec{E}_{\perp}\right)}_{=0}$$

Therefore, we get that

$$\vec{E}'\cdot\vec{B}'=\vec{E}_{//}\cdot\vec{B}_{//}+\gamma^2\left(\vec{E}_{\perp}\cdot\vec{B}_{\perp}\right)\left(1-\frac{V^2}{c^2}\right)=\vec{E}_{//}\cdot\vec{B}_{//}+\vec{E}_{\perp}\cdot\vec{B}_{\perp}$$

$$=\left(\vec{E}_{//}+\vec{E}_{\perp}\right)\cdot\left(\vec{B}_{//}+\vec{B}_{\perp}\right)$$

$$=\vec{E}\cdot\vec{B}$$

In this way we prove that $\vec{E}\cdot\vec{B}$ is an invariant quantity under a Lorentz transformation.

EXAMPLE 4.14

Is the Poynting vector corresponding to a plane monochromatic electromagnetic wave a Lorentz invariant?

SOLUTION

Let us consider the specific case of a monochromatic electromagnetic wave propagating in a lossless and isotropic medium with magnetic permeability μ. In example 4.13 we have shown that $\vec{E}\cdot\vec{B}$ is a Lorentz invariant quantity.

Does the same occur for the Poynting vector given by $\vec{\wp}=\vec{E}\times\vec{B}/\mu$? This is analyzed as follows:

Let us consider the specific case of a plane and monochromatic electromagnetic wave of angular frequency ω propagating in a direction parallel to the x-axis with linear polarization in the y-direction. In this case the corresponding electric and magnetic field vectors are

$$\vec{E}\left(z,t\right)=\hat{y}E_o sin\left(kx-\omega t\right)=\hat{y}E_o sin\omega\left(\frac{x}{c}-t\right) \quad (4.85)$$

$$\vec{H}(z,t) = \hat{z}H_o sin(kx - \omega t) = \hat{z}H_o sin\omega\left(\frac{x}{c} - t\right) \tag{4.86}$$

Using Lorentz transformations for the electromagnetic field and for this specific case, the corresponding electric and magnetic vectors for an observer S' moving in the x direction with velocity V are

$$E'_y(x,t) = \gamma\,(E_y - VB_z) \tag{4.87}$$

$$B'_z(x,t) = \gamma\left(B_z - \frac{V}{c^2}E_y\right) \tag{4.88}$$

The situation is sketched in Figure 4.20 where the traveling of an electromagnetic wave is observed from reference frames S and S', this latter moving with velocity vector $\vec{v}$ parallel to the x direction.

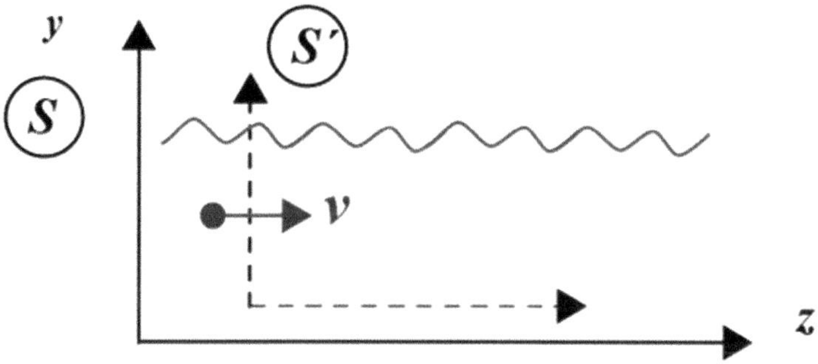

FIGURE 4.20 Plane electromagnetic observed from reference frames S and S'

Therefore for observer S' the corresponding magnitude of the Poynting vector is

$$|\wp|' = \frac{E'_y B'_z}{\mu_o} = \frac{\gamma^2}{\mu_o}\left\{E_o B_o - VB_o^2 - \frac{V}{c^2}E_o^2 + \frac{V^2}{c^2}E_o B_o\right\} sin^2\omega'\left(\frac{x'}{c} - t'\right)$$

$$= \gamma^2 \frac{E_o B_o}{\mu_o}(1-\beta)^2\, sin^2\omega'\left(\frac{z'}{c} - t'\right) = \left(\frac{1-\beta}{1+\beta}\right)\frac{E_o B_o}{\mu_o} sin^2\omega'\left(\frac{z'}{c} - t'\right) \tag{4.89}$$

Therefore, the Poynting vector is not invariant if one applies the Lorentz transformations to the electromagnetic field components $\vec{E}$ and $\vec{H}$ and we have the following relation between its magnitudes as observed in S and S':

$$|\vec{\wp}|' = \left(\frac{1-\beta}{1+\beta}\right)|\vec{\wp}| \tag{4.90}$$

EXAMPLE 4.15

As a complement to the previous example, let us analyze whether or not the energy density is invariant under Lorentz transformation.

SOLUTION

The electromagnetic energy density for the same wave of Example 4.15 is given by

$$w_{em} = \frac{\varepsilon_o}{2} E_y^2 + \frac{B_z^2}{2\mu_o} \tag{4.91}$$

As known, for a plane electromagnetic wave propagating through the vacuum $\vec{E} = c\vec{B}$ and $1/\mu_o = c^2\varepsilon_o$. Therefore, eq 4.90 can be written as

$$w_{em} = c^2\varepsilon_o B_z^2 \tag{4.92}$$

From eqs 4.87 to 4.88, we have that for observer S' the electromagnetic energy density is

$$w'_{em} = \frac{\varepsilon_o}{2} \gamma^2 \left(E_y - VB_z\right)^2 + c^2 \frac{\varepsilon_o}{2} \gamma^2 \left(B_z - \frac{V}{c^2}\right)^2 = \gamma^2 c^2 \varepsilon_o B_z^2 \left(1 - \frac{V}{c}\right)^2 \tag{4.93}$$

Since $\gamma^2 = 1/(1 - \beta^2)$ with $\beta = V/c$ the above equation can be rewritten as

$$w'_{em} = \left(\frac{1-\beta}{1+\beta}\right) w_{em} \tag{4.94}$$

EXAMPLE 4.16

As known from courses on classical electrodynamics, in a lossless medium the Poynting vector and the electromagnetic energy density satisfy the so called continuity equation for the energy

$$\nabla \cdot \vec{\wp} + \frac{\partial}{\partial t} w_{em} = 0 \tag{4.95}$$

What can we say about the behavior under time reversal symmetry for the electromagnetic wave considered in the two above examples, both for observers S and S'?

SOLUTION

The Poynting vector is given by $\vec{\wp} = \vec{E} \times \vec{H}$. As discussed in Chapter 2 the electric field $\vec{E}$ is even under time reversal but the magnetic field $\vec{H}$ is odd, that is $\vec{H} \to -\vec{H}$. Therefore, under time reversal $\vec{\wp} \to -\vec{\wp}$. Since under this operation t changes its sign but the electromagnetic energy density is kept (it depends on the square of $\vec{E}$ and $\vec{H}$), eq 4.95 is time reversal invariant.

From eqs 4.90 to 4.94 we can see that under time reversal for observer S' exists the magnitude of the Poynting vector is related with the electromagnetic energy density by a factor c, that is $|\vec{\wp}|' = cw_{em}$ as equally occurs for observer S.

Therefore, this example shows that although under Lorentz transformation of the vectors $\vec{E}$ and $\vec{H}$, the Poynting vector and energy density are not Lorentz invariant, the way they are related does not change with time reversal.

4.12.3 LORENTZ TRANSFORMATION OF ENERGY AND TIME

As shown in several texts about Special Relativity Theory, a free particle of mass m moving with speed υ has a total energy

$$E = \gamma mc^2 \tag{4.96}$$

where $\gamma = 1/\sqrt{1-\upsilon^2/c^2}$

As we have seen in Section 4.12.1 energy and momentum form a 4-vector given by

$$p^{\mu} = \left(\frac{E}{c}, p_x, p_y, p_z\right) \tag{4.97}$$

where $E^2 - c^2p_x^2 - c^2p_y^2 - c^2p_z^2$ is a Lorentz invariant quantity. As seen in Section 4.12.1, for a relative motion between two reference frames along the x direction, with speed V, Lorentz transformations for energy and momentum are expressed by the following equations:

$$E' = \gamma(E - Vp_x) \tag{4.98}$$

$$p'_x = \gamma\left(p_x - \frac{V}{c^2}E\right) \tag{4.99}$$

$$p'_y = p_y \tag{4.100}$$

$$p'_z = p_z \tag{4.101}$$

Let us compare the form of these transformations with those corresponding to time and position for the same case of relative motion along the x direction

$$x' = \gamma(x - Vt) \tag{4.102}$$

$$t' = \gamma\left(t - \frac{V}{c^2}x\right) \tag{4.103}$$

$$y' = y \tag{4.104}$$

$$z' = z \tag{4.105}$$

The usual form of making this comparison is to consider that the quantity E/c (which has units of momentum) is transformed in a similar way as the quantity ct (which has units of distance). In fact, if we divide each side of eq 4.98 by c and multiply each side of eq 4.103 by c we would get

$$\frac{E'}{c} = \gamma\left(\frac{E}{c} - \frac{V}{c}p_x\right) \tag{4.106}$$

$$ct' = \gamma\left(ct - \frac{V}{c}x\right) \tag{4.107}$$

These last two equations have the same form, even though we are dealing with different physical quantities, if we set the correspondence $\frac{E'}{c} \Leftrightarrow ct'$. Perhaps we would be more peaceful if we divide eqs 4.106 and 4.107 by known physical constants. For instance, we could use the Compton momentum and Compton wavelength of the electron, defined as $p_{eC} \equiv m_{eC}$ and $\lambda_{eC} \equiv h/m_{eC}$, respectively. In this case we would obtain the following equations corresponding to the Lorentz transformations of coordinates in space time and in the energy-momentum space:

$$\tilde{t}' = \gamma\left(\tilde{t} - \beta\tilde{x}\right) \tag{4.108}$$

$$\tilde{E}' = \gamma\left(\tilde{E} - \beta\tilde{p}_x\right) \tag{4.109}$$

where $\tilde{t} \equiv ct / \lambda_{eC}, \tilde{E} \equiv E / m_e c^2; \tilde{x} \equiv x / \lambda_{eC}, \tilde{p}_x \equiv p_x / p_{eC}$

These last two equations have a similar structure, so that $\tilde{t}$ and $\tilde{E}$ which are dimensionless quantities related with time and energy, respectively, have similar Lorentz transformations.

EXAMPLE 4.17

This example has the objective to analyze the similarities between the results of applying Lorentz transformations for time and energy in the case of the decaying of a particle of mass M where a massless particle results. We illustrate this case with the example of the decaying of a pi-meson in a mu-meson and a massless neutrino, as proposed in problem 11.19 of Ref. [4].

Let M be the mass of the decaying particle, m_1 the mass of the mu-meson and let us assume that the mass m_2 of the neutrino can be considered equal to zero. The velocities of these resulting particles can be obtained applying the principles of conservation of energy and momentum, as follows.

Let us consider that after the decay, for an observer at rest in the laboratory frame, the resulting particles have energies and momenta ($\vec{E}_1$, $\vec{p}_1$) and ($\vec{E}_2$, $\vec{p}_2$), respectively, so that by conservation of energy and momentum we have $E_2 = Mc^2 - E_1$ and $\vec{p}_2 = -\vec{p}_1$.

For each one of the resulting particles, from the Lorentz invariance of the quadrivectors $\left(\frac{E_j}{c}, \vec{p}_j\right)$ for j=1,2, and with $p = |\vec{p}_1| = |\vec{p}_2|$ we can write

$$E_2^2 - c^2p^2 = E_1^2 - c^2p^2 = \left(Mc^2 - E_1\right)^2 - c^2p^2 \tag{4.110}$$

Additionally, for each particle the relativistic relation between energy and momentum is $E_j = \sqrt{(cp)^2 + (m_jc)^2}$ for j=1,2. Using this relation in eq 4.110 and after some algebra, we obtain that for $m_2 = 0$ the corresponding energies of the resulting particles are

$$E_1 = \frac{M^2 - m_1^2}{2M}c^2 \tag{4.111}$$

$$E_2 = \frac{M^2 + m_1^2}{2M}c^2 \tag{4.112}$$

From eq 4.111 we have

$$E_1 = \frac{M^2 - m_1^2}{2M}c^2 = \gamma_1 m_1 c^2 \tag{4.113}$$

with $\gamma_1 = 1/\sqrt{1-\beta_1^2}$ and $\beta_1 = \upsilon_1/c$.

Solving eq 4.113 we obtain

$$\beta_1 = \frac{\upsilon_1}{c} = \frac{M^2 - m_1^2}{M^2 + m_1^2} \tag{4.114}$$

Since particle 2 is a massless particle we have that

$$E_2 = cp = c\gamma_2 m_2 \upsilon_2 = \gamma_2 m_2 c^2 \Rightarrow \beta_2 = \frac{\upsilon_2}{c} = 1 \tag{4.115}$$

Let us consider that at a distance D from the position where this reaction occurs, there is a system that detects the mu-meson and its corresponding energy as schematized in Figure 4.21.

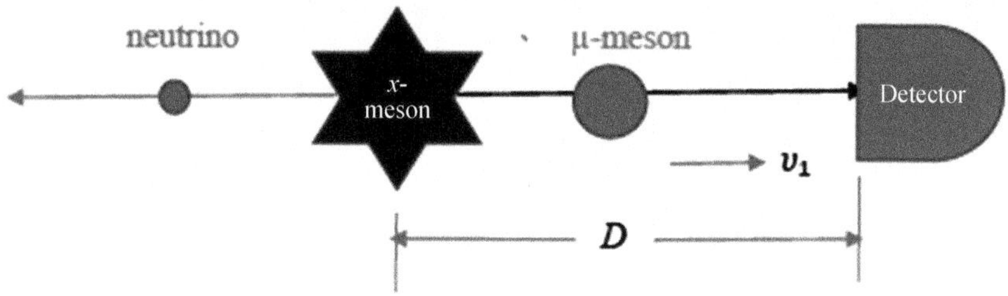

FIGURE 4.21 Decaying of a π-meson in a μ-meson and a massless neutrino.

From a reference frame S fixed to the detector, the time interval for detection of the mu-meson after the reaction is $\Delta t = t_d = D/\upsilon_1$ and the corresponding energy of the particle incident to the detector is that given by eq 4.114.

From a reference frame S' fixed to the mu-meson the time detection is

$$t'_d = \gamma_1\left(t_d - \frac{\upsilon_1}{c^2}D\right) = \gamma_1\left(t_d - \frac{\upsilon_1}{c^2}\upsilon_1 t_d\right) = t_d \frac{1}{\sqrt{1-\beta_1^2}}\left(1-\beta_1^2\right) = t_d\sqrt{1-\beta_1^2} = t_d/\gamma_1 \tag{4.116}$$

and the corresponding energy of this particle referred to S' is

$$E'_1 = \gamma_1\left(E_1 - \upsilon_1 p_1\right) = \gamma_1\left(\gamma_1 m_1 c^2 - \upsilon_1\gamma_1 m_1 \upsilon_1\right) = m_1 c^2 = E_1/\gamma_1 \tag{4.117}$$

From eqs 4.116 to 4.117 we notice that in system S' the detection time of the mu-meson behaves as a proper time and we observe an analogy with its energy that can be considered a "proper energy."

4.13 A NOTE ABOUT SUPERLUMINAL PARTICLES

In a paper published in 1967, Gerald Feinberg considered the possibility of describing, within the theoretical frame of the special relativity theory, particles moving with velocity greater than that of light in vacuum.[14] He noted that what was really not allowed by the special theory of relativity, was not really motion faster than the speed of light, but rather the acceleration of particles with nonzero mass up to such speeds, going through the value c, where the Lorentz transformations become meaningless and the energy of the particles with mass becomes infinity. Admitting the existence of a class of particles always moving at speeds greater than c, he proposed for them the name of ***tachyon*** and proposed that its energy as a function of its mass m_t and speed u might be given by the expression $E_t = m_t c^2 / \sqrt{\beta^2 - 1}$ where $\beta = u/c > 1$. According to that suggestion the energy of a tachyon becomes infinite if its speed decreases to a value equal to that corresponding to light. This is analogous to the behavior of particles known so far, whose energy has an infinite value when their speed increases up to reaching the value c. In this way, just as the up to now known particles cannot move with a speed exceeding c (subluminal particles), if tachyons exist, they must move faster than c (superluminal particles).

To postulate the existence of a particle such as the tachyon seemed an interesting idea at that time, but as was pointed out by G. Feinberg, it originated paradoxes because by definition, tachyons move faster than light, so their world lines are space like. Therefore, as discussed in Section 4.7 of this chapter, the sequence of events along its world lines is not the same for all inertial reference frames. This implies that tachyons give rise to the same type of paradoxes as those associated with travel into the past. If tachyons existed, it would not be possible for human beings to travel with superluminal speed. However, the existence of these superluminal particles gives the possibility of using them to send information faster than c, which implies the possibility of sending information into the past, which in turn gives rise to paradoxes related to the principle of causality.

To exemplify the above, let us consider a situation in which a device produces tachyons that are emitted with speed u>c with respect to a reference frame S fixed to the ground. Let us suppose that these tachyons are produced at an instant chosen as t=0 and that they move in a direction parallel to the positive x-axis, with respect to the reference system S. Since these tachyons are superluminal particles, in space time the point (t_d, χ_d) is outside the cone of light. Therefore, as discussed earlier in this chapter, the sign of t_d is not the same for all inertial reference frames.

For an observer fixed to a reference frame S' moving with velocity V parallel to the x-axis with respect to frame S, each tachyon will be detected at the instant $t' = t_d'$ and position $x' = x_d'$ (the respective clocks are assumed to be adjusted so that $t = t' = 0$ when the origins of the two reference frames S and S' coincide). Using Lorentz transformations, we obtain that the observer in the reference frame S' will detect a tachyon at the following time and position:

$$t_d' = \gamma\left(t_d - Vx_d / c^2\right) \tag{4.118}$$

$$x_d' = \gamma\left(x_d - Vt_d\right) \tag{4.119}$$

If we substitute $x_d = ut_d$ into the second of these last two equations, we get the following relation:

$$t_d' = \gamma t_d\left(1 - Vu / c^2\right) \tag{4.120}$$

It should be noted that in these equations, V is the speed with which the system S' is moving with respect to S, such that $V<c$. From eq 4.120 we can see that if $u > c^2/V$ we obtain $t_d' < 0$. That is, if it is possible to generate a tachyon moving subluminally fast enough, it will travel backward in time according to the reference frame S' so that for an observer in that frame, it will be detected before it is produced. Even though the speed of this tachyon is not high enough to satisfy this condition, it is always possible to find an inertial reference frame for which its speed V, being high, is less than c. For this system, the tachyon will travel backward in time. For instance, the tachyon could have a speed slightly greater than *c*, but it is possible to find a value of V lower than c, as for example, $u = 1.1\ c$ and $V = 0.95\ c$, values for which the condition $u > c^2/V$ is satisfied.

The mere fact that $t_d' < 0$ it does not by itself give rise to possible paradoxes. This would only happen if it were possible to send a return signal from the event corresponding to the tachyon detection, so that it reaches the point x = 0 before t = 0. If that were the case, the receiver of the return signal could block the transmission of the original tachyon from x = 0 and then we would have the paradox that the tachyon is sent if and only if it is not sent.

Since the tachyon detection event occurred outside the light cone with apex at the origin, the return signal would have to propagate faster than the speed of light and therefore would require a second tachyon for such a paradox to occur. In the reference frame S', the second tachyon would travel into the future, traveling the distance x_d' back to the origin in a positive time less than $-t_d'$ (remember that $t_d' < 0$). Hence, the speed of the return tachyon must satisfy $u_r > x_d'/(-t_d')$ which implies that $u_r > u$ as can be seen from eqs 4.118 to 4.120. Therefore, the tachyon used for the return signal must be moving somewhat faster than the original tachyon. From the relativity theory, it follows that if it were possible for an observer fixed on the earth (reference frame S) to produce tachyons traveling into the future, then it should be possible for observers in frame S' to produce such tachyons, and thus send the return signal. Therefore, the application of the theory of relativity to superluminal particles, such as tachyons, gives rise to paradoxes. One of these paradoxes is concerned with the energy of a particle as explained in what follows:

For ordinary particles, according to the Lorentz transformations the sign of the energy of a particle is the same in all inertial frames. For all inertial observers, the energy of a subluminal particle is positive, despite differences in its corresponding value. However, from eq 4.98 and with $p_x = Eu/c^2$ (assuming that the motions of the ship and of the tachyons are in the x direction), we obtain that if a tachyon has an energy E in the earth's frame, for an observer in the frame S' (ship) the energy is given by $E' = \gamma E(1 - Vu/c^2)$. From this last expression and eq 4.118 we can see that when t_d' is negative the same happens with E'. Therefore, a tachyon traveling backward in time has negative energy. That means that according to frame S' at t_d' the detector absorbs a tachyon coming from the future and bringing negative energy. But the absorption of negative energy is equivalent to losing positive energy. In this way, the detector has apparently lost energy by emitting a tachyon with positive energy at t_d' and continues present, appearing with movement toward the future together with the absorber, up to time $t' = t = 0$, instant at which the tachyon was initially emitted backward in time with energy E', according to an observer at S'. If this observer lives forward in time (supposedly like all of us), for him it is as if the tachyon disappeared in the device that produced it. This observer will then conclude that the device, rather than having emitted a negative-energy tachyon traveling backward in time, has absorbed a positive-energy tachyon traveling forward in time and so coming from the past. Both processes are physically equivalent.

Arguments of this type have been proposed as a possible solution to the problem of negative energy of tachyons, applying the "principle of reinterpretation," which tries to avoid concepts not yet clear, such as travel into the past and negative energy, considering that the absorption of a particle with negative energy is equivalent to the creation of a particle with positive energy, as well as the creation of a particle with negative energy is equivalent to the absorption of a particle with positive energy.

The above arguments have been refuted in reference[16] where the authors prove that the principle of reinterpretation does not eliminate the possibility of communication with the past if the existence of tachyons were possible.

In addition, if the existence of tachyons was possible it would be necessary a generalization of Lorentz transformations to relate the coordinates of events in the corresponding space time of superluminal and subluminal reference systems.

To obtain that relation, Parker has proposed a theory which considers, as usual, a temporal dimension, but only one spatial dimension.[17] Nevertheless, the fact that no known physical law requires to choose a specific spatial direction as the preferred puts this theory in contradiction with the accepted principle of symmetry of the space under rotations, which in turn leads to the conservation of the angular momentum. Therefore, a theory of tachyon involving a preferred direction in space could be experimentally feasible only if the coupling of tachyons with matter is extremely weak.

4.14 CONCLUSIONS

In this chapter, we have studied the implications of the SRT on the conception of time and on the results of measurements of time intervals. One of the great changes introduced by this theory was that time must be considered in an equal foot as the spatial coordinates and combined with them, to form a frame called space time, where events occur differently for observers in relative motion with speed comparable with light velocity.

One very important effect is that the result of measurements of time interval between two events is not the same for these observers and even more, its sequence of occurrence can be different if they are not causally connected. For events causally connected, the possibility of changing its sequence would imply that an effect can occur before than its corresponding cause, requiring velocities greater than the light velocity, what is precluded by relativity theory, as formulated by Einstein. This is expressed by means of the causality principle which is associated with the postulate that it is not possible that the propagation of signals is faster than the speed of light in vacuum. This follows from the special theory of relativity and is valid for all observers moving relative to each other with constant speed (inertial observers).

Our analysis has included a review of the possibility of describing, within the theoretical frame of the special relativity theory, particles moving with velocity greater than that of light in vacuum (these particles are known as ***tachyons***). According to the references here quoted, what is really not allowed by special relativity is not actually travel faster than the speed of light, but rather, the acceleration of matter up to such speeds, going through the value c, where the Lorentz transformations become meaningless and the energy of the particles with mass becomes infinity. The possibility of existence of such particles gives rise to the same kind of unsolved paradoxes as those associated with backward time travel, mainly the communication with the past if tachyon existed. Another paradox is concerned with the fact that a tachyon traveling backward in time would have negative energy for some observers. Up to now there is not any experimental evidence of the existence of this kind of particles.

KEYWORDS

- **inertial observer**
- **Lorentz transformation**
- **Lorentz invariance**
- **time dilation**
- **Doppler effect**
- **light cone**
- **simultaneity**

REFERENCES

1. Einstein, A. Relativity the Special and General Relativity, Digital Reprint of the Book First Published in 1920, PDF ebook CREATED by José Menéndez.
2. Norton, J. D. Einstein's Special Theory of Relativity and the Problems in the Electrodynamics of Moving Bodies that Led Him To It, Department of History and Philosophy of Science; University of Pittsburgh, Cambridge University Press, 2004.
3. Aranoff, S. Postulates of Special Relativity. *Phys. Essays* **2013,** *26*, 110–112.
4. Jackson, J. D. *Classical Electrodynamics*, 3rd edn; Wiley: New York, 1998.
5. Levy, J-M. A Simple Derivation of the Lorentz Transformation and of The Related Velocity and Acceleration Formulae. *Am. J. Phys., Am. Assoc. Phys. Teachers* **2007,** *75* (7), 615–618.
6. Susskind, L. Art Friedman, Special Relativity and Classical Field Theory: The Theoretical Minimum, Published September 26th 2017 by Basic Books.
7. Hill, J. M.; Cox, B. J. Einstein's Special Relativity Beyond the Speed of Light. *Proc. R. Soc. A* **2012,** *468*, 4174–4192.
8. Hassani, M. E. Foundations of Superluminal Relativistic Mechanics. *Commun. Phys.* **2014,** *24* (4), 313–332.
9. Grushka, Y. I. Notes on Extended Lorentz Transformations for Superluminal Reference Frames. *Progress Phys. 13*, 200–201.
10. Einstein, A. Jahrbuch der Radioaktivitat und Elektronik **1907,** *4*, 411.
11. Huang, Y-S.; Lu, K-H. Formulation of the Classical and the Relativistic Doppler Effect by a Systematic Method. *Can. J. Phys.* **2004,** *82* (11).

12. Zhang, J.; Zhang, K.; Grenfell, R.; Deakin, R. On the Relativistic Doppler Effect for Precise Velocity Determination Using GPS. *J. Geodesy* **2006,** *80*, 104–110.
13. Griffiths, D. J. *Introduction to Electrodynamics*, Chapter 12, 3rd ed.; Prentice Hall: Upper Saddle River, 1999.
14. Feinberg, Possibility of Faster-Than-Light Particles. *Phys. Rev.* **1967,** *159*, 1089.
15. Bilaniuk, O. M.P.; Deshpande, V. K.; Sudarshan, E. C. G. "Meta" Relativity. *Am. J. Phys.* **1962,** *30*, 718.
16. Benfoord, G. A.; Book, D. L.; Newcomb, W. A. The Tachyonic Antitelephone. *Phys. Rev. D* **1970,** *2*,263.
17. Parker, L. Faster-Than-Light Inertial Frames and Tachyons. *Phys. Rev.* **1969,** *188*, 2287.

CHAPTER 5

Time in General Relativity Theory

ABSTRACT

The relativity theory formulated by Albert Einstein during the first years of the 20th century radically changed the notions about time and space accepted up to then in physics and philosophy. A first step toward this revolutionary change was given by the special relativity theory, having as one of its main ideas the postulates that the laws of physics are the same for all inertial observers and that there does not exist an absolute time. The latter implies that time intervals are different for inertial observers in relative motion with a velocity comparable with light velocity. In this theory, events take place in a space of four dimensions called the space time (time is considered as a fourth coordinate additional to the three coordinates which specify the position in space). A second step in this change was represented by the general relativity theory where gravitational effects are included, postulating that the presence of massive bodies curves the space time in a way determined by the Einstein field equations. One of the predictions of this theory is that these effects are such that time goes slower as the lowest is the distance of an observer to a massive body. Another important prediction of this theory is the existence of black holes, which corresponds to regions of space with a very high density of mass so that gravity is so strong that even light cannot escape from the interior of the corresponding volume and that space time is distorted so that time looks as frozen at points over the so called "event horizon."

In this chapter, we analyze these effects from the perspective of time, starting from the equivalence principle, which allows to understand that gravity is best described not as an external interaction, but rather as a way in which the geometry and curvature of space time are manifested. This equivalence can explain the bending and the red shift of light due to the action of a gravitational field, and the time differences between clocks placed

Looking at Time from a Physics Perspective. Patricio Robles (Author)

at different distances from a massive body. In particular, we analyze the behavior of time in the presence of a black hole and discuss the possibility of time-reversal symmetry in the evolution of our universe.

5.1 INTRODUCTION

The special relativity theory (SRT) is adequate for describing situations where there is a relative motion between inertial frames. Observers fixed to such frames cannot say from measurements made in their own frame, if they are moving with constant velocity. However, they can detect if its movement has acceleration with respect to a certain inertial reference system.

Albert Einstein wondered by the above dichotomy and also why Newton's law of gravitation is not the same in all inertial frames, on the contrary to Maxwell equations, which are the same for all inertial reference systems (they are invariant under a Lorentz's transformation). Newton's law of gravitation is an action at distance since the interaction between two masses is instantaneous, so that the motion of one of these masses has an immediate effect over the other. This is in contradiction with the predictions obtained from the SRT, which states that no influence or signal can propagate faster than light.

As previously discussed in Chapter 4, one of the basic postulates of SRT is that for all inertial reference frames, the laws of physics are the same. This theory does not consider the distortion of the space time due to the effect of the gravity.

At an atomic scale, gravity is one of the weakest interaction forces, but its effect is relevant in astrophysics where cosmological dimensions are involved. In addition, it must be considered that gravity has relevant effects on technological applications in our current society, such as the precision of a global positioning system (GPS), as will be discussed in this chapter.

Immediately after his work on SRT, Einstein began to think about gravity and how to give it a relativistically invariant formulation. Between 1915 and 1916, he presented his general relativity theory whose key insight is now known as the Einstein equivalence principle, the equivalence of gravitation and inertia. This ultimately led him to think that gravity is most adequately described and understood not as an external force, but as a manifestation of the geometry and curvature of the space time itself. The principle of equivalence determines how particles must move in a space time of given curvature, but it does not determine how this is itself curved by mass. That was a much more difficult limitation, which for Einstein took several more years to overcome.

The eventual solution was given by Einstein's equations, which he presented to the Prussian Academy of Sciences at the end of 1915.

The basic concepts on which the general theory of relativity is built can be summarized as follows:

A. Equivalence principle: In a bounded region of the space and during a finite time interval, there is no experiment capable of distinguishing between the effects of a gravitational field and the acceleration of a reference frame.
B. Gravity produces a curvature of the space time.
C. This curvature is due to massive objects, and considering the equivalence between mass and energy, any form of energy produces this curvature.
D. In presence of massive objects, the shorter distance between two points is not a straight line but a curve called geodesic.

In this chapter, we analyze the implications of these principles over the form as the notion of time is perceived by observers placed at different points with respect to a body producing a gravitational field.

5.2 EQUIVALENCE PRINCIPLE

The equivalence of gravitation and energy led Einstein to realize that gravity is best described not as a physical external force, but rather as a manifestation of the geometry and curvature of space time.[1] As discussed in Section 5.6 of this chapter, this equivalence can explain the increase in the wavelength of light (redshift) due to a gravitational field and the time differences between clocks placed at different distances from a massive body. These considerations make adequate to start this chapter by analyzing the equivalence principle.

There are two properties that must be considered to describe the concept of mass of an object. One is the *inertial mass*, which represents the way in which an object responds to a force, that is, a push or a pull. More precisely, inertial mass represents the way an object appears to resist change in its state of motion, that is, its resistance to being accelerated or decelerated.

In addition to inertial mass, it must be considered that the state of motion of any object is affected by a gravitational field, which is associated with the concept of *gravitational mass*.

Although the properties of inertial mass and gravitational mass are not associated with the same concept, they are identified under a single name, which is the ***mass*** of an object.

Yet, it turns out that these two kinds of mass are equal to one another: ***the inertial mass of an object is equal to its gravitational mass.*** As a result, when one writes down the expression for the acceleration of an object subject to the gravitational force, the mass of the object cancels out from the two sides of the equation ***Force= mass· acceleration*** and we obtain an expression for the acceleration that it is independent of the mass of the object considered, as shown below:

As known, Newton's second law relates the inertial mass m of an object to the acceleration a that it experiences due to a net external force F acting on it according to the expression $F = ma$.

In addition, according to Newton's law of gravitation, the gravitational force F_g felt by an object with mass m near to another body of mass M is given by $F_g = -\frac{G_N Mm}{r^2}$, where r denotes the distance between the masses and G_N is Newton's gravitational constant.

The negative sign in the above equation indicates that the force of gravity is always attractive. If the net external force to which the mass m is subjected is F_g, so that $F = F_g$, then we have that

$$ma = -\frac{G_N Mm}{r^2} \tag{5.1}$$

We can see that the term m present on both sides of the above equation cancels out so that the acceleration due to the gravitational field produced by the mass M is simply given by

$$a = -\frac{G_N M}{r^2} \tag{5.2}$$

If, for example, the mass M corresponds to the mass of the Earth, then this equation expresses that the acceleration acquired by an object of mass m, due to the Earth's gravity, is independent of m. Therefore, it follows that the action of gravity produces the same acceleration of all objects.

Einstein argued that the equivalence of inertial and gravitational mass was a deep feature of nature, which he elevated to the "Principle of Equivalence." He argued that a man falling to the ground from a certain height would not feel his own weight during the fall. That led Einstein to conceive the following experiment:

Let us consider a person who is in a spaceship in a region where there is no gravitational field, and that the engines of this ship are switched off. If the person takes out an object like a watch and releases it, this object will remain in the same position—the watch, the person, and everything in the spaceship will "float" with respect to any point inside the ship.

Now, let us assume that a person is inside an elevator car in a building and that the elevator cable has snapped. Such a person will also float with respect to the elevator walls.

Moreover, if he takes out a watch and lets it go, it will stay there—it will "float" relative to the person and the walls of the car. That occurs because the watch is falling as fast as the person in the elevator. So, during the fall of the elevator the person inside it will have the same feeling as if gravity has been "turned off." This is schematically shown in Figure 5.1. Einstein's argument was that no experiment carried out inside the elevator could determine whether the car is in space infinitely removed from the influence of any gravitational field or in free fall near the earth's surface.

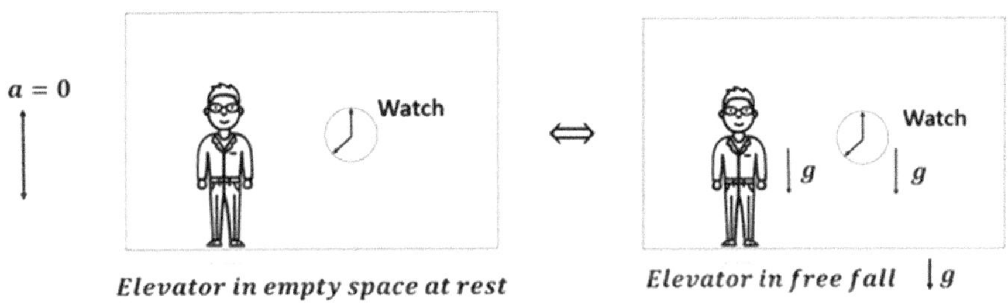

FIGURE 5.1 Principle of equivalence: The behavior of objects inside an elevator in free fall is indistinguishable from that in an identical elevator car far away from all gravitating bodies.

Einstein further extended his thought experiment in the following way. Suppose an elevator car out in empty space is subjected to an upward acceleration of value 9.8 m/s^2 (equal to g, the acceleration acquired by objects close to the Earth in free fall). The person in the elevator takes off his watch and drops it. He sees it fall to the floor of the elevator with an acceleration of 9.8 m/s^2. As seen from inside the car, the watch and the rest of the objects in the elevator will behave as if it were at rest on the surface of the earth (see Figure 5.2).

An observer, not shown in the figure and at rest with respect to the earth, looking at the situation in the left half of Figure 5.2 would see the watch momentarily floating in space and the elevator floor speeding up to meet it.

Einstein's conclusion was that "locally," that is, within the closed elevator car, an observer could not tell whether the elevator car was near the surface of the earth, or out in empty space far away from all gravitating bodies and accelerating upward at a rate of 1 g. Here, we assume that the car is "small" and falls for a "suitably short" period of time, so that apart from gravity, there is no other interaction or effect not considered in the situation here described.

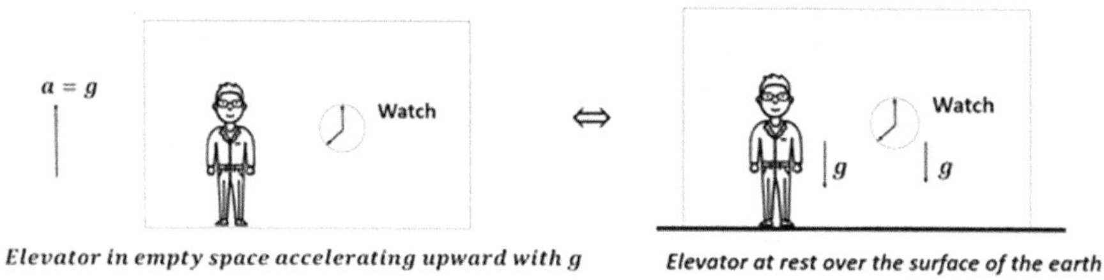

FIGURE 5.2 Principle of equivalence: Objects inside an elevator car accelerating upward (left side) behave the same way as inside a small elevator car on the surface of the Earth (right side).

5.3 GRAVITY AND LIGHT

Einstein considered the behavior of a beam of light inside the elevator car in each of the above two cases. Suppose a tiny horizontally mounted laser attached to the wall of the car emits a light beam. We describe the subsequent motion of the beam as seen by an observer inside the closed elevator car. On the left of Figure 5.3, the elevator car is not accelerated in empty space, so the beam simply travels in a horizontal line across the car. On the right, the elevator car is in free fall.

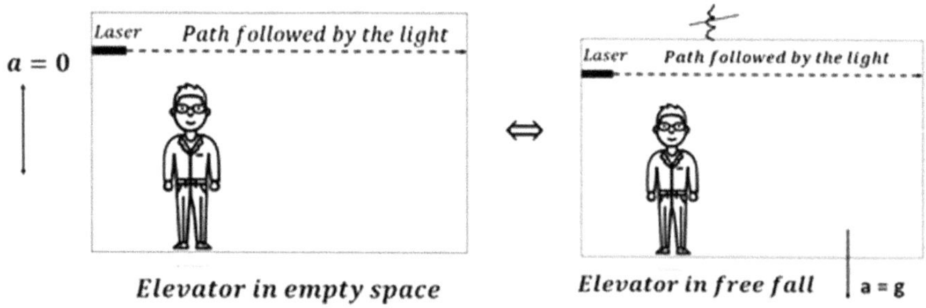

FIGURE 5.3 Principle of equivalence. The light ray moves in a straight line for observers at the left and at the right. On the left, the elevator is in empty space without acceleration. On the right, the elevator is freely falling.

Since an observer inside the car cannot tell whether he is in free fall or drifting in space, everything inside the car—including the light beam as well as the observer—falls at the same rate. So, again the observer will see the light beam move across the car in a horizontal line.

On the left of Figure 5.4, the elevator is shown accelerating upward with an acceleration equal to g. Because of the upward acceleration of the car, everything inside the car, including the light beam, appears to observers in the elevator moving downward with acceleration equal to g. As a result, the beam does not travel in a horizontal line relative to the car but appears to bend downward, striking the opposite wall at a point below the point of impact of the initial horizontal path. Einstein concluded that no experiment done inside the elevator can determine whether the car is being accelerated in empty space or sitting at rest on the surface of the earth. Therefore, in the second case (illustrated in Figure 5.4), the path of a beam of light must also bend in the earth's gravitational field.

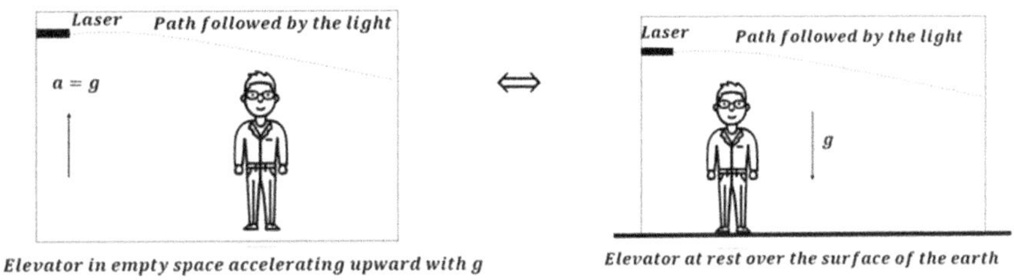

FIGURE 5.4 Principle of equivalence: According to an observer inside an elevator accelerating upward with g, the light ray emitted by a laser bends (figure at the left). The same effect is seen by an observer inside an elevator at rest over the surface of the earth (figure at the right). Both observers see that the light ray bends.

Hence, the principle of equivalence implies that "gravity bends light." On the left, the elevator is accelerating upward with g. On the right, the elevator is freely falling.

This effect appears as a curvature of the light's path due to the action of gravity. One could understand it considering that light is a form of energy, and energy has mass (according to the expression $E = mc^2$), and mass is affected by gravity. Therefore, one would be led to think that in this case light behaves as having mass that should be affected by gravity as well. However, this argument needs careful consideration, as it is not entirely correct as it stands if we consider that the reason why the light path bends when passing near objects with high mass is because the gravity warps the space time through which the photons travel.

Since the gravitational force, unlike the electromagnetic force, affects all objects equally Einstein reasoned that it might therefore be more appropriate to describe gravity in terms of the geometry of space and time rather than as a separate force acting in space and time. Furthermore, he argued that all reference frames, both inertial and non-inertial, should be equally adequate for describing the laws of physics. This idea was called by Einstein as the "general principle of covariance."

5.4 GEOMETRY OF SPACE TIME

As we have seen in Relativity Theory, events are specified by the set formed by the three spatial coordinates (x, y, z) and the corresponding time t of occurrence, which appears as a fourth dimension. Therefore, a point in this four-dimensional space may be considered to represent an "event' (t, x, y, z) in this space, which is called the space time.

In his general relativity or gravitational theory, einstein postulated that gravity can be described as curvature or warping of the geometry of space time. Without gravity and any other interacting force, particles and light move in straight lines but in presence of gravity they move along the closest analogs to straight lines, called geodesics.

In general relativity, we have the additional feature that in the presence of gravity, space time becomes curved. In places within our solar system, the curvature will in general be slight but, in some circumstances, for example, near a black hole, it may become highly curved.

The corresponding geometry is specified by means of a metric that allows to determine the distance between two points in space time. This metric is obtained from Einstein field equations relating the curvature of space time with the amount of energy and momentum possessed by the matter. As will be discussed in Section 5.9 of this chapter, this is expressed by the following expression equivalent to ten differential equations:

$$G_{\mu\nu} = 8\pi G_N T_{\mu\nu}/c^2 \tag{5.3}$$

where $G_{\mu\nu}$ is a tensor which describes the curvature of space time; $T_{\mu\nu}$ is a tensor describing the effect of the matter; G_N is the Newton constant describing the strength of the effect of the matter over the curvature of space time, and c is the speed with which light travels in a vacuum.

If there are no effects of gravity, events in the space time are adequately described by special relativity theory according to Minkowski metric as discussed in Chapter 4. Relevant milestones in the development of the general theory of relativity were Einstein's field equations, their solution known as the Schwarzschild solution, and their corresponding consequences that have been observed and measured.

The Schwarzschild solution makes possible to describe what space time is like in the vicinity of a static mass and with spherical symmetry, as the Sun can be considered in a first approximation, and observed astronomical effects such as the perihelion of planetary orbits (particularly the orbit of the planet Mercury and the deviation of the path of a beam of light due to a gravitational field).

A more recent verification of the predictions made by the general theory of relativity is the so-called delay in the echo of a signal sent from a planet (e.g., the earth) and reflected back toward the source.

A further verification of the general theory of relativity is the gravitational redshift of light, which can be deduced from the equivalence principle. The successful results that have been obtained from this theory by these verifications have established it as an adequate theory of gravity. A feature of the Schwarzschild solution is the concept of the "Schwarzschild" radius, which is a parameter characteristic for describing how objects move near a black hole.

In Figure 5.5, we show a spherical massive object of mass M. The position of a point in space time can be specified by spherical coordinates as $(\vec{r}, t)$ and according to the Schwarzschild solution of Einstein equations, the distance between two points in space time infinitesimally separated is given by

$$ds^2 = c^2 dt^2 \left(1 - \frac{2G_N M}{rc^2}\right) - \frac{dr^2}{1 - \frac{2G_N M}{rc^2}} - r^2 \left(d\theta^2 + sin^2\theta \, d\phi^2\right) \tag{5.4}$$

It must be noted that along this chapter, terms like dt^2, dr^2, and so on, must not be confused with second-order differential quantities. They are the square of infinitesimal intervals of time or spatial separation.

Equation 5.4 can be written in terms of the Schwarzschild radius $r_s = 2G_N M/c^2$, so that

$$ds^2 = c^2 dt^2 \left(1 - \frac{r_S}{r}\right) - \frac{dr^2}{1 - \frac{r_S}{r}} - r^2 \left(d\theta^2 + sin^2\theta \, d\phi^2\right) \tag{5.5}$$

In some texts about general relativity, the function $\sqrt{1 - \frac{r_S}{r}}$ is called the *lapse function* which in a colloquial way can be associated to the rhythm of time flow.[2] The gravitational dilation of time is inversely proportional to the lapse function as shown in Example 5.1.

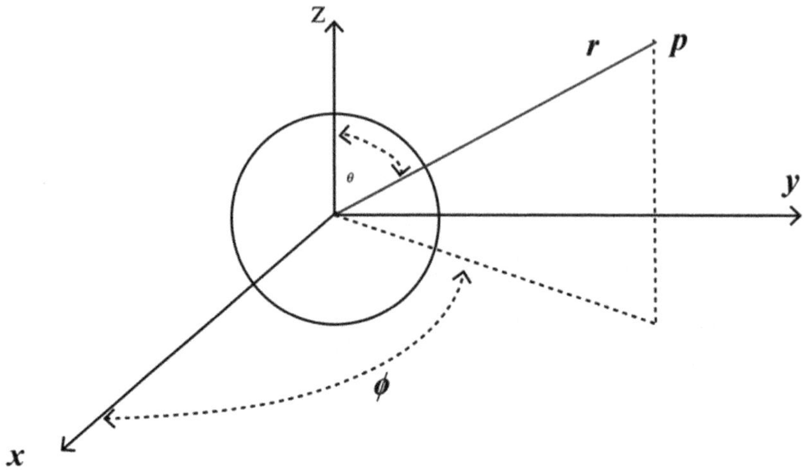

FIGURE 5.5 Position vector of a point P with respect to a spherical massive object of mass M which produces a gravitational field.

EXAMPLE 5.1

Consider a light source placed very far away from a massive body of mass M so that $r \to \infty$ in eq 5.5. This source emits periodically light pulses with a time interval $\Delta\tau$ between two successive pulses. Which is the corresponding period measured by an observer A at rest near the surface of the massive body?

SOLUTION

From eq 5.5 and assuming that the positions of the source and the observer A are radially separated as shown in Figure 5.6, we can write the following relation between his proper time interval $\Delta\tau$ and a coordinate time interval Δt (interval of time measured by a clock at a laboratory placed at a position where the gravitational potential is zero, $r \to \infty$ in this case):

$$\Delta s = c\Delta\tau = \sqrt{1-\frac{r_S}{R}}\; c\Delta t,$$

where R is the distance from the point A to the center of the gravitational body.

Therefore, we obtain that for $R > r_s$, the time interval between the events measured by the observer A is lower than that measured by the distant clock.

This effect is referred to as gravitational time dilation, and it is interpreted as that due to gravity the hands of a clock move slower. For $R = r_s$, $\Delta \rightarrow 0$ and time seems as frozen for the observer A. For $R < r_s$, this observer would measure an imaginary time interval, an effect that will be examined in Section 5.10 in the context of black holes.

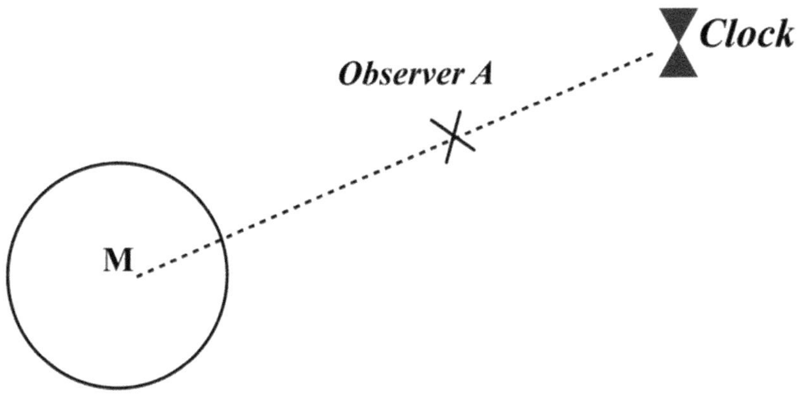

FIGURE 5.6 Time dilation due to the gravitational field produced by a body of mass M. An observer A measures a smaller time interval than a clock very far away from the body.

5.5 PLAUSIBILITY ARGUMENTS FOR SCHWARZSCHILD METRIC

A rigorous deduction of the Schwarzschild metric requires the full mathematical formalism of general relativity, considering that the curvature or warping of the geometry of space time due to gravity must be formally taken into account.

In a way similar to that of Refs. [3, 4], we give here plausibility arguments for understanding this metric using concepts of classical mechanics, such as Newton's law of gravitation and SRT (specifically, time dilation and contraction of lengths).

As we have previously seen, the metric of space time in SRT corresponds to the Minkowski metric. According to this metric, if we consider two events separated spatially by an infinitesimal distance and temporally by an infinitesimal interval dt, in space time (which is a four-dimensional space), the interval between these two events is given by the following expression corresponding to a Lorentz invariant:

$$ds^2 = c^2dt^2 - dx^2 - dy^2 - dz^2 \tag{5.6}$$

Using curvilinear spherical coordinates, this equation can be recast as

$$ds^2 = c^2dt^2 - dr^2 - r^2 - (d\theta^2 + sin^2\, d\phi^2) \tag{5.7}$$

Since gravity curves the space time, this metric is not valid in the presence of massive objects except for regions very far away from these bodies. In 1916, the German physicist Karl Schwarzschild gave the first and still the most important exact solution of the Einstein vacuum field equations. This solution represents the field outside a spherically symmetric mass in otherwise empty space. Schwarzschild's solution allows the exact calculation of several of the post-Newtonian' effects of general relativity, such as the precession of planetary orbits, the bending of light around the sun, and the exact gravitational frequency shift, among others.

Let us consider a local reference frame system *SA*, having a clock and a length measurement device (ruler), both at rest and subject to a radial acceleration $-\vec{g}$ toward a spherical massive object of mass *M*. According to the Einstein's equivalence principle, the influence of gravitation on phenomena observed from a local reference frame that is at rest in the gravitational field is equivalent to the influence produced by the accelerated motion of another local reference frame, denoted as *SB*, in which phenomena are described in the absence of gravitation, moving away from *M* with radial acceleration $\vec{g}$. We also consider a clock and a ruler, both identical to those of *SA* and fixed to the reference frame *SB*. It is assumed that *SB* is in a gravitation-free region, so that the metric given by eqs 5.6 or 5.7 is also valid.

Let us suppose that the clock in *SA* has a downward speed υ_A with respect to *SB* at the instant when the two reference frames coincide. According to special relativity, the relation between time intervals measured by clocks at *SA* and *SB* is

$$dt_A = dt_B\sqrt{1-\frac{\upsilon_A^2}{c^2}} \tag{5.8}$$

Similarly, the lengths measurements in the radial direction made by rulers in *SA* and *SB* are related by

$$dr_A = dr_B \,/\, \sqrt{1-\frac{\upsilon_A^2}{c^2}} \tag{5.9}$$

Therefore, for system *SA*, the interval between two events is

$$ds_A^2 = c^2dt_A^2 - dr_A^2 - r^2\left(d\theta_A^2 + sin^2\theta\, d\Delta\phi^2\right)$$

$$= c^2dt_B^2\left(1-\frac{\upsilon_A^2}{c^2}\right) - \frac{dr_B^2}{1-\frac{\upsilon_A^2}{c^2}} - r^2\left(d\theta_A^2 + sin^2\theta\, d\phi^2\right) \tag{5.10}$$

We will assume that the magnitude of υ_A is equal to the speed that an object must possess to escape the action of a gravitational field if it is initially at a distance *r* from the center of a massive object (considered as a spherical body). This assumption is based on the form that the gravitational potential appears in the first two terms at the right of eq 5.4 given by $2G_N M/r$. This last quantity is proportional to the energy that must be supplied to a mass *m* by an external agent, so that it could be able to escape from the attractive gravitational field produced by the body of mass *M*. Under this assumption and considering conservation of energy, we have

$$\frac{1}{2}m\upsilon_A^2 + \left(-\frac{G_N Mm}{r}\right) = 0 \Rightarrow \upsilon_A^2 = 2G_N M/r \tag{5.11}$$

Combining this last expression with eq 5.8 and making $dt_B = dt$ and $dr_B = dr$, we obtain a relation identical to that corresponding to the Schwarzschild metric:

$$ds^2 = c^2 dt^2\left(1 - \frac{2G_N M}{rc^2}\right) - \frac{dr^2}{1 - \frac{2G_N M}{rc^2}} - r^2\left(d\theta^2 + sin^2\theta\, d\phi^2\right) \tag{5.12}$$

Of course, this argument is not intended to be a deduction but only an attempt to better understand the structure of the Schwarzschild metric.

5.6 GRAVITY AND TIME

In a paper published in 1911, Einstein analyzed how the light velocity is affected by a gravitational field.[5] Using the equivalence principle and classical laws of motion, he concluded that two clocks placed at different positions in a gravitational field do not give equal results for the respective intervals of time they are measuring.

This result can be obtained using arguments based on SRT and also directly from Schwarzschild metric as follows.

Gravitational Time Dilation from SRT

This can be analyzed considering the Doppler effect in a freely falling system finding that the frequency of an atom (clock) increases with its height with respect to the surface of the Earth. In what follows, we explain this fact.

In Figure 5.7, we show an atom denoted as *S1* at rest near the Earth surface (according to a reference frame *K*), which emits a photon upward so that it is absorbed by a second atom denoted as *S2* and also at rest, as also seen from reference frame *K* . The emitted photon corresponds to radiation of period T_1. Both atoms are identical and *S2* is at a height *H* with respect to *S1*. If we consider that according to the principle of equivalence formulated by Einstein,[5] there is a local equivalence between the behavior of a physical system in a gravitational field and in a reference frame with accelerated motion, we can look at this process of emission-absorption from an elevator freely falling (reference frame *K'*).

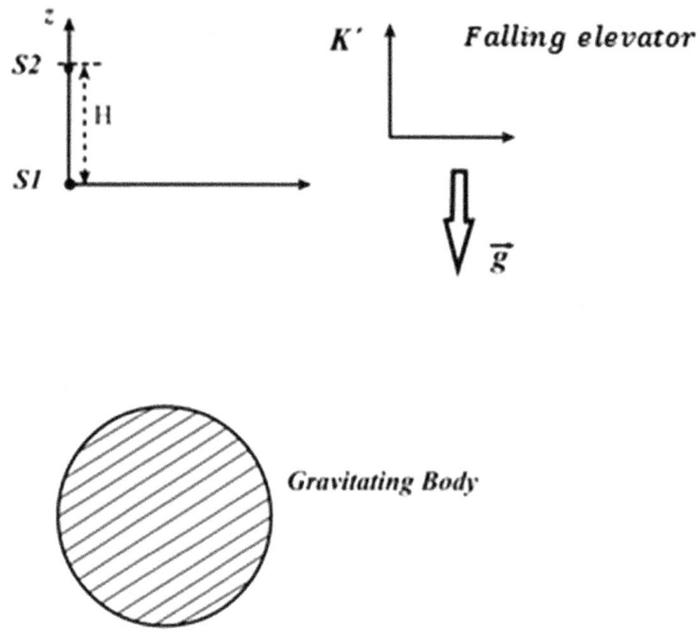

FIGURE 5.7 Clocks S1 and S2 are subjected to a homogeneous gravitational field with gravity acceleration *g*. The clock S1 is programmed to emit light pulses with period T_1 toward the clock S2 which measures the period T_2 of the received pulses. The same events are seen from a free falling elevator (system *K'*).

We will assume that at the moment of emission of the photon by *S1* ($t = 0$) the velocity of the elevator is zero. Then, at the instant $t = H/c$, when the photon reaches the upper atom, the velocity of the latter with respect to the elevator is $\upsilon = gH/c$ and directed upward, so that the atom "runs away" from the photon. Therefore, for an observer at the elevator, there is a Doppler effect since the situation is equivalent to a source moving away, so that according to eq 4.55 of Chapter 4, the period detected by clock S2 is given by

$$T_2 = T_1 \left\{ \frac{1+\beta}{1-\beta} \right\}^{1/2} \tag{5.13}$$

where $\beta = \frac{\upsilon}{c} = gH/c^2$ and $T_1 = H/c$ since this is the time interval between two consecutives photons emitted by S1, according to an observer inside the elevator (S2 would receive the pulses at the same rate as S1 emits them). We can consider that $\beta \ll 1$, so that from eq 5.13 we obtain that

$$T_2 \approx T_1 \left(1 + \frac{gH}{c^2} \right) \tag{5.14}$$

If we express this last result in terms of frequency, we obtain that the photon reddens since

$$\frac{\Delta \nu}{\nu_1} = \frac{\nu_2 - \nu_1}{\nu_1} \approx -\frac{gH}{c^2} \tag{5.15}$$

From eqs 5.13 to 5.14, it is concluded that the clock closer to the body, which produces the gravitational field, runs slower than the more distant clock, so it's like *"gravity slows down time."*

Gravitational Time Dilation According to the Schwarzschild Metric

The same previous conclusion can be obtained from the Schwarzschild solution to Einstein's equations as follows:

Let us consider two twins, John and Patrick, both infinitely distant from a massive body of mass M as shown in Figure 5.6, so that they do not feel the gravitation due to this mass. Patrick begins to move radially toward this body and when is at a distance r_p from the body, he and his brother measures the time interval between two events.

Based on the Schwarzschild metric specified by eq 5.4, with $dr = d\theta = d\phi = 0$ we have that the relation between the interval of proper time $d\tau = ds/c$ and the interval of coordinate time dt (time interval measured by a clock at the laboratory, where the gravitational potential is taken as zero) is given by

$$d\tau = \left(1 - \frac{2G_N M}{c^2 r} \right)^{1/2} dt \tag{5.16}$$

This last equation indicates that the clock runs slower in the laboratory, which is deeper in the gravitational potential (closer to the body of mass M). Therefore, when Patrick is at rest at a distance r_p from the mass M, his proper time interval $d\tau_p$ is given by

$$d\tau_P = \left(1 - \frac{2G_N M}{c^2 r_P}\right)^{1/2} dt \qquad (5.17)$$

Since John keeps infinitely distant from the mass M, his proper time interval $d\tau_J$ is simply given by $d\tau_J = cdt$.

For a static gravitational field, the integration of these last two relations give the following relation between the measured time intervals of both observers:

$$\Delta t_P = \Delta t_J \sqrt{1 - \frac{2G_N M}{c^2 r_P}} \qquad (5.18)$$

Therefore, for $r > r_s = 2G_N M/c^2$, the time interval between the considered events measured by Patrick is lower than the corresponding value measured by John, confirming our previous conclusion that it seems that gravity "slows down" time.

Is There Some Relation between the Gravitational Red-shift of the Frequency and the Gravitation of Energy?

According to a paper of Einstein published in 1911,[5] the answer is ***Yes***. Here, we give arguments for justifying this answer considering that as expressed by Einstein in this paper ***"energy must therefore possess a gravitational mass which is equal to its inertial mass" (extracted verbatim from Ref. [5]).*** For that purpose, we consider Figure 5.8 where a clock emits radiation of energy E_2 toward an observer placed at point A.

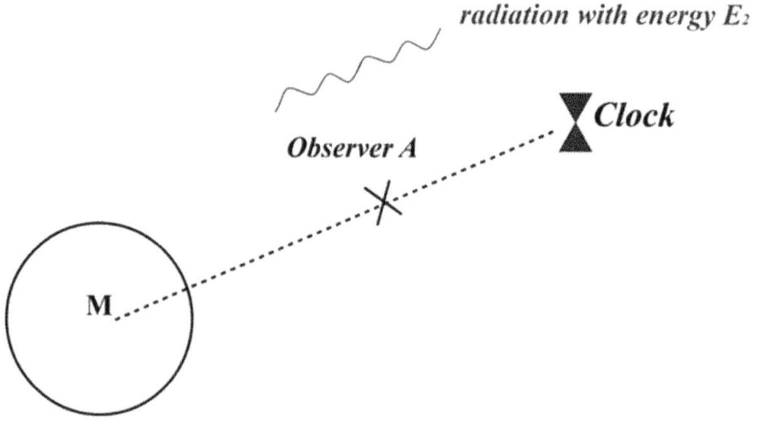

FIGURE 5.8 Clock emitting radiation with energy E_2 toward the observer placed at point A, in presence of a gravitational field produced by the mass M.

Let r_C and r_A be the respective distances from the clock and observer A to the center of the body of mass M. If we consider that this energy has an associate gravitational mass $m_{rad} = E^2/c^2$, as postulated by Einstein,[5] and if there is not loss of energy we can write the following equation relating the total energies at the points where the clock and observer A are placed:

$$E_2 - \frac{G_N M m_{rad}}{r_C} = E_1 - \frac{G_N M m_{rad}}{r_A} \tag{5.19}$$

where E_1 is the energy received by the observer at point A of Figure 5.8, which can be expressed as

$$\begin{aligned} E_1 &= E_2 + G_N M m_{rad}\left(\frac{1}{r_A} - \frac{1}{r_C}\right) = E_2 + G_N M m_{rad}\left(\frac{r_C - r_A}{r_A r_C}\right) \\ &\approx E_2 + m_{rad} M\left(\frac{G_N H}{r^2}\right) H = E_2 + m_{rad}\, gH \end{aligned} \tag{5.20}$$

In this last equation, we have used the approximation $1/r_A r_C \approx 1/r^2$ at great distances from the surface of the Earth, with $g = G_N M/r^2$ and with $H = r_C - r_A$ being the difference of distances between the positions of observer A and that of the clock.

Let us consider that the radiation emitted by the clock corresponds to a monochromatic beam of light of frequency energy ν_2. In terms of photons, each emitted photon by the clock has an energy $h\nu_2$ and a gravitational mass $m_{rad} = h\nu_2/c^2$ whereas each photon received at the position of the detector has an energy $h\nu_1$ and an associated gravitational mass $m_1 = h\nu_1/c^2$. Therefore, we can interpret eq 5.20 as follows: In the passage from the clock to the observer A, each emitted photon losses a potential energy $m_{rad}\, gH = \frac{h\nu_2}{c^2} gH$.

From eq 5.20, we obtain that the frequency of the radiation received by the detector at point A has a value greater than that corresponding to the radiation emitted by the clock, according to the following relation

$$\nu_1 = \nu_2\left(1 + \frac{gH}{c^2}\right) \tag{5.21}$$

A similar relation was deduced by Einstein in Ref. [5], which he used to predict that the spectral lines corresponding to light emitted by the sun, compared to the spectral lines corresponding to light sources on earth, are displaced toward the red (gravitational red shift). However, it is essential to consider the arguments given by Einstein in this quoted paper and expressed verbatim as follows:

"We cannot regard v_2 or respectively v_1 simply as frequencies (as the number of periods per second) since we have not determined a time in system K. What v_2 denotes is the number of periods per second with reference to the time-unit of the clock U at S2, while v_1 denotes the number of periods per second with reference to the identical clock at S1." This implies that if we make time measurements in the reference frame S1 with a clock denoted as U, then it is necessary measure the time in the reference frame S2 with a clock that advances $1 + gH/c^2$ times slower than the clock U if compared simultaneously with that clock at the same location.

It is worth mentioning that the effect of gravitational potential differences on electromagnetic radiation frequency measurements was found by Pound and Rebka to agree with Einstein's predictions based on his principle of equivalence. They made measurements for the apparent frequency of γ rays getting an adequate concordance between the experimental results and the theoretical predictions.[6]

5.7 EFFECTS OF GRAVITATIONAL FIELD DUE TO EARTH ON TIME MEASUREMENTS

In Section 5.6, we have seen that according to the general relativity theory, measurements of time intervals made at different positions in a gravitational field give different values.

In this section, we discuss in more detail this effect for the case of measurements of time intervals under the influence of the gravitational field produced by the Earth and analyze the corrections that should be considered.

Let us first consider a light source fixed with respect to some observer which emits periodic pulses, with duration very low in comparison with its period. Using the Schwarzschild metric given by eq 5.5 and considering that the distances from the emitting source and the detector to the center of the gravitational body of mass M are much greater than its Schwarzschild radius r_S, we obtain the following relation between the period T_2 of the signal emitted from a source placed at $P_2 = (R_2, \theta_2, \phi_2)$ and the corresponding period T_1 measured at $P_1 = (R_1, \theta_1, \phi_1 = \theta_2, \phi_1, \phi_2)$:

$$\frac{T_1}{T_2} = \frac{\sqrt{1-\dfrac{r_S}{R_1}}}{\sqrt{1-\dfrac{r_S}{R_2}}} \approx \frac{1-\dfrac{r_S}{2R_1}}{1-\dfrac{r_S}{2R_2}} \approx \left(1-\frac{r_S}{2R_1}\right)\left(1+\frac{r_S}{2R_2}\right) \approx 1-\frac{r_S}{2}\left(\frac{1}{R_1}-\frac{1}{R_2}\right) \tag{5.22}$$

If the point P_2 is more distant from the gravitational body than the point P_1, so that $R_2 > R_1$, the period T_1 of the signal measured in this last point is lower than that of the signal emitted.

If we consider that the point P_1 is over the surface of the Earth while P_2 is at an altitude h over this surface, eq 5.22 leads to the following difference between these periods:

$$\frac{T_1 - T_2}{T_2} \approx -\frac{r_S}{2}\frac{h}{R^2} \tag{5.23}$$

where R represents the radius of the Earth.

EXAMPLE 5.2

Consider that for the case corresponding to eq 5.23, the emitting source is at an altitude of 10 km with respect to the surface of the earth. Evaluate the difference between the periods measured by an observer at the earth and the corresponding values measured at the point of emission.

SOLUTION

Considering that the Schwarzschild radius of the Earth is approximately equal to 9 mm with a radius R of about 6400 km, from eq 5.23, we obtain that this difference is given by a fraction of the order of 10^{-12} with respect to the period of the emitted signal.

Differences between Time Measurements of a Clock in Orbit and Another Fixed to the Earth

We will analyze the difference between proper times of two identical clocks A and B which are initially synchronized when they are in contact and at rest on the Earth's surface. Clock A is kept over the Earth's surface and clock B travels inside an aircraft making a circular orbit at a height h with respect to the Earth's surface and with relative speed υ relative to the ground (see Figure 5.9). For making this analysis, we will use the Schwarzschild metric given by eq 5.10 considering the effect of gravity on the measurements of time for two clocks placed at different spatial points in the gravitational field produced by the Earth. We will consider that with respect to an inertial frame the Earth and its atmosphere rotate with angular velocity Ω (of the order of $7\cdot10^{-5}$ rad/s), which implies that the tangential velocity of the clock A is ΩR_{earth}, whereas, if we neglect relativistic correction to the velocity addition formula, the speed of the clock B with respect to the same inertial system is $\upsilon + (R_{earth} + h)\Omega$.[7]. In this last expression, the sign of υ is positive if B rotates in the same sense as the earth rotates and negative on the contrary.

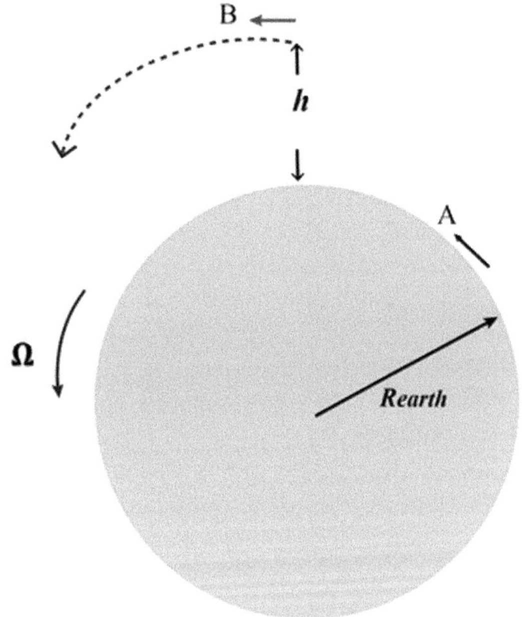

FIGURE 5.9 Airplane describing a circular orbit at an altitude h from the Earth's surface and with relative velocity υ.

After completing the journey, time measurements of clocks A and B are compared. That means a comparison between proper times τ_A and τ_B and for this purpose the Schwarzschild metric can be used. For clock A, its proper time is determined from eq 5.10 making $r = R_{earth}$, $\Delta r = 0$, $\Delta\theta = 0$, and assuming $\theta = \pi/2$, we get

$$\tau_A = \int dt\sqrt{1-\frac{2G_N M}{c^2 R_{\text{earth}}}-\frac{R_{\text{earth}}^{\;2}}{c^2}\left(\frac{d\phi}{dt}\right)^2} = \Delta t\sqrt{1-\frac{2G_N M}{c^2 R_{\text{earth}}}-\frac{R_{\text{earth}}^{\;2}}{c^2}\Omega^2} \qquad (5.24)$$

Similarly, for clock B, its proper time is given by

$$\tau_B = \int dt \sqrt{1 - \frac{2G_N M}{c^2 (R_{\text{earth}+h})} - \frac{(R_{\text{earth}} + h)^2}{c^2}\left(\frac{d\phi}{dt}\right)^2}$$

$$= \Delta t \sqrt{1 - \frac{2G_N M}{c^2 (R_{\text{earth}} + h)} - \frac{[(R_{\text{earth}} + h)\Omega + \upsilon]^2}{c^2}} \tag{5.25}$$

In eq 5.25, it has been considered that $(R_{\text{earth}} + h)\frac{d\phi}{dt} \approx \upsilon + (R_{\text{earth}} + h)\Omega$ as previously explained.

The mismatch factor between the proper times measured by clocks A and B is defined as

$$\delta = \frac{\tau_B - \tau_A}{\tau_A} \tag{5.26}$$

Taking into account that $\frac{G_N M}{R_{\text{earth}} c^2} = \frac{g}{c^2} \ll 1, \frac{h}{R_{\text{earth}}} \ll 1, \frac{\upsilon^2}{c^2} \ll 1$ and $\frac{h\Omega}{\upsilon} \ll 1$ this factor is given approximately by

$$\delta \approx \frac{gh}{c^2} - (2R_{\text{earth}}\Omega + \upsilon)\upsilon / 2c^2 \tag{5.27}$$

Considering typical values of parameters for aircraft flights, if we assume $h = 10$ [*km*] and $\upsilon = 300$ [*m/s*], we obtain $\delta(\upsilon < 0) = 2\cdot10^{-12}$ and $\delta(\upsilon > 0) = 2\cdot10^{-12}$. Taking into account that the precision of atomic clocks is $\sim10^{-13}$, this mismatch cannot be neglected.

How can the effects of gravity and time dilation be combined?

We have seen in Chapter 4 that the lapse of time between the occurrence of two events is not the same for two inertial observers with relative motion. Consider two events which according to an inertial system K occur within a time interval dt. For another observer fixed to a coordinate system K' and moving with velocity $\vec{\upsilon}$ with respect to K, the interval $d\tau$ between the same two events is

$$d\tau = dt\sqrt{1 - \beta^2} \tag{5.28}$$

where $\beta = \upsilon/c$.

This last equation describes time dilation due to relativistic velocity effect discussed in Chapter 4 and indicates that if a time interval is measured by an observer 2 moving relative to an observer 1, the corresponding value is lower for observer 2.

Let us consider now a clock 1 describing a circular orbit of radius r around the Earth and a clock 2 at rest and very far away from the Earth, so that it is free from its gravitational field. From the Schwarzschild's metric and with $dr = 0$, we can obtain the relation between the proper time interval measured by the clock 1 ($d\tau_1$) with respect to the proper time interval measured by the clock 2, as follows:

$$c^2 d\tau_1{}^2 = c^2 dt^2 \left(1 - \frac{2G_N M}{rc^2}\right) - r^2 \left(d\theta^2 + sin^2\theta \, d\phi^2\right)$$

$$= c^2 dt^2 \left\{1 - \frac{2G_N M}{rc^2} - \frac{\vec{\upsilon}_{tang}{}^2}{c^2}\right\} \tag{5.29}$$

where $\vec{\upsilon}_{tang}$ is the tangential components of the velocity of the clock 1, measured with respect to the observer 2 and dt is the interval of coordinate time. If the clock 2 is infinitely distant from the Earth, we have ($dt = d\tau_2$).

Therefore, we can define the time dilation factor for this case as:

$$\gamma = \frac{d\tau_1}{d\tau_2} = \sqrt{1 - \frac{2G_N M}{rc^2} - \frac{\vec{\upsilon}_{tang}{}^2}{c^2}} \tag{5.30}$$

From this last equation, we can appreciate that if the relative speed between the clocks is very small with respect to the speed at which light travels, the lapse of time between two events measured by the clock subjected to a gravitational potential is lower than that corresponding to a very distant clock not affected by this potential. This means that in this case the gravitational time dilation predominates with respect to the relativistic time dilation by speed.

5.8 APPLICATION OF GENERAL RELATIVITY THEORY TO GLOBAL POSITIONING SYSTEM

5.8.1 WHAT IS A GLOBAL POSITIONING SYSTEM?

According to Ref. [8], this system has 24 satellites describing orbits corresponding to an altitude of about 2000 [km] with respect to the Earth's surface and with a rotation period of 11 h and 58 min. These satellites are distributed in six orbital planes with equal angular separation between them. Each satellite carries an operating atomic clock (and several backup clocks) emitting timed signals to which a code is assigned which gives information about its location. By analyzing signals from at least four of these satellites, a receiver on the surface of Earth can display with a built-in microprocessor the location of the receiver (latitude and altitude).

The principle of operation of a global positioning system (GPS) is based on one of the postulates of the special theory of relativity which establishes that for all inertial frames the light velocity is the same. With reference to Figure 5.10, let us consider a receptor located nearby to the Earth's surface receiving simultaneously pulses emitted at instants t_i from four satellites located at known positions determined by position vectors $\vec{r}_i$ $(i=1,2,3,4)$. The knowledge of the emission instants t_i, allows to obtain the spatial coordinates (x, y, z) corresponding to the vector position $\vec{r}$ of the receptor at any instant t solving the set of four equations

$$|\vec{r}-\vec{r}_i| = c(t-t_i) \tag{5.31}$$

This is a non-linear system and according to Ref. [9] the explicit equations that must be solved by an iterative algorithm are:

$$(x-x_i)^2+(y-y_i)^2+(z-z_i)^2 = c^2(t-t_i)^2, i=1,2,3,4 \tag{5.32}$$

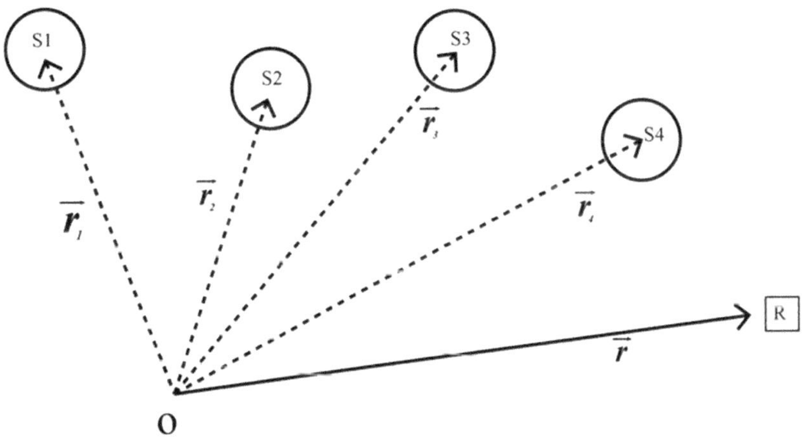

FIGURE 5.10 The receptor R receives signals from transmitters placed at each one of the four satellites. The reference frame has its origin at the center of mass of the planet (ECI).

Even though the relativity theory allows to choose the coordinate system, the easiest approach is to use an approached solution of field equations in order to determine the corrections to Minkowski metric of special relativity that must be included to take into account the effect of the mass of the Earth. Therefore, a coordinate system whose origin is the center of mass of the Earth, denoted as ECI, is chosen.

Synchronization of clocks is fundamental for technological applications such as communication systems, navigation, and engineering. For example, in the case of electrical power systems, the technology based on synchro phasors is a widely used system for monitoring the operation of a transmission network, transmitting to a control center the instantaneous values of the phase angles of the voltages at each bus at the same instant of time as required.[10] If the corresponding clocks associated to this systems were not synchronized for a high-voltage transmission network of length 1000 km, considering a frequency of 50 Hz we would have a phase difference of up to the order of 1° between the wave voltages at different points of the electrical network, what could be relevant from the point of view of operational information.

In most of cases, synchronization must be made between clocks in relative motion and placed at points with different gravitational potential, as is the cases of satellites at different altitudes. Therefore, for these cases, corrections due to gravitational effects must be considered. In particular for the synchronization of clocks of satellites in the ECI reference system, correction factors are applied considering relativistic effects for errors in the frequency originated by differences of gravitational potential and relativistic time dilation.[11]

5.8.2 RELATIVISTIC EFFECTS AFFECTING THE ACCURACY OF GPS

The GPS allows both a global position and time determination, by means of atomic clocks. These clocks are subjected to relativistic effects, mainly due to time dilation and gravitational frequency shifts, which if they are not considered would lead to timing and positioning errors in the accuracy of a GPS that cannot be ignored. Here is an analysis of these effects.

First. Gravitational Blue Shift Effect

Clocks advance faster as they move away from the center of gravitational attraction, which in the case of a global positioning system is the center of the Earth. This implies a blue shift of the frequency of the signal sent toward the earth, due to gravitational effects that depend on the distance to the center of the earth.

Considering that clocks at the satellites are subject to a gravitational potential different from clocks at the Earth, atomic clocks at satellites seem to oscillate faster. In fact, according to eq 5.22, the period of oscillation of the atomic clocks at one of the satellites of a GPS is higher than the corresponding value measured at Earth. In terms of frequency of oscillations, we can see that as observed on the Earth, the difference in gravitational potential causes the oscillators of GPS atomic clocks to run faster, with a fractional deviation given by

$$\frac{\Delta\nu}{\nu} = \frac{\Phi_1 - \Phi_2}{c^2} \tag{5.33}$$

in which Φ_1 and Φ_2 represent the gravitational potentials at the site of the satellite and at the surface of the earth, respectively. For GPS, the fractional deviation is $\Delta\nu/\nu \approx 5.3 \cdot 10^{-10}$. In eq 5.33:

$$\Phi_1 - \Phi_2 = -\frac{GM_E}{r_{SE}} - \left(-\frac{GM_E}{r_E}\right) = \frac{GM_E}{r_E} - \frac{GM_E}{r_{SE}} \tag{5.34}$$

where M_E is the mass of the earth and r_E its radius, while r_{SE} is the distance between the satellite and the center of the Earth. Since $r_E < r_{SE}$, $\Delta\nu/\nu$ is positive meaning that the frequency ν_1 measured at the Earth is higher than ν_2 (the frequency measured at the satellite).

EXAMPLE 5.3

Calculate the error in the measurement of time in a GPS during a day.

SOLUTION

In terms of differences in measurements of time, from the relation between frequency and period $T = \frac{1}{\nu}$, we have

$\Delta T = \frac{dT}{d\nu}\Delta\nu = -\frac{1}{\nu_o^2}\Delta\nu$, where $\Delta\nu$ is the deviation of the frequency with respect to a reference value ν_o.

The above expression can be rewritten as

$$\Delta T = -\frac{1}{\nu_o}\frac{\Delta\nu}{\nu_o} = -T_o\frac{\Delta\nu}{\nu_o}$$

Therefore, the fractional deviation between the period measured at the satellite of a GPS and the period measured at the Earth is

$$\frac{\Delta T}{T_o} = -\frac{\Delta \nu}{\nu_o} = -5.3 \cdot 10^{-10}$$

According to this result, the period $\Delta T < 0$ meaning that the clock in the satellite runs faster compared with a clock on the Earth's surface. For a day, this implies a difference of

$$24 \cdot 3600 \cdot 5.3 \cdot 10^{-10} = 45.6\ \mu s/day$$

Second. Time Dilation According to the Special Theory of Relativit

The motion of both the satellites and the receptors located on the surface of the earth should be considered. Since the clocks located on the satellites advance faster than the clocks on the Earth's surface this gives rise to the special relativistic Doppler effect of order υ_s^2/c^2 in the satellite speed due to the relativistic effect of time dilation as seen in Chapter 4. Consequently, this effect produces a red shift for a signal received at the Earth from the satellites.

Furthermore, the effect of the gravitational field affects the measurement of time, so that a clock closer to a massive body as the Earth goes slower (it measures a smaller interval of time between two events) than another clock more distant.

Considering eq 5.30, the effects produced by relativistic time dilation and gravitational potential can be combined according to the following expression

$$\frac{\Delta\tau_S}{\Delta\tau_E} = \frac{\sqrt{1 - \dfrac{2G_N M}{c^2 r_{SE}} - \dfrac{\upsilon_S^{\,2}}{c^2}}}{\sqrt{1 - \dfrac{2G_N M}{c^2 r_E} - \dfrac{\upsilon_E^{\,2}}{c^2}}} \approx 1 + \frac{G_N M}{c^2}\left(\frac{1}{r_E} - \frac{1}{r_{SE}}\right) - \frac{1}{2}\frac{\upsilon_S^{\,2} - \upsilon_E^{\,2}}{c^2} \tag{5.35}$$

where r_E and r_{SE} are the radial coordinates of the clock near to the surface of the Earth and the clock at the satellite, respectively, with υ_E and υ_S the corresponding tangential velocities.

If the time dilation effect due to velocity can be neglected, the last term at the right side of this equation tends to be zero, and considering that the second term is positive since $r_E < r_{SE}$, we confirm that due to gravitation, the clock closer to the body, which produces the gravitational field, runs slower than a more distant clock. On the other hand, if the term containing velocities in this equation cannot be neglected and it is positive, as expected, the time dilation effect due to velocity compete with the gravitational time dilation.

THIRD. SAGNAC EFFECT

The rotation of the earth must be considered for the following reasons: Firstly, because of the kinematic Sagnac effect which is of order c^{-2} and gives rise to a maximum error that has an order of 207.4 ns/d in the GPS. Secondly, the effect of the gravitomagnetic field (GM) generated by the proper rotation of the Sun, which has influence on the orbit of the earth, and which in turns has its proper rotation. The effect of this GM field is of the order 0 (c^{-3}), that is, picoseconds, and hence can be considered negligible for the accuracy presently required.

During signal propagation from transmitter to receiver, the latter is most likely in motion. Even if the receiver remains at rest on the earth's surface, the rotation of the Earth will carry the receiver into a different position while the signal propagates to the receiver. From Ref. [12], in the rotating frame one may account for this to sufficient accuracy by writing the navigation equations as

$$t = t_j + \frac{\left|\vec{r}(t) - \vec{r}_j\right|}{c} = t_j + \frac{\left|\vec{r}(t_j) + \vec{\upsilon}(t - t_j) - \vec{r}_j\right|}{c} \tag{5.36}$$

where $\vec{r}(t)$ is the receiver position at time t and $\vec{\upsilon}$ is its velocity of the receiver at the instant t_j where the transmission event begins from a point placed at position $\vec{r}_j$. Since receiver velocity is very small with respect to c, the term containing the velocity at the right of eq 5.36 can be neglected and the equation can be solved by iteration. We define the distance from transmitter to receiver at the transmission time as

$$\vec{R} = \vec{r}(t_j) - \vec{r}_j \tag{5.37}$$

With the above-mentioned approximation, eq 5.36 can be rewritten as

$$t = t_j + \frac{\left|\vec{r}(t_j) - \vec{r}_j\right|}{c} = t_j + \frac{R}{c} \tag{5.38}$$

Combining these equations and expanding to first order in $\vec{\upsilon}$, we get

$$t = t_j + \frac{\left|\vec{R} + \vec{\upsilon}(t - t_j)\right|}{c} = t_j + \frac{\sqrt{R^2 + 2\vec{R}\cdot\vec{\upsilon}(t - t_j)}}{c} \approx t_j + \frac{R}{c} + \frac{\vec{R}\cdot\vec{\upsilon}}{c^2} \tag{5.39}$$

If the speed with which the receiver moves is due only to the Earth rotation (angular velocity $\vec{\omega}$), then

$$\vec{\upsilon} = \vec{\omega} \times \vec{r}(t_j) \tag{5.40}$$

Therefore, the Sagnac correction term is given by

$$\Delta t_{\text{Sagnac}} = \frac{\vec{R}\cdot\vec{\upsilon}}{c^2} = \frac{\vec{R}\cdot\{\vec{\omega}\times\vec{r}(t_j)\}}{c^2} \tag{5.41}$$

5.9 AN OVERVIEW OF EINSTEIN'S FIELD EQUATIONS

From the general theory of relativity formulated by Einstein, it follows that the matter or energy distorts the geometrical structure of space time, so that particles and light follow geodesics, the straightest possible paths available to them. The famous physicist John Wheeler said that "space time tells matter how to move; matter tells space time how to curve." In the book of Allen Everett and Thomas Roman, *Time Travel and Warp Drives*,[1] this is summarized as a conceptual equation

$$\textit{geometry} \triangleq \textit{matter and energy}$$

Of course, this is a highly simplified form of describing the general relativity theory and does not give the full content of the Einstein's equations. First of all, stresses and pressures in the matter, as well as matter flows, contribute to the right-hand side. Second, only a part of the curvature of space time is contained in the left-hand side. There are solutions of Einstein's field equations of gravitation which are "vacuum solutions," representing the warping of empty space. One example is the curvature of the region of empty space outside a massive body, such as a star, as can be obtained from Schwarzschild metric which is one of one of the solutions obtained from Einstein's equations.

In this section, we will try to present these equations in a summarized way as follows, starting with the definition of a tensor, which roughly can be described as an array of numbers, or functions, that transforms following established rules for coordinates. It is like a vector field except that it can have two or more index components, for example,

$$A_{\mu\nu}, B^{\mu\nu}, C^{\mu}{}_{\nu\rho}, R^{\rho\tau}{}_{\kappa\alpha}$$

A vector field V is a differentiable function defined on certain region of space time whose values are vectors "tangent" to space time, in a form similar to the case of the components of a vectors which are tangential to a spherical surface when one analyzes, for example, the electric and magnetic field in the radiation zone of a dipolar antenna. A natural way to describe V is by means of its components V^{μ} relative to a given coordinate system denoted as K. When suprascripts are used the components of such a vector field are called contravariant, because for expressing them with respect to another coordinate system K' the transformation is determined by the rule

$$V'^{\nu} = \sum_{\mu} \frac{\partial x'^{\nu}}{\partial x^{\mu}} V^{\mu} \equiv \frac{\partial x'^{\nu}}{\partial x^{\mu}} V^{\mu} \tag{5.42}$$

In this last equation, we have considered the rule that when an upper index is repeated as a lower index, this automatically is equivalent to sum over that index.

Covariant vector fields are written using subscripts V_{μ} and they transform as

$$V'_{\nu} = \frac{\partial x^{\mu}}{\partial x'^{\nu}} V_{\mu} \tag{5.43}$$

THE METRIC TENSOR

The general relativity is based on the metric tensor $g_{\mu\nu}$, that is transformed under changes of coordinates so that the interval ds between two events with coordinates x^{μ} and $x^{\mu} + dx^{\mu}$ remains unchanged, that is

$$ds^2 = g_{\mu\nu} dx^{\mu} dx^{\nu} \tag{5.44}$$

With $dx^1 = dx^2 = dx^3 = 0$, we obtain the relation between the interval of ***proper time*** $d\tau = ds/c$ and the so-called interval of ***world time,*** $dt = dx^0/c$ for an observer at rest, so that

$$d\tau = \sqrt{g_{00}}\, dt \tag{5.45}$$

where according to eq 5.5 in the metric of Schwarzschild $g_{00} = 1 - r_S/r$.

According to Einstein's equivalence principle that we quote verbatim: "for an observer falling free *from the roof of a house there exists*—at least in his immediate surroundings—no gravitational field." This means that space time is locally inertial and at each point P there is a coordinate system where the metric tensor $g_{\mu\nu}$ gives exactly the Minkowski metric corresponding to the tensor $\eta_{\mu\nu}$:

$$\eta_{\mu\nu} = \begin{bmatrix} 1 & 0 & 0 & 0 \\ 0 & 1 & 0 & 0 \\ 0 & 0 & 1 & 0 \\ 0 & 0 & 0 & -1 \end{bmatrix} \tag{5.46}$$

The Minkowski metric is that used in special relativity theory where the space time interval is

$$ds^2 = \eta_{\mu\nu} dx^\mu dx^\nu = \mathrm{d}x^2 + \mathrm{d}y^2 + dz^2 - c^2 \mathrm{d}t^2 = -c^2 d\tau^2 \tag{5.47}$$

In absence of gravitation as it is considered in special relativity theory or in the limit of a weak gravitational field $g_{\mu\nu} \rightarrow \eta_{\mu\nu}$.

According to the general principle of covariance of the general theory of relativity, the laws of physics for systems where gravitational effects cannot be ignored, must be expressed in covariant form by means of tensors. This allows that these laws have the same form regardless of which coordinate system (observer) is used.

Covariant Derivative

Since space time is generally curved, the notion of the derivative becomes curved also, becoming in what is called the covariant derivative. For a covariant vector field A_μ, the covariant derivative is expressed as

$$A_{\mu;\nu} = \frac{\partial A^\mu}{\partial x^\nu} - \Gamma^\lambda_{\mu\nu} A_\lambda \tag{5.48}$$

where

$$\Gamma^\lambda_{\mu\nu} = \frac{1}{2} g^{\lambda\rho} \left[\frac{\partial g_{\rho\nu}}{\partial x^\mu} + \frac{\partial g_{\rho\mu}}{\partial x^\nu} - \frac{\partial g_{\mu\nu}}{\partial x^\rho} \right] \tag{5.49}$$

is called the Christoffel symbol.

RIEMANN–CHRISTOFFEL CURVATURE TENSOR

This is a rank 4 tensor that determines the curvature of space time and is given by

$$R^\lambda_{\mu\nu\kappa} = \frac{\partial}{\partial x^\kappa} \Gamma^\lambda_{\mu\nu} - \frac{\partial}{\partial x^\nu} \Gamma^\lambda_{\mu\kappa} + \Gamma^\eta_{\mu\nu} \Gamma^\lambda_{\kappa\eta} - \Gamma^\eta_{\mu\kappa} \Gamma^\lambda_{\nu\eta} \tag{5.50}$$

The Riemann–Christoffel tensor has two important contractions. One of these contractions corresponds to the so-called Ricci tensor, which is expressed as

$$R_{\mu\kappa} = R^\lambda_{\mu\lambda\kappa} \tag{5.51}$$

A second contraction of the Riemann–Christoffel tensor leads to the scalar curvature tensor

$$R = R^\mu_\mu \tag{5.52}$$

With this elements of tensor calculus which have been briefly presented, we discuss now the Einstein's field equation which in the presence of matter and radiation is

$$G_{\mu\nu} = R_{\mu\nu} - \frac{1}{2} g_{\mu\nu} R = \frac{8\pi G_N}{c^2} T_{\mu\nu} \tag{5.53}$$

This last equation is called the Einstein Field Equations where in the right-hand side, matter and energy are represented by the tensor $T_{\mu\nu}$ called the stress–energy–momentum tensor, which contains information respect to the density, energy flow, and momenta of matter particles and/or radiation. In the left-hand side, $G_{\mu\nu}$ is known as the Einstein tensor, $R_{\mu\nu}$ is the Ricci tensor, and R is the scalar that represents the curvature of the space time.

Let's now consider a special case. Just as in Maxwell's equations, there are solutions which involve no sources, the same situation is true here. So, let's consider the case with either no sources or when we are in a region of space where there are no sources, so that $T_{\mu\nu} = 0$ in the right-hand side of equation (5.60).

We will show that, in this case, the Ricci tensor $R_{\mu\nu}$ vanishes. The proof is as follows:

If $T_{\mu\nu} = 0$, from eq 5.53 we have

$$R_{\mu\nu} = \frac{1}{2} g_{\mu\nu} R$$

Multiplying both members of this equation by $g^{\mu\nu}$, we have

$$g^{\mu\nu} R_{\mu\nu} = \frac{1}{2} g^{\mu\nu} g_{\mu\nu} R$$

This leads to

$$R = \frac{1}{2} \delta^{\nu}_{\mu} R = \frac{1}{2} R$$

Therefore, R must be zero which implies that

$$R_{\mu\nu} = 0 \tag{5.54}$$

Equation 5.54 is called the Einstein equation in vacuum. Note, as we have already said, that it doesn't imply that the Riemann curvature tensor be zero. The solutions to eq 5.54 are nontrivial. They contain gravitational waves with no sources, just like there can be electromagnetic waves even in space without electric charges.

The Schwarzschild metric provides another example. It is roughly speaking analogous to a point mass. Outside the point mass there is no matter, nothing. So just like Newton's equations, everywhere outside the mass is the same as for empty space, the Schwarzschild metric outside the singularity, constitutes a solution to Einstein's equation in a vacuum.

Metrics for which the Ricci tensor is zero are called Ricci flat. So anywhere except at the singularity—where everything is crazy and meaningless anyway, just like for Newton's point masses—the Schwarzschild metric is of this type. It is not the same as flat.

So gravitational waves satisfy eq 5.54. Schwarzschild metric also satisfies it, except at the singularity. And that is one of the basic facts about Einstein field equations.

5.10 GENERAL RELATIVITY AND BLACKHOLES

5.10.1 WHAT IS A BLACKHOLE?

Black holes have been studied in various books and scientific publications, being one of the most extreme objects known in the universe. The strange phenomena that occur around black holes are difficult to understand with our basic conceptions of space, time, determinism, irreversibility, information, and causality. For that reason, when the concept of black holes predicted by general relativity theory was formulated, it had a great impact on Physics and Philosophy.[13]

The simplest way to describe a black hole is as an object subjected to such intense gravity that not even light can escape from its interior. It is known as such because its surface neither reflects nor emits light, at least according to classical general relativity based on the differential geometry of the solutions to Einstein's equations. According to Stephen Hawking, if quantum effects are considered, the black holes emit a thermal flux of radiation that for black holes of a few solar masses is very low (equivalent to a temperature lower than 3 K).

In this section, we will not consider these effects and our discussion about time is based on classical general relativity, so that a black hole can be represented as an enormous mass density resulting from the gravitational collapse of an object like an star, producing a deformation of the space time.

Let us consider a star like the sun which gradually is exhausting all its nuclear combustible, so that at some moment the heat produced by nuclear reactions will not be enough for holding the star up against its own gravitational pull, and it will collapse into a "black hole." What this means is that the star collapses so that a ray of light emitted from the surface of the star gets immediately dragged back in. The star becomes shrouded by an "event horizon." Just as a ship on earth that passes below the horizon cannot be seen, the region inside of the event horizon is cut off from the outside universe. This is because inside the horizon, an object or light ray would need to travel at a speed greater than c to escape to the exterior universe.

The German physicist Karl Schwarzschild was one of the pioneers in research on the concentration of matter in a mathematical point, giving rise to the concept of black hole.

These objects present among their characteristics, a geometric property called "event horizon," which mathematically separates the internal and external regions of the black hole. In these two regions, space time behaves in a radically different way. The region outside the event horizon contains what we know about our universe and its behavior, but inside this horizon space and time behave in a very different way, as we will discuss in this section.

Matter and light cannot escape from inside a black hole. However, this horizon has no relation to the Newtonian concept of escape velocity of a particle from the gravitational field produced by a very massive body, as discussed in relation with eq 5.11.

5.10.2 WHAT HAPPENS WITH TIME AT AN EVENT HORIZON OF A BLACK HOLE?

As we have seen in Section 5.4 of this chapter, Schwarzschild metric which is one of the solutions of Einstein's equations in a region around a massive body of mass M, predicts the following relation between time intervals measured by two observers one of them placed at a distance R from its center (observer 1) and the other infinitely distant (Observer 2):

$$\Delta t_{\text{observer 1}} = \Delta t_{\text{observer 2}} \sqrt{1 - \frac{2G_N M}{c^2 R}} = \Delta t_{\text{observer 2}} \sqrt{1 - \frac{r_S}{R}} \tag{5.55}$$

where $r_S = 2G_N M/c^2$ is the Schwarzschild radius (some approximated values of this parameter are: 9 millimeters for the Earth, 2.8 m for Jupiter, 3 km for the Sun).

Let us consider that initially both observers are infinitely distant from the body of mass M and that observer 1 begins to move closer to this body, so that when he is at a distance $r = r_S$ time seems to become stopped for him and becomes an imaginary number if he continues his approaching so that $r < r_S$. A common interpretation of this result is that inside a black hole time runs in a new dimension.

In particular, from the Schwarzschild metric given by eq 5.12, we can deduce that for a light beam emitted from a source infinitely distant from the considered body and moving radially towards its center (without considering the effect of light deflection in a gravitational field), the velocity of propagation is given by

$$\upsilon_r = \frac{dr}{dt} = c\left(1 - \frac{r_S}{r}\right) \tag{5.56}$$

In the above equation, the time t is called the coordinate time, or world time, established by clocks infinitely far from the gravitating body. Therefore this velocity is called the coordinate or world velocity. The decrease of this coordinate velocity as r decreases up to r_S leads to two types of effects both of which are well known. One of them is the famous deflection of light by the sun, predicted by Einstein[5] and first observed during 1919 solar eclipse. The other effect is the delay of radar echo from relatively closer planets, such as Venus or Mercury, whose measurement was proposed and realized by I. I. Shapiro.[14,15] This effect consists in that for a radar pulse transmitted from the earth and reflected by another planet, the increase in the round-trip delay, attributable to the gravitational influence of the sun on the propagation, can take values of the order of some hundred microseconds.

Figure 5.11 shows schematically how the orientation of light cones changes near a black hole. Away from this object, the light cones on the left look like their flat space time counterparts without gravity. As we get closer to the Schwarzschild radius, it can be seen that the cones gradually begin to tip inward. (Here, r represents the radial coordinate.) Right at the Schwarzschild radius, it can be observed that the outgoing leg of the light cone (i.e., the one pointing away from the black hole) is vertical in the diagram. This indicates that a ray of light emitted right at the horizon would require an infinite amount of time to escape, what is consistent with eq 5.56 which gives a null velocity of light for $= r_S = 2G_N M/c^2$. Inside the horizon, for $r < r_S$, both the ingoing and outgoing parts of the light cone point inward. Since any observer's world line must always lie inside the local light cone, this implies that, once inside the horizon, all observers must inevitably fall toward smaller values of r.

EXAMPLE 5.4

There are two values of the radial coordinate r for which the Schwarzschild metric has troubles: $r = 0$ and $r = r_S = 2G_N M/c^2$. At each of these values, one of the components of this metric diverges. At the point $r = 0$ there is a ***singularity***, and the theory of general relativity breaks down as we get close to this point, being necessary to have a quantum theory of space time as required by the spatial dimensions involved.

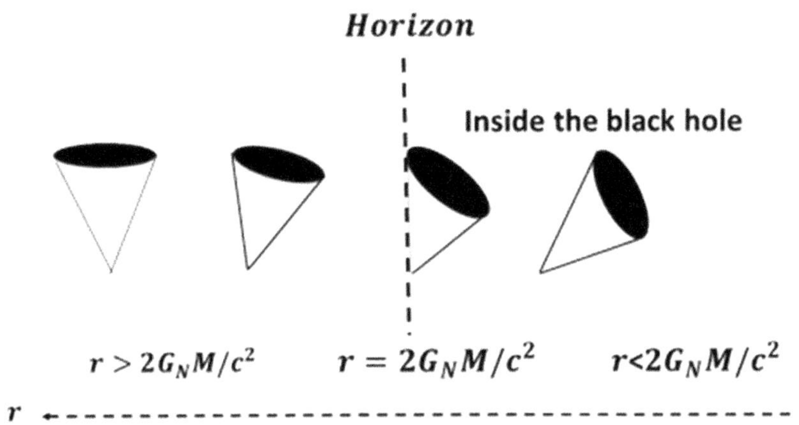

FIGURE 5.11 Light cones in the vicinity of a black hole. At the horizon, the outgoing leg of the light cone is parallel to that horizon. Inside the black hole, all the light cones point inward toward the singularity.

In contrast, the divergence at the surface $r = r_S$ is not dramatic and can be solved by a suitable coordinate transformation as we shall discuss in this example. This surface is referred to as the ***event horizon.***

We will develop this example step-by-step considering a light ray moving radially toward a black hole, according to the following guideline:

a) Obtain the slope dt/dt and discuss your result as we approach to the event horizon.
b) Obtain an explicit expression for the relation between the time t and the radial coordinate r, and also for the time that a photon requires to propagate from an external point up to the event horizon of the black hole.
3) Using your previous result, show that the divergence at $r = r_S$ disappears if the Schwarzschild metric is written in terms of two new variables r^* and u given by:

$$r^* = r + r_S \, ln\left[\frac{r}{r_S} - 1\right] \tag{5.57}$$

$$u = ct + r^* \tag{5.58}$$

SOLUTION

a) From eq 5.12, we have that for a light ray ($ds^2 = 0$) and if its motion is in the radial direction, we get

$$\frac{dt}{dr} = \pm\frac{1}{c}\left(1-\frac{r_S}{r}\right)^{-1} \tag{5.59}$$

where the + (-) sign corresponds to photons moving away (approaching) to the black hole.

For large r, the slope dt/dr tends to $\pm 1/c$, as it would be in the flat space time of the special relativity theory, while as we approach the event horizon $r \to r_S$ and therefore, dt/dr tends to ∞. Thus, at least in this coordinate system, a light ray which approaches $r = r_S$ never seems to reach this point.

b) By integration of eq 5.59, we get $ct = \pm r^* + constant$, where r^* is given by eq 5.57. Therefore, for photons approaching the black hole (incoming photons), we obtain the following expression for the time t required for each photon to propagate up to a distance ηr_S from the center of the black hole if the emitter is located at a distance d (in units of the Schwarzschild radius r_S) from that center:

$$t = d - \eta + ln\left[\frac{d-1}{\eta-1}\right] \tag{5.60}$$

As can be shown in Figure 5.12, with η slightly greater than 1, it seems that this horizon is never reached. Nevertheless, this is an apparent result, as we shall see in what follows of this example.

c) In terms of t and r^* the Schwarszchild metric becomes

$$ds^2 = \left(1-\frac{r_S}{r}\right)\left(dr^{*2} - c^2dt^2\right) + r^2d\Omega^2 \tag{5.61}$$

where

$$d\Omega^2 = d\theta^2 + sin^2\theta\, d\phi^2 \tag{5.62}$$

Schwarzschild metric written in the form of eq 5.61 does not diverge for $r = r_S$ as occurs in the standard form given by eq 5.12, which we repeat here as

$$ds^2 = \left(1-\frac{r_S}{r}\right)c^2dt^2 - \left(1-\frac{r_S}{r}\right)^{-1} dr^2 - r^2d\Omega^2$$

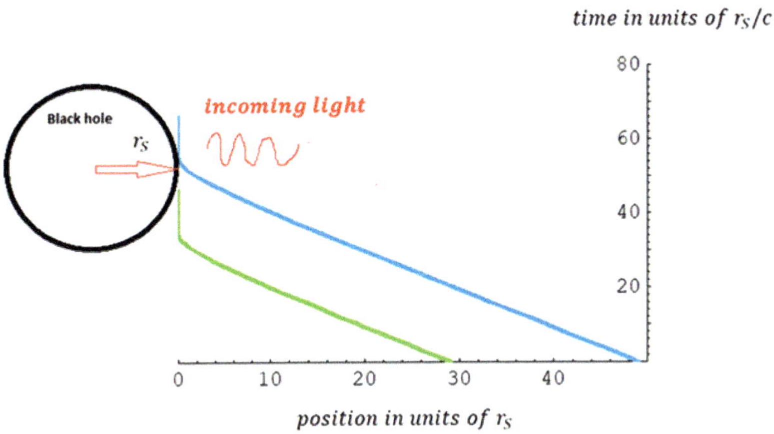

FIGURE 5.12 Time required for a light ray to reach and traverse the event horizon. Blue (green) curve corresponds to an emitter placed at a distance d=50 r_S (30 r_S) from the center of the black hole.

Nevertheless, for $r \to \infty$ the metric given by eq 5.61 diverges, and therefore, it is better to write the Schwarzschild metric in terms of the coordinates (u, r) which are naturally adapted to the null geodesics, as follows:

From eqs 5.57 to 5.58, we have for dr^* and cdt, respectively:

$$dr^* = \frac{dr}{1-\frac{r_S}{r}} \tag{5.63}$$

$$cdt = du - dr^* \tag{5.64}$$

Using these two last equations in 5.61, we obtain that Schwarzschild metric expressed in terms of coordinates (u, r^*) becomes

$$ds^2 = -\left(1-\frac{r_S}{r}\right)du^2 + 2dudr + r^2 d\Omega^2 \tag{5.65}$$

From the above equation, we deduce that du/dr can take the values:

$$\frac{du}{dr} = \begin{cases} 0 & \text{infalling photons} \\ 2\left(1-\frac{r_S}{r}\right)^{-1} & \text{outgoing photons} \end{cases} \tag{5.66}$$

From eqs 5.58 to 5.59 combined with eq 5.66, we obtain

$$\frac{dr}{dt} = \pm c\left(1-\frac{r_S}{r}\right) \tag{5.67}$$

where the + (-) sign corresponds to outgoing (infalling) photons. This is the same equations as eq 5.56 obtained from the standard form of Schwarzschild metric. Therefore, if $r < r_S$ we have $dr/dt < 0$ for outgoing photons, meaning that all future-directed paths are in the sense of decreasing r.

EXAMPLE 5.5

This example is concerned with time measurements for a particle with falling radial motion in a Schwarzschild field produced by a gravitational body, comparing the time seen by a distant observer with respect to that measured by the clock of another observer fixed to the particle.

SOLUTION

From eqs 5.57 to 5.58, the time required for a particle to fall from a distance r_1 up to $R \ll r_1$ is given by

$$\Delta t(r_1 \to R) = \frac{r_1 - R}{c} + \frac{r_S}{c} ln\left[\frac{r_1 - r_S}{R - r_S}\right]$$

As R becomes closer to the Schwarzschild radius, the time interval calculated with the above expression tends to infinity. Therefore, the radial fall of a particle approaching a black hole down to the Schwarzschild radius lasts for infinite time when observed from a safe distance.

For the time of fall of a particle measured by its proper clock, it is required to calculate the proper time for the particle to approach up to a distance $R = r_S$, from the center of the black hole, and therefore we need to know the velocity $dr/d\tau$.This is made as follows:

In several books about general relativity theory, it is explained that the motion of a particle subject only to a gravitational field (see, e.g., Ref. [16]), obeys the geodesic equation

$$\frac{d^2 x^\mu}{ds^2} + \Gamma^\mu_{\nu\rho} \frac{dx^\nu}{ds} \frac{dx^\rho}{ds} = 0$$

where $ds^2 = c^2 d\tau^2$ with τ being the proper time and where $\Gamma^\mu_{\nu\rho}$ corresponds to the Christoffel symbol given by eq 5.49. For $\mu = 0$ the geodesic equation becomes

$$\frac{d^2 t}{d\tau^2} + \frac{r_S}{r(r - r_S)} \frac{dt}{d\tau} \frac{dr}{d\tau} = 0 \Rightarrow \frac{d}{ds}\left[\left(1 - \frac{r_S}{r}\right)\frac{dt}{d\tau}\right] = 0 \Rightarrow \left(1 - \frac{r_S}{r}\right)\frac{dt}{d\tau} = \text{constant}$$

This last relation implies that for any r:

$$\left(1-\frac{r_S}{r}\right)\frac{dt}{d\tau}=\left(1-\frac{r_S}{r}\right)\left(\frac{dt}{d\tau}\right)_{r=r_1}$$

From the Schwarzschild metric expressed by eq 5.10 and assuming that initially the particle was at rest at $r = r_1$, we get that

$$\left(\frac{dt}{d\tau}\right)_{r=r_1}=\left(\frac{r_1}{r_1-r_S}\right)^{1/2}$$

Combining the two preceding equations, we get

$$\text{constant}=\left(\frac{r_1-r_S}{r_1}\right)^{1/2}$$

Finally, we have that

$$\frac{dt}{d\tau}=\left(\frac{r}{r-r_S}\right)\left(\frac{r_1-r_S}{r_1}\right)^{1/2}$$

$$\frac{dr}{d\tau}=\frac{dr}{dt}\frac{dt}{d\tau}=-c\left(\frac{r_S}{r_1}\right)^{1/2}\left(\frac{r_1}{r}-1\right)^{1/2}$$

Therefore, if the particle starts from a point placed at a distance r_1 from the gravitational body, the time required to reach another point at a distance r< r_1 from this body is given by

$$\tau=-\frac{1}{c}\sqrt{\frac{r_1}{r_S}}\int_{r_1}^{r}\frac{dr'}{\left(\frac{r_1}{r'}-1\right)^{\frac{1}{2}}}=\frac{1}{c}\sqrt{\frac{r_1^3}{r_S}}\left[\sqrt{\rho(1-\rho)}+cos^{-1}\sqrt{\rho}\right]$$

where $\rho = r/r_1$.

This proper time is finite for all r, even $r = 0$. In fact, the proper time to reach this point is $\tau_o=\frac{\pi}{2c}\sqrt{\frac{r_1^3}{r_S}}$.

As an example, if initially the particle is at rest at a distance $r_1 = 100r_S$ from the center of the star, this proper time takes a value of the order of $5\mu s$ seconds to cross its surface and reach its center.

In summary, we have that an object falling into a collapsed star, viewed from the outside, it may still be seen even after an infinite time, as it approaches the Schwarzschild radius. Measured on its own clock, however, the object only exists for a time given by the last equation. It looks as if nothing strange happens when the Schwarzschild surface is crossed. This is an excellent example where the General Relativity Theory gives us an illustration about the relativity of time.

5.11 TIME REVERSAL SYMMETRY IN AN EXPANDING UNIVERSE MODEL

In 1917, Albert Einstein (1879–1955) put forward the first relativistic cosmological model, that is, a model based in the General Relativity static model. This model is finite, has spherical spatial symmetry and considers that all the objects are at rest.

At this time, the hypothesis of a static universe was quite reasonable. The observations by Edwin Hubble (1889–1953) consistent with non-static solutions, had not yet been realized.

Almost immediately after Einstein's proposition, W. de Sitter (1872–1934), A. Friedmann (1888–1925), and G. Lemaıtre (1894–1966) proposed alternative models to Einstein's static model, also based on the General Relativity Theory. These models represent expanding universes. The light emitted by any galaxy arrives at an observer located in a distant galaxy with its wavelength redshifted. In other words, light arrives with a wavelength larger than the wavelength at emission. In Einstein's model, this behavior is not present because it represents a static universe.

In this section, we briefly discuss the time reversal symmetry in one of these models, that based on Friedmann's equations, where a homogeneous and isotropic universe filled with one perfect fluid is assumed. Such a universe is described by the Robertson–Walker Metric (RWM), which allows the curvature radius to be a function of time. For a flat universe model and using Cartesian coordinates, this frame is given by

$$ds^2 = -c^2 dt^2 + \left(a(t)\right)^2 \left(dx^2 + dy^2 + dz^2\right) \tag{5.68}$$

Here, $a(t)$ is a dimensionless parameter defined so that $a(t_0) = 1$ where instant t_0 is taken as the present. At other times, the curvature radius is $r(t) = a(t)\, R_0$ with R_0 corresponding to the initial curvature.

It must be considered that this metric depends on the coordinate frame of the observer. Even a homogeneous and isotropic universe only appears so to a special set of freely falling observers, called comoving observers, and the proper distance between them increases in proportion to $a(t)$.

From the metric given by eq 5.68 and taking $c \equiv 1$, we find that the covariant and contravariant metric tensors are:

$$g_{\mu\nu} = \text{diag}\left(-1, a^2, a^2, a^2\right) \tag{5.69}$$

$$g^{\mu\nu} = \text{diag}\left(-1, 1/a^2, 1/a^2, 1/a^2\right) \tag{5.70}$$

EXAMPLE 5.6

Show that the component G_{00} of the Einstein's tensor $G_{\mu\nu}$ is given by

$$G_{00} = 3\left(\frac{\dot{a}}{a}\right)^2 \tag{5.71}$$

where $\dot{a} = da/dt$ gives the variation of the curvature with respect to time.

SOLUTION

From eq 5.53

$$G_{00} = R_{oo} - \frac{1}{2} g_{00} R$$

From eqs 5.50 to 5.51, we can write the Ricci Tensor as

$$R_{\mu\nu} = \Gamma^{\alpha}_{\mu\nu,\alpha} - \Gamma^{\alpha}_{\mu\alpha,\nu} + \Gamma^{\alpha}_{\beta\alpha}\Gamma^{\beta}_{\mu\nu} - \Gamma^{\alpha}_{\beta\nu}\Gamma^{\beta}_{\mu\alpha} \tag{5.72}$$

For $\mu = \nu = 0$, we have

$$R_{00} = \Gamma^{\alpha}_{00,\alpha} - \Gamma^{\alpha}_{0\alpha,0} + \Gamma^{\alpha}_{\beta\alpha}\Gamma^{\beta}_{00} - \Gamma^{\alpha}_{\beta 0}\Gamma^{\beta}_{0\alpha} \tag{5.73}$$

eq 5.49 can be recast as

$$\Gamma^{\mu}_{\alpha\beta} = \frac{g^{\mu\nu}}{2}\left(g_{\alpha\nu,\beta} + g_{\beta\nu,\alpha} - g_{\alpha\beta,\nu}\right) \tag{5.74}$$

For $\mu = 0$, this last equation becomes

$$\Gamma^{0}_{\alpha\beta} = \frac{g^{0\nu}}{2}\left(g_{\alpha\nu,\beta} + g_{\beta\nu,\alpha} - g_{\alpha\beta,\nu}\right) \tag{5.75}$$

Considering that from eq 5.70 $g^{0\nu} = -\delta_{0\nu}$, this last equation is recast as

$$\Gamma^{0}_{\alpha\beta} = \frac{g^{0\nu}}{2}\left(g_{\alpha\nu,\beta} + g_{\beta\nu,\alpha} - g_{\alpha\beta,\nu}\right) = -\frac{1}{2}\left(g_{\alpha 0,\beta} + g_{\beta 0,\alpha} - g_{\alpha\beta,0}\right) \tag{5.76}$$

From eq 5.69, we obtain that $g_{\alpha 0,\beta} = g_{\beta 0,\alpha} = 0$ so that

$$\Gamma^{0}_{\alpha\beta} = \frac{1}{2} g_{\alpha\beta,0} \tag{5.77}$$

This last equation implies that $\Gamma^{0}_{00} = \Gamma^{0}_{0i} = \Gamma^{0}_{i0} = 0$ and

$$\Gamma^{0}_{ii} = a\dot{a}\,\delta_{ij} \tag{5.78}$$

For $\mu = i$ eq 5.74 implies that

$$\Gamma^{i}_{\alpha\beta} = \frac{g^{iv}}{2}\left(g_{\alpha v,\beta} + g_{\beta v,\alpha} - g_{\alpha\beta,}\right) = \frac{1}{2a^2}\left(g_{\alpha i,\beta} + g_{\beta i,\alpha} - g_{\alpha\beta,i}\right)$$

$$= \frac{1}{2a^2}\left(g_{\alpha i,\beta} + g_{\beta i,\alpha}\right) \tag{5.79}$$

From eq 5.74 we get:

$$\Gamma^{i}_{00} = \frac{1}{2a^2}\left(g_{0i,0} + g_{0i,0}\right) = 0 \tag{5.80}$$

$$\Gamma^{i}_{0j} = \frac{1}{2a^2}\left(g_{0i,j} + g_{ji,0}\right) = \frac{1}{2a^2} g_{ji,0} = \frac{1}{2a^2}\delta_{ij} 2a\dot{a} = \frac{\dot{a}}{a}\delta_{ij} \tag{5.81}$$

and $\Gamma^{i}_{jk} = \frac{1}{2a^2}\left(g_{ji,k} + g_{ki,j}\right) = 0$ since the components of tensor $g_{\mu v}$ given by eq 5.69 do not depend on the spatial components.

Therefore, from these results we obtain that for $\mu = v = 0$ the Ricci tensor is

$$R_{00} = \Gamma^{\alpha}_{00,\alpha} - \Gamma^{\alpha}_{0\alpha,0} + \Gamma^{\alpha}_{\beta\alpha}\Gamma^{\beta}_{00} - \Gamma^{\alpha}_{\beta 0}\Gamma^{\beta}_{0\alpha} = -\Gamma^{i}_{0i,0} - \Gamma^{i}_{j0}\Gamma^{j}_{0i}$$

$$= \sum_{i} \delta_{ii}\frac{\partial}{\partial t}\frac{\dot{a}}{a} - \sum_{i,j}\delta_{ij}\frac{\dot{a}}{a}\delta_{ij}\frac{\dot{a}}{a}$$

$$= -3\frac{\dot{a}}{a} \tag{5.82}$$

Now, we need to calculate the Ricci Scalar R. This is made as follows:

$$R = g^{\mu v} R_{\mu v} = -R_{00} + \frac{1}{a^2}\sum_{i} R_{ii} \tag{5.83}$$

Use of the results for Christoffel symbol previously obtained leads to

$$R_{0i} = \Gamma^{\alpha}_{0i,\alpha} - \Gamma^{\alpha}_{0\alpha,i} + \Gamma^{\alpha}_{\beta\alpha}\Gamma^{\beta}_{0i} - \Gamma^{j}_{\beta i}\Gamma^{\beta}_{0\alpha} = 0 \tag{5.84}$$

$$R_{ij} = \Gamma^{\alpha}_{ij,\alpha} - \Gamma^{\alpha}_{i\alpha,j} + \Gamma^{\alpha}_{\beta\alpha}\Gamma^{\beta}_{ij} - \Gamma^{\alpha}_{\beta j}\Gamma^{\beta}_{i\alpha} = \Gamma^{0}_{ij,0} - 0 + \Gamma^{\alpha}_{0\alpha}\Gamma^{0}_{ij} - \Gamma^{0}_{\beta j}\Gamma^{\beta}_{i0} - \Gamma^{k}_{\beta j}\Gamma^{\beta}_{ik}$$

$$= \delta_{ij}\left(\dot{a}^2 + a\ddot{a}\right) + 3\delta_{ij}\dot{a}^2 - \delta_{ij}\dot{a}^2 - \delta_{ij}\dot{a}^2 = \delta_{ij}\left(2\dot{a}^2 + a\ddot{a}\right) \tag{5.85}$$

Therefore,

$$R = -R_{00} + \frac{1}{a^2}\sum_i R_{ii} = 3\frac{\ddot{a}}{a} + \frac{1}{a^2}\sum_i \delta_{ii}\left(2\dot{a}^2 + a\ddot{a}\right) = 6\left\{\frac{\ddot{a}}{a} + \left(\frac{\dot{a}}{a}\right)^2\right\} \tag{5.86}$$

Using eqs 5.82 and 5.86 we get that the component G_{00} of the Einstein's tensor $G_{\mu\nu}$ is

$$G_{00} = R_{oo} - \frac{1}{2}g_{00}R = -3\frac{\dot{a}}{a} + \frac{1}{2}6\left\{\frac{\ddot{a}}{a} + \left(\frac{\dot{a}}{a}\right)^2\right\} = 3\left(\frac{\dot{a}}{a}\right)^2 \tag{5.87}$$

The result given by eq 5.87 combined with eq 5.53 (Einstein's Field Equation) allows to relate $\dot{a}/a$ with the density of matter filling the universe. In fact, from eq 5.53 we can write

$$G_{00} = \frac{8\pi G_N}{3}T_{00} \tag{5.88}$$

where T_{00} is the time-like component of the energy-momentun tensor which for the Friedmann model is

$$T^{\mu\nu} = \begin{pmatrix} \rho & 0 & 0 & 0 \\ 0 & P/a^2 & 0 & 0 \\ 0 & 0 & P/a^2 & 0 \\ 0 & 0 & 0 & P/a^2 \end{pmatrix} \tag{5.89}$$

where ρ is the mass density of the fluid filling the universe and P is the corresponding pressure.

From eq 5.89, we get

$$T_{00} = g_{0\mu}g_{\sigma\nu}T^{\mu\nu} = T^{00} = \rho \tag{5.90}$$

Therefore, inserting eqs 5.87 and 5.90 in eq 5.88, we get

$$\left(\frac{\dot{a}}{a}\right)^2 = \frac{8\pi G_N}{3}\rho \tag{5.91}$$

EXAMPLE 5.7

The general theory of relativity assumes that there is local energy–momentum conservation, which translates into saying that the energy–momentum tensor satisfies the covariant form of the continuity equation:

$$T^{\mu\nu}{}_{;\nu} = \frac{\partial T^{\mu\nu}}{\partial x^{\nu}} + \Gamma^{\mu}_{\nu\beta} T^{\nu\beta} + \Gamma^{\beta}_{\nu\beta} T^{\mu\nu} = 0 \tag{5.92}$$

Show that for $\mu = 0$ eq 5.92 leads to

$$\frac{\partial \rho}{\partial t} + 3\frac{\dot{a}}{a}(P + \rho) = 0 \tag{5.93}$$

SOLUTION

This result is obtained from (5.92) which gives $\Gamma^{0\nu}_{;\nu} = 0$.

Using the Christoffel symbol given by eq 5.49, considering the covariant and contravariant metric tensors given by eqs 5.69 and 5.70, and making $\mu = 0$ in eq 5.92 we have that

$$\Gamma^{0\nu}_{,\nu} = \frac{\partial \rho}{\partial t} + \Gamma^{0}_{\beta\nu} T^{\nu\beta} + \Gamma^{\nu}_{\beta\nu} T^{00} = 0$$

From eqs 5.75 to 5.79, it can be verified that the only non-zero Christoffel symbols are

$$\Gamma^{0}_{ij} = a\dot{a}\delta_{ij} \ ; \Gamma^{i}_{0j} = \Gamma^{i}_{j0} = \delta_{ij}\frac{\dot{a}}{a}$$

Therefore, one gets

$$\Gamma^{0}_{ij} T^{ij} = \sum_i \sum_j \delta_{ij} (a\dot{a}) \delta_{ij} \frac{p}{a^2} = 3\frac{\dot{a}}{a} p$$

Combining these results, we obtain

$$\Gamma^{0\nu}_{;\nu} = \frac{\partial \rho}{\partial t} + 3a\dot{a}\frac{P}{a^2} + 3\frac{\dot{a}}{a}\rho = \frac{\partial \rho}{\partial t} + 3\frac{\dot{a}}{a}(\rho + P) = 0 \tag{5.94}$$

It is worth to consider that eq 5.91 can be interpreted by means of a classical model of a fluid with homogeneous mass density ρ filling all the space (Ref. [17]), as shown in Figure 5.13. Let us imagine that a portion of this fluid is contained in a sphere of radius r and mass $M = \frac{\rho 4\pi r^3}{3}$ so that it produces a gravitational field at its surface equal to $-G_N M/r^2$.

From Newton's law, the corresponding acceleration of an object placed over its surface is given by

$$\frac{d^2 r}{dt^2} = -\frac{G_N M}{r^2} \tag{5.95}$$

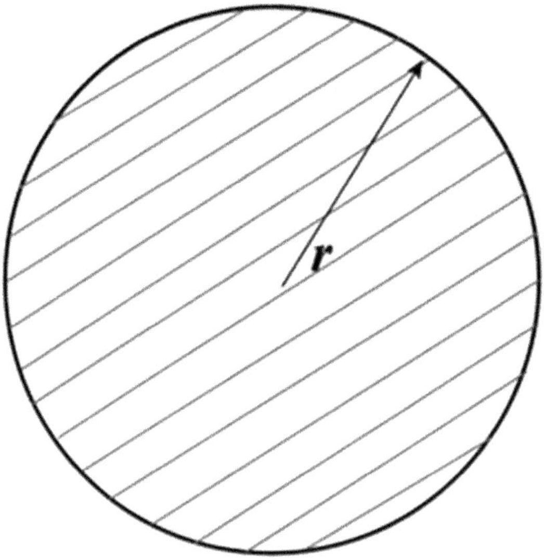

FIGURE 5.13 Spherical portion of a fluid with mass density ρ

Multiplying both sides of this equation by dr/dt gives

$$\underbrace{\frac{d^2r}{dt^2}\frac{dr}{dt}}_{=\frac{1}{2}\frac{d}{dt}\left(\frac{dr}{dt}\right)^2} = -\frac{G_N M}{r^2}\frac{dr}{dt} \tag{5.96}$$

After integration with respect to r this leads to

$$\left(\frac{dr}{dt}\right)^2 - 2G_N M / r = K \tag{5.97}$$

with K a constant resulting from integration.

This last equation can be rewritten as

$$\frac{\left(\frac{dr}{dt}\right)^2}{r^2} = \frac{8\pi G_N \rho}{3} + \frac{K}{r^2} \tag{5.98}$$

If we assume that the integration constant K is zero, we can see that eq 5.98 is of the form corresponding to eq 5.91 obtained using the mathematical formalism of general relativity theory.

Furthermore, eq 5.93 can be interpreted by means of the following argument based on the before mentioned reference, where the work over the external medium made by the sphere of Figure 5.13 is given by

$$\frac{dE}{dt} = -P\frac{dV}{dt} \tag{5.99}$$

where P denotes the pressure of the fluid, $V = 4\pi r^3/3$ and $\mathrm{E} = Mc^2$ are, respectively, the volume and the energy contained in the spherical portion shown in Figure 5.13.

The left-hand side of eq 5.99 is

$$\frac{dE}{dt} = c^2\frac{dM}{dt} = c^2\frac{d(\rho V)}{dt} = c^2 V\frac{d\rho}{dt} + c^2\rho\frac{dV}{dt} = c^2\left\{\frac{4}{3}\pi r^3\frac{d\rho}{dt} + \rho 4\pi r^2\frac{dr}{dt}\right\} \tag{5.100}$$

Combining eqs 5.99 and 5.100 and making a bit of algebra we obtain

$$\frac{\partial\rho}{\partial t} + 3\frac{\dot{r}}{r}\left(\rho + \frac{P}{c^2}\right) = 0 \tag{5.101}$$

which is similar to eq 5.93 with $r = a$ and $c \equiv 1$.

It can be readily seen that eqs 5.93 and 5.101 are time reversal symmetric so that according to this model of the universe at some moment the expansion could be stopped and there would be a contraction leading to what is named as a big-crush.

Let us come back to eq 5.98 and consider now that the integration constant K of its right-hand is different from zero. If we define the term in its left-hand side as

$$H(t) = \frac{\frac{dr}{dt}}{r(t)} = \frac{\dot{a}}{a(t)} \tag{5.102}$$

and in terms of the constant of integration K of the right-hand side we define a new constant B as

$$B = K\left(\frac{a}{r}\right)^2 \tag{5.103}$$

after making some algebra we can rewrite eq 5.98 as:

$$H^2 = \frac{8\pi G}{3}\rho + \frac{H_0^2(1-\Omega_0)}{a^2} \tag{5.104}$$

where $H_0 = H(t = 0)$ and Ω_0 is a parameter defined in terms of Ω_0 and the matter density ρ_o at $t = 0$ as

$$\Omega_0 = \frac{8\pi G_N \rho_0}{3H_0^2} \tag{5.105}$$

With a little more of algebra eq 5.104 can be recast as

$$\left(\frac{H}{H_0}\right)^2 = \frac{\Omega_0}{a^3} + \frac{1-\Omega_0}{a^2} \tag{5.106}$$

The case $\Omega_0 = 1$ corresponds to a matter-dominated universe, so that from eqs 5.102 to 5.106 we can obtain that:

$$a(t) = \left(\frac{3}{2}H_o t\right)^{2/3} \tag{5.107}$$

EXAMPLE 5.8

Show that this Newtonian derivation of the Friedmann model can be used for obtaining the temporal evolution of the density of matter and size parameter *a(t)* under the assumption of a flat universe containing only matter ($\Omega_0 = 1$ in eq 5.106). Estimate the age of this universe.

SOLUTION

From eq 5.106, in this case

$$\frac{H}{H_0} = \pm \frac{1}{a^{3/2}}$$

With $H = \frac{1}{a}\frac{da}{dt}$ and considering the positive sign since we are interested in the evolution of the universe for $t > 0$, we obtain

$$\frac{1}{a}\frac{da}{dt} = \frac{H_0}{a^{3/2}} \Rightarrow a(t) = \left(9H_0^2\, t^2/4\right)^{1/3} \tag{5.108}$$

If we assume that the mass of the expanding sphere is constant we have that the evolution of the density of matter is given by:

$$\rho(t) = \frac{\rho_0}{a^3} = \frac{3H_0^2}{8\pi G_N}\frac{4}{9H_0^2 t^2} = \frac{1}{6\pi G_N t^2} \tag{5.109}$$

From eq 5.108 the age of our universe can be estimated by considering $a(t) = 1 \Rightarrow t_U = \frac{2}{3}\frac{1}{H_o}$.

H_o can be associated with the Hubble constant which allows us to describe the speed with which the universe is expanding when observed from different distances with respect to a specific point in space.

At the present time, this constant has a value of approximately $7 \cdot 10^{-11}$ [1/year] so that according to this model, it can be estimated that the age of the universe is approximately $9 \cdot 10^{9}$ [years].

5.12 TIME REVERSAL SYMMETRY IN COSMOLOGY

Cosmology is an empirical discipline which studies the properties of the universe on a large scale, being usual to associate the expansion of the "radius of the universe" with a so called ***"cosmological arrow of time."*** However, there is no well-founded reason to privilege this "radius" to define the arrow of time without considering other properties of space time.

On the other hand, discussions regarding the arrow of time generally consider the concept of entropy. In cosmology, the past-future direction is generally related to the direction determined by the gradient of the entropy function of the universe and the problem is conceived in terms of "ordered" or "disordered" boundary conditions. Nevertheless, it must be taken into account that entropy is a thermodynamic magnitude: a given entropy value is compatible with several different configurations of the system. The question that then arises is whether there is a well-founded way to define an arrow of time for the universe as a whole.

In a paper of M. Castagnino, O. Lombardi, and L. Lara[18] it is shown that if certain conditions are fulfilled, it is possible to define for the universe a cosmological arrow of time, and that moreover, current models of cosmology satisfy such conditions. One of these conditions is the possibility of an asymmetry of universe that would make possible to distinguish between the two temporal senses, implying an asymmetry of the geometric properties that space time presents in its dimension corresponding to time.

The other condition is related with the possibility of describing an asymmetric universe with time-reversal invariant laws.

S. J. Robles-Pérez,[19] has shown that the evolution of the universe can be considered as the world line formed by the scale factor, which for formal purposes represents a time-like variable, while the scalar fields play the role of spatial components. Positively oriented trajectories with respect to scale factor in the minisuperspace entail an increasing value of the scale factor, for which they can be associated with expanding universes (the concept of minisuperspace in quantum cosmology corresponds to an approach by which it is possible to infer initial conditions for dynamical systems which can adequately represent observable and non-observable universes). Conversely, world lines that are oriented negatively with respect to the scale factor can represent contracting universes. The names "expansion" and "contraction" can be interchanged, so that we would end up with four symmetric solutions—two oriented forward and two oriented backwards. The former represent two expanding universes with their time variables reversely related and the latter represent two virtual universes that rapidly return to the gravitational vacuum from which they emerged.

These concepts are illustrated by A. E. Fisher in Ref. [20] where he mathematically shows that the universe did not start with the big-bang at $t = 0$ in a state with infinite density, but rather occurred at the instant $t = -\infty$ with the universe in a state of infinite dilution. He also shows that considering a generic quadratic Friedmann's equation of the form $\dot{a}^2 = F(a)$, it is possible to extrapolate classical cosmological models into the infinite past by passing through the big-bang.

EXAMPLE 5.9

Consider the general form used for the Friedmann's equation[20,21]:

$$\dot{a}^2 = H_0^{\ 2}\left\{\frac{\Omega_{m,0}}{a} + \frac{\Omega_{r,0}}{a^2} + \Omega_{\Lambda,0}a^2 + \Omega_{k,0}\right\} \tag{5.110}$$

where $\Omega_{m,0}$, $\Omega_{r,0}$, $\Omega_{\Lambda,0}$, and $\Omega_{k,0}$ are cosmological constants such that $\Omega_{m,0} + \Omega_{r,0} + \Omega_{\Lambda,0} + \Omega_{k,0} = 1$. If the Big Bang is represented by means of the radiation-dominated model ($\Omega_{r,0} = 1$) determine the temporal evolution of this universe and analyze the extrapolation to $t = -\infty$.

SOLUTION

According to this model the early universe was dominated by radiation composed mainly of photons and neutrinos. With $\Omega_{r,0} = 1$ in eq 5.110 and all the others cosmological constants equal to zero, we have

$$\dot{a}^2 = \left(\frac{H_0}{a}\right)^2 \tag{5.111}$$

which implies

$$\dot{a} = \pm \frac{H_0}{a} \tag{5.112}$$

Considering the positive sign in the above equation and the big bang initial condition $a(0) = 0$, we get

$$a(t) = (2H_0 t)^{1/2} \tag{5.113}$$

This solution is defined in the semi-infinite domain $t \in [0,\infty]$ and cannot be extended to the negative domain epoch corresponding to $\in [0,-\infty]$. However, considering eq 5.111 we observe that if $a(t)$ is a solution of eq 5.112 then $a^{TR}(t) = a(-t)$ is also a solution in the domain $t \in [0,-\infty]$, where TR means time reversal. In fact,

$$\frac{da^{TR}(t)}{dt} = \frac{da(-t)}{dt} = -\frac{da(-t)}{d(-t)} = -\sqrt{F(a(-t))} = -\sqrt{F(a^{T}(t))} \tag{5.114}$$

where F denotes the Friedmann's equation in its native quadratic form $\dot{a}^2 = F(a)$.

Therefore, $a(t)$ satisfies the positive square root Friedmann equation on the interval $t \in [0,\infty]$ and $a^{TR}(t)$ satisfies the negative square root Friedmann equation on the domain $t \in [0,-\infty]$, so that the join $a_{global}(t) = a(t) \vee a^{TR}(t)$ satisfies the native quadratic on the entire global domain $(-\infty, 0) \cup (0, \infty)$. Nevertheless, the union does not satisfy either the positive or negative signed square root of the Friedmann equation in the entire global domain $(-\infty, 0) \cup (0, \infty)$. Under the additional assumption that $F(a) \to \infty$ as $t \to 0^+$, $\dot{a}(t) \to \infty$ as $t \to 0^+$ and $a^{TR}(t) \to \infty$ as $t \to 0^-$, $a_{global}(t) = a(t) \vee a^{T}(t)$ although is continuous at $t = 0$, has a cusp singularity at this instant which corresponds to the ***Big Bang.*** This is shown in Figure 5.14 which is adapted from Refs. [20, 21].

5.13 TIME TRAVELING THROUGH WORMHOLES: IS IT POSSIBLE?

The equations of general relativity describe the geometry of space time as we have seen and besides black holes they also lead to predict the existence of wormholes, which are shortcuts between distant regions of space time and join two black holes. Traversable wormholes are space time tunnels connecting universes or distant parts of the same universe, through which transit of mass and energy is possible.[22] They were proposed by Ellis[23] and later by Morris, Thorne, and Yurtsever.[24]

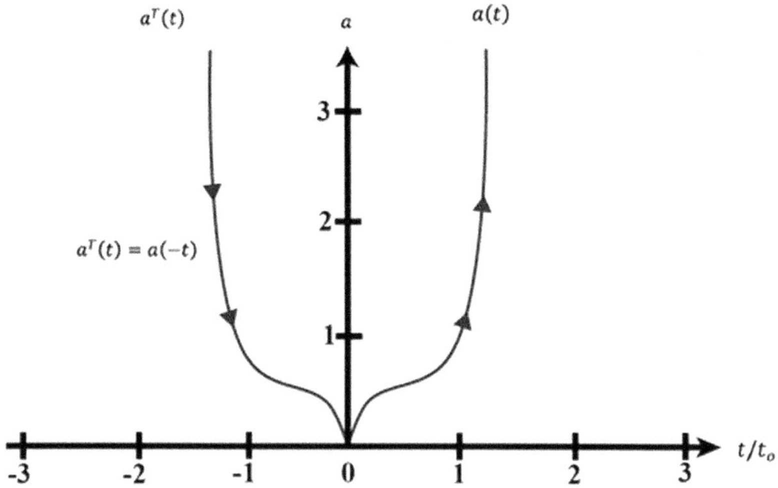

FIGURE 5.14 Time reversal symmetry in the evolution of the universe according to Friedmann's equation.[20] The time axis is expressed in terms of t_o which is the time elapsed from the big-bang to the present, solving $a(t_o) = 1$.

Source: Adapted from Ref. [20] and Ref. [21]

From the outside, they appear as separate entities sitting at distant locations, even though they share an interior. In 1935, Einstein and Rosen anticipated that this shared interior corresponds to a wormhole.[25] For this reason, wormholes are also called ER-bridges. Figure 5.15 shows a scheme where two regions of space time are connected by a wormhole.

The wormhole in Schwarzschild solution does not contain matter, only curved space time. This wormhole varies with time: it elongates and becomes thinner as time progresses. Meanwhile the two black hole horizons, which at one point touches, separate so rapidly that such a wormhole cannot be used for travelling from one exterior to the other. This is consistent with the laws of general relativity which states that it is not possible to travel faster than the speed of light.

The shortcut through space provided by a wormhole could be theoretically a means of faster-than-light travel. Imagine the wormhole connecting the earth to a star 10 light-years away. A beam of light, traveling in the space outside the wormhole, would require 10 years to make the trip from the earth to the star. By contrast, an observer could make the trip in a much shorter time by traveling through the wormhole. This would appear to violate the special relativistic speed limit, which prohibits exceeding the velocity of light. A person going through the wormhole could arrive at the star before a light beam traveling the outside route. However, it must be considered that when space time is curved, according to the special relativistic speed limit it is not possible to exceed the speed of light relative to the immediate surroundings. Furthermore, even if we admit this possibility, as Fuller and Wheeler realized,[26] this type of wormhole is unstable—the throat collapses in on itself so rapidly that even a beam of light does not have time to travel through it. Light or anything else falling into such a wormhole gets caught in the "pinch-off" of the throat, where the curvature of space becomes infinite. If that was not bad enough, this kind of wormhole is like a black hole. It has an "event horizon." This implies that an infinite amount of time would be required, as measured by outside observers, for one to fall into the wormhole. And once inside, one could never escape. All seems to indicate that it is not possible to go through such wormhole. Nevertheless, in papers published around 1988,[27,24] Thorne and his collaborators argued that not only it is theoretically possible the instantaneous travel between two points of space (for instance, two stars) but it would also be possible to travel in time. Morris and Thorne found that to hold a wormhole open it is necessary that it has at its throat matter with negative energy density. In Ref. [24] it is argued that if the laws of physics allow that an advanced civilization to create and maintain a wormhole in space for interstellar travel, then it is possible to have a time machine with which causality could be violated.

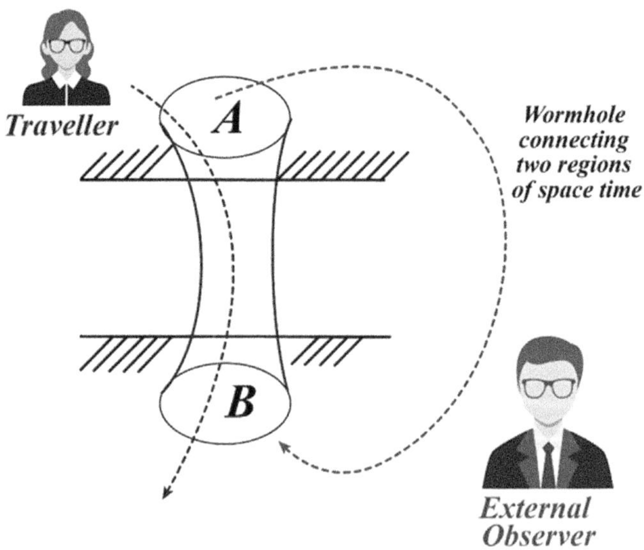

FIGURE 5.15 Two regions of space time connected by a wormhole.

In Chapter 6, Section 6.8, we will come back to the subject of time travel through wormholes, and closed time like curves as that which exist in the Gödel's rotating model of the universe.[28]

5.14 QUANTUM ENTANGLEMENT ACCORDING TO GENERAL RELATIVITY

In 1974, Stephen Hawking demonstrated the existence of radiation emitted by black holes, meaning that they have a temperature associated.[29] Since 19th century, it is known that the temperature of an object is determined by the movement of its constituent particles, with various micro-states being possible.

Considering that black holes must be considered as quantum systems, they should be subject to the laws of quantum mechanics which predict the Entanglement effect. Let us remember that quantum entanglement is a physical effect that occurs when group of particles are created at very close points and interact with each other in such a way that the quantum state of each of them is correlated with the state of the others, a correlation which is maintained even when the particles are separated by a great distance.

Maldacena argues that two black holes with entangled microstates would produce a space time where a wormhole links the interior of these black holes.[30] That is, quantum entanglement establishes a geometric connection between these black holes. According to Maldacena, this predicted result would be surprising, considering that in general *quantum entanglement* is understood as a correlation without a physical connection. But in this case the two black holes are connected through the wormhole. Maldacena and Susskind[31] call the equivalence between wormholes and quantum entanglement as "$ER \triangleq EPR$" because it relates two articles written by Einstein, Rosen, and Podolsky in 1935. One of these articles (Ref. [32]) is very critical with the correlation at distance involved in quantum entanglement (EPR paradox) and the other formulates the possibility of a bridge between the two black holes (ER-bridge,[25]). Maldacena and Susskind believe that this equivalence could be a clue for a quantum description of space time and an important step toward the achievement of unifying the theory of general relativity with quantum mechanics.

5.15 MEASUREMENT OF INTERVALS BETWEEN EVENTS IN THE SPACE TIME

In ordinary quantum mechanics, the distance between two space time points is given in terms of their coordinates, which are assumed to have a physical meaning independent of observations (no prescription for the measurement of these coordinates is to be given). In general relativity, instead, space time points can only be meaningfully identified by events.

Suppose we want to measure the length of an object and for that purpose we use a luminous clock and a mirror both fixed to an inertial reference frame S. The clock is put at one end of the object and is set to zero when emits a light signal towards the other end where a mirror is placed reflecting the light signal back to the clock. From the time interval Δt measured by the clock between the emission events of the light signal and its reception after being reflected by the mirror, one deduces that the length of the object is given by

$$l = \frac{c\Delta t}{2}$$

Nevertheless, all this analysis is fully classic and does not consider that during the emission of the light signal and the reception of the light reflected by the mirror, the clock acquires a momentum that cannot be ignored as its dimensions decreases at values where quantum effects can affect in a not negligible way the accuracy of the measurement. As shown below, using the uncertainty principle of quantum mechanics it can be established that the uncertainty in the measurement of a length l using this procedure based on light signals is given by

$$(\delta l)^2 \gtrsim \hbar l / mc$$

DEMONSTRATION

Considering that since the clock acquires a momentum from the emitted and back received light it has a spread in its velocity which can be obtained from position-momentum uncertainty principle:

$$\Delta l \cdot \Delta p \gtrsim \hbar / 2 \Rightarrow \Delta \upsilon = \frac{\Delta p}{m} = \frac{\hbar}{2m\Delta l}$$

where m is the mass of the clock and $\Delta l = \Delta x$ is the uncertainty in the length of the clock at the moment of the emission of light towards the mirror.

At time t the uncertainty in this measurement will be

$$\Delta \upsilon \cdot t = \frac{\hbar}{2m\Delta l} t$$

Since $t = 2l/c$ we obtain

$$\Delta l(t) = \Delta \upsilon \cdot t = \frac{\hbar}{2m\Delta l(0)} \frac{2l}{c} = \frac{\hbar l}{mc\Delta l(0)}$$

Therefore, we can estimate that the uncertainty in the measurement of the length has a value of the order of

$$\Delta l \gtrsim \sqrt{\frac{\hbar l}{mc}}$$

If we are measuring the length of a thin rod with l of the order of 4 cm and m~40 g we could estimate an uncertainty Δl of approximately $4 \cdot 10^{-19}$ cm which for all practical purposes is negligible, if for example, we compare this value with the radius of Bohr which corresponds to the smallest electronic orbit for the hydrogen atom with a value of the order of $5 \cdot 10^{-11}$ m.

From the above expression, one could think that the uncertainty in the measurement of length could be arbitrarily decreased and eventually could be near zero as the mass m increases. Nevertheless, a very massive clock would disturb the curvature of space time. Large values of the mass of the clock + light-gun +detector system necessarily lead to great distortions of the geometry, and well before the limit $m \to \infty$ the measurement procedure can no longer be followed. If we assume the clock to be spherically symmetric with a radius r from the Schwarzschild metric we can deduce the following limit for the mass m:

$$r > \frac{2G_N m}{c^2} \Rightarrow m < M_{max} = \frac{rc^2}{2G_N}$$

where $r_S = \dfrac{2G_N m}{c^2}$ is the Schwarzschild radius.

For values of mass greater than $M_{\max}$ a horizon forms around the center of mass of the clock+light-gun+detector system and it is not possible to send a light signal to the mirror positioned on the other side of the object whose length is being measured.

The replacement of m by $M_{\max}$ in the above expression for Δl leads to

$$\Delta l \gtrsim \sqrt{\frac{2l\, l_p^2}{r}}$$

where l_p is the Plank length defined as $l_p = \left(\frac{\hbar G_N}{c^3}\right)^{1/2} = 1.62 \cdot 10^{-35}\, m$ which is a lower bound to physical proper length in any space time and at present it is imposible to build an apparatus for measuring length scales smaller that this value.[33,34] According to V. Gharibyan,[35] quantum models predict unusual optical properties of space time at very short distances, such as behavior as a crystal, which at present cannot be experimentally verified since the smallest distance experimentally accessible is about $10^{-19}\, m$ at the large hadron collider (LHC).

If we consider that the dimensions of the clock are comparable with the length of the object to be measured we obtain that $\Delta \gtrsim l_p$. Since the Plank length is so short, even for a massive body as our Earth ($m = 6 \cdot 10^{24}\, Kg$, $r = 6371\, Km$) and this procedure of luminous signals allows to conclude that the distances between the south and north poles would be $l = 2r$ with an uncertainty not lower than $2l_p$.

Associated with Plank length there exists a Plank time defined as $t_p = \frac{l_p}{c} = 5.4 \cdot 10^{-44}\, s$. Up to now, the shortest duration events that can be directly created, controlled, and measured are pulses of light that occur on a time scale of the order of femtoseconds (1 femtosecond equals 10^{-15} s). According to more recent measurements, these pulses can last only some attoseconds (one attosecond is equal to 10^{-18} second) [www.scientificamerican.com/article/what-is-the-fastest-event/].

The meaning or interpretation of these two quantities, Plank Length and Time Length, have been discussed in several publications related with quantization of the space time and quantum theory of gravity, which up to now are theories in development. For example, P. A. Zizzi has assumed that the space time at the Plank scale is discrete and quantized in Plank units, so that each pixel corresponding to a Plank area is associated to a qubit and in this form the space of time can be conceived as a quantum computer.[36]

5.16 GENERAL RELATIVITY THEORY AND QUANTUM MECHANICS

The history of physics had two great revolutionary theories during the 20th century, which are relativity and quantum mechanics. However, what has been advanced so far in particle physics does not simultaneously comply with the principles of both theories. One of the main problems is to obtain an explanation for the symmetry breaking in the proton–antiproton collision experiments and why more matter than antimatter particles are created. A theory that can explain these phenomena should answer questions such as how matter was at the beginning of the universe and what is the space time structure of the universe.[37]

The standard formulation of quantum field theory considers a fixed space time metric to determine the localization and the causal order of events. The corresponding metric is generally implicit in the way an operation is defined and determines the causal relationship in the space time in which the operations are carried out.

In the general theory of relativity, the metric is affected by matter, which makes a description of its behavior using quantum mechanics impossible. According to this theory, the space time metric is obtained by solving Einstein's equations. A solution of these equations depends on the way matter is distributed, which is assumed to be classically described. An essential aspect to consider in a quantum theory of gravity is to understand how a description of matter considering the principles of quantum mechanics can be used in Einstein's equations.[38] At the moment, there is not a fully satisfactory and accepted theory of quantum gravity, and it is not clear how the concepts of quantum mechanics and general relativity would be modified in such a theory. Not having a quantum theory of gravity poses a huge challenge in making predictions when quantum systems act as gravitational field sources, giving rise to several conceptual difficulties, such as an indefinite space time metric.[39,40]

As analyzed by R. Hedrich in Ref. [41], the following issues are relevant to solve the incompatibility that exists between quantum mechanics and general relativity:

1) General relativity, which is currently the best theory of gravity as well as of space time, considers the gravitational field as a classical dynamic field, which is represented by the metric of space time. However, according to quantum mechanics, dynamic fields have quantum properties, so it seems reasonable to consider the need to quantize the gravitational field.
2) The theory of general relativity describes the gravitational field by means of the space time metric. Therefore, quantizing the gravitational field requires quantizing the metric of space time. A quantum description of the gravitational field that considers its dynamics corresponds to a dynamic space time that must be described using the

laws of quantum mechanics. But, it must be considered that quantum field theories assume a fixed and non-dynamic space time, which is in contradiction with point (1).

3) General relativity considers time as a component of space time, which is dynamically included in the interaction between matter and energy. The time can only be defined locally and there is no global time. For its part, quantum mechanics considers time as a global parameter and not as a physical observable represented by an operator.

5.17 CONCLUSIONS

In this chapter, we have analyzed how according to general relativity theory, gravity affects the space time and in particular time intervals measured by observers placed at different distances from a massive body such as our earth. An important effect predicted by this theory is the existence of black holes, corresponding to regions of space with a very high density of mass so that gravity is so strong that even light cannot escape from the interior of the corresponding volume, and space time is distorted so that time looks as frozen at points over the so called "event horizon."

Other important prediction of general relativity theory concerned with time and causality is the existence of wormholes which are shortcuts between distant regions of space time and joining two black holes. These shortcuts could theoretically provide a mean of faster than light travel. This would appear to violate the special relativistic speed limit, which prohibits exceeding the velocity of light. A person going through the wormhole could arrive at a distant star before a light beam traveling the outside route. Nevertheless, wormholes are unstable since the throat collapses in itself so rapidly that even a beam of light does not have time to travel through it.

It must be considered that there are phenomena in which both gravitational as well as quantum effects play a crucial role. This is the case of the conditions which would have existed in the very early universe, as also in the case of black holes where high energy densities are combined with strong gravitational fields. A theory based on a combination of general relativity with quantum mechanics to analyze these kind of phenomena gives rise to several points of conflict between these two theoretical structures, and one of them is the treatment of time. Quantum mechanics treats time as if it were a parameter presumed by, and hence independent of dynamics. This is in conflict with general relativity considering that time is closely related with space as members of a frame called the space time which is distorted by gravity. Almost a century has passed since the formulation of both theories and it is still not clear how these theories should be unified into one that could be accepted as a consistent theory. The generally used approach is to keep the principles on which quantum theory was founded, such as the principle of superposition, but modify the theory of general relativity. This is known as quantization of gravity to obtain a quantum theory of gravity. The opposite approach is "gravitizing quantum theory" in which it is tried to maintain the principles of the general relativity, such as the principle of equivalence, and to analyze the modifications required in the quantum theory.

Although the unification of quantum theory and general relativity is still a matter of intense research and there are several publications with proposals, till date this is a pending challenge.

KEYWORDS

- **space time**
- **geodesic**
- **gravity**
- **equivalence principle**
- **covariance**
- **Einstein tensor**
- **black hole**

REFERENCES

1. Everett, A.; Roman, T. *Time Travel and Warp Drives*; University Chicago Press: Chicago and London, 2012.
2. Gubser, S. S.; Pretorius, F. *The Little Book of Black Holes*; Princeton University Press, 2017.
3. Schiff, L. I. On Experimental Tests of the General Relativity Theory. *Am. J. Phys.* **1960,** *28*, 340.
4. Czerniawski, J. The Possibility of a Simple Derivation of the Schwarzschild Metric, 2016. arXiv:gr-qc/0611104v2
5. Einstein, A. On the Influence of Gravitation on the Propagation of Light. *Annalen der Phys.* **1911,** *35*, 898–908.
6. Pound, R. V.; Snider, J. L. Effect of Gravity on Nuclear Resonance. *Phys. Rev. Lett.* **1964,** *13*, 539–540.

7. Berry, M. *Principles of Cosmology and Gravitation*; Adam Hilger: Bristol and Philadelphia, 1989.
8. Ashby, N. Relativity and the Global Positioning System. *Physics Today*, May 2002.
9. Kalman, D. An Undetermined Linear System for GPS. *Collegue Math. J.* **2002,** *33*, 384–390.
10. Robles, P. Some Issues Concerned with PMU-Based Security Control of Electric Interconnected Systems. *Adv. Robot. Mech. Eng.* **2019.** DOI: 10.32474/ARME.2019.01.000119.
11. Bahder, T. B. Relativity of GPS Measurement. *Phys. Rev. D* **2003,** *68*, 063005-1.
12. Ashby, N. Relativistic Effects in the Global Positioning System. *J. Syst. Eng. Electron.* **1995,** *6* (4).
13. Romero, G. E. *Philosophical Issues About Black Holes*, Chapter 1; Nova Science Publishers Inc., 2014.
14. Shapiro, I. I. The Fourth Test of General Relativity. *Phys. Rev. Lett.* **1964,** *13*, 789–791.
15. Okun, L. B. Photons and Static Gravity. *Mod. Phys. Lett.* **2000,** *A15*, 1941.
16. Ryder, L. *Introduction to General Relativity*; Cambridge University Press: Cambridge, 2009.
17. Reeves, H. *Last News from Cosmos, Towards the First Second*; Stoddart: Atlanta, 1997.
18. Castagnino, M.; Lombardi, O.; Lara, L. The Arrow of Time in Cosmology, 2002. http://philsci-archive.pitt.edu/id/eprint/800
19. Robles-Pérez, S. J. Time Reversal Symmetry in Cosmology and the Creation of a Universe–Antiuniverse Pair. *Universe* **2019,** *5*, 150.
20. Fisher, A. E. Friedmann's Equation and the Creation of the Universe. *Int. J. Modern Phys. D27 Gravitation; Astrophys. Cosmol.* **2018,** *14*, 1847013.
21. Fischer, A. A Simple All-Time Model for the Birth, Big Bang, and Death of the Universe. *Int. J. Modern Phys. D* **2017,** *26* (12).
22. Wielgus, M.; Horak, J.; Vincent, F.; Abramowicz, M. Reflection-Asymmetric Wormholes and Their Double Shadows. *Phys. Rev. D* **2020,** *102*, 084044.
23. Ellis, H. G. Ether Flow Through a Drainhole: A Particle Model in General Relativity. *J. Math. Phys.* **1973,** *14*, 104.
24. Morris, M. S.; Thorne, K. S.; Yurtsever, U. Wormholes, Time Machines, and the Weak Energy Condition. *Phys. Rev. Lett.* **1988,** *61*, 1446.
25. Einstein, A.; Rosen, N. The Particle Problem in the General Theory of Relativity. *Phys. Rev.* **1935,** *48*, 73–77.
26. Fuller, R. W.; Wheeler, J. A. Causality and Multiply Connected Space-Time. *Phys. Rev.* **1962,** *128*, 919.
27. Morris, M. S.; Thorne, K. P. Wormholes in Spacetime and Their Use for Interstellar Travel: A Tool for Teaching General Relativity. *Am. J. Phys.* **1988,** *56*, 395–416.
28. Gödel, K. An Example of a New Type of Cosmological Solutions of Einstein's Field Equations of Gravitation. *Rev. Mod. Phys.* **1949,** *21*, 447.
29. Hawking, S. W. Black Hole Expansions? *Nature* **1974,** *248*, 30–31.
30. Maldacena, J. Black Holes, Wormholes and the Secrets of Quantum Space Time. *Sci. Am.* **2016,** *315*, 26–31.
31. Maldacena, J.; Susskind, S. Cool Horizons for Entangled Blak Holes. *Fortsch. Phys.* **2013,** *61*, 781–811.
32. Einstein, A.; Podolsky, B.; Rosen, N. Can Quantum-Mechanical Description of Quantum Reality Be Considered Complete. *Phys. Rev.* **1935,** *47*, 777–780.
33. Camelia, G. A.; Stachel, J. Measurement of the Space–Time Interval Between Two Events Using the Retarded and Advanced Times of Each Event with Respect to a Time-Like World-Line. *Gen. Relativ. Gravit.* **2009,** *41*, 1107–1124.

34. Padmanabhan, T. Physical Significance of Plank Length. *Ann. Phys.* **1985,** *165*, 38–58.
35. Gharibyan, V. Testing Planck-Scale Gravity with Accelerators. *Phys. Rev. Lett.* **2012,** *109*, 141103.
36. Zizzi, P. Spacetime at the Planck Scale: The Quantum Computer View, Feb 2005. https://arxiv.org/abs/gr-qc/0304032v2
37. Mamedov, A. *Unification of Quantum Mechanics with the Relativity Theory*, Chapter 4; Intech, 2015.
38. Kiefer, C. Conceptual Problems in Quantum Gravity and Quantum Cosmology. *Hindawi Pub. Corp. ISRN Math. Phys.* **2013,** *2013*, 17. Article ID 509316.
39. Castro, E.; Giacomini, F.; Belenchia, A.; Bruckner, C. Time Reference Frames and Gravitating Quantum Clocks **2019**. arXiv: Quantum Physics.
40. Carlip, S. Is Quantum Gravity Necessary? *Classical Quant. Gravity* **2008,** *25* (15), 6. Article ID 154010.
41. Hedrich, R. Has Spacetime Quantum Properties? 2009. arXiv:0902.0190v1 [gr-qc]

CHAPTER 6

A Glance at Some Recent Theories About Time in Physics

ABSTRACT

In the previous chapters, we have made an imaginary travel in a ship which we might call "time" for analyzing how this concept is present in the physical theories which describe our world and the diverse interactions which make possible its existence as we perceive it under our notion of reality. We have traveled from the submicroscopic scale—requiring a quantum description—up to the cosmological dimensions where gravity acquires a crucial role distorting the space-time of the special relativity theory.

In this chapter, we discuss some of the most recent publications related to physical theories about time, starting with the cellular automation interpretation of quantum mechanics (QM) formulated by Gerard't Hooft, where information loss is considered as the cause of the time-reversal asymmetry. Then we continue with the thermal time hypothesis of Carlo Rovelli, who asserts that it is possible to formulate classical mechanics in a way in which the time variable is treated on equal footing as other physical variables, and not considered as a special and independent variable.

Afterward, we analyze theories which propose quantization of time, and then continue with the discussion of proposed time operators for non-relativistic and relativistic QM. At the end of that section, we discuss theories related with time in quantum gravity and then we briefly explore in the physics of time travel.

6.1 INTRODUCTION

As we have seen through previous chapters of this book, the universe as we know can be characterized by a frame of reference called space–time and

Looking at Time from a Physics Perspective. Patricio Robles (Author)

in which all events take place. Events are first described by the location at which they occur (where?) and then by the instants at which they occur (when?), all of which are specified in terms of coordinates. The number of coordinates required for this specification determines the dimension of space-time. For example, we have seen that in special relativity theory an event occurring at a given point in space-time is specified by means of three spatial coordinates (x, y, z) indicating its location and a fourth coordinate corresponding to the quantity t indicating its instant of occurrence, traditionally called *time*. In almost all of the current accepted physical theories, this last coordinate is the only coordinate of the *space-time* which meaningfully allows to establish an order in the sequence of occurrence of the different events occurring in our world. Based on this order, an orientation of time can be established, which is generally called the arrow of time. Several theories about the physics of time explain the apparent lack of symmetry in time reversal considering concepts of thermodynamics and entropy giving rise to what is called the ***thermodynamic arrow of time,*** related to the concept that processes always occur in a sense such that the disorder increases, and this makes the difference between *past* and *future*.

Before the formulation of relativity theory, the notion of time played a clear and widely accepted role in physics. In the Newton's physics, time was seen as one of the fundamental ideas in terms of which the laws of the nature were built. Time was assumed to "flow," and mechanics was understood as the science that allows describing the laws of change of state of physical systems with respect to time. For the description of macroscopic processes, that point of view is still implicitly accepted in the laws of classical mechanics and classical electrodynamics. In these theories, the notion of time is described by a real and continuous variable t. If the state of any physical system can be described by a set of physical variables, then the evolution of the system is described by a set of functions $a_n(t)$ obtained by solving the corresponding dynamical equations given by Newton's laws in classical mechanics and Schrödinger's equation in quantum mechanics (QM).

We can appreciate that there is an asymmetry between time and space by considering that physical systems evolve over time, whereas there are not spontaneous displacements through the space. This asymmetry can be seen in the equations of motion and in the conservation laws, which behave differently with respect to spatial coordinates and time.

Using the sum-over paths formalism (formalism which can be conceived as a discrete version of Feynman's notion of path-integral), Joan A. Vaccaro has shown that the absence of time-reversal symmetry (T) might be the cause.[1] This absence of T-symmetry produces states localized in space and localized without limit in time, which creates an asymmetry between space and time.

According to the research made by Carlo Rovelli,[2] it is possible to formulate classical and QM treating the time variable in a similar way as the other physical variables, and not treated as a special and independent variable. The same formulation would be possible for QM and this would be a suitable formalism for a quantum theory of gravity.

In this chapter, we discuss some of the most recent publications related with physical theories about time. We start with the analysis of the cellular automaton interpretation of QM formulated by Gerard Hooft, where he postulates that what today is known as QM can result from an analysis of vector spaces for a classical system. This researcher concludes that information loss is considered as a source that gives rise to time reversal asymmetry.

Then we continue with the thermal time hypothesis of Carlo Rovelli, who asserts that it is possible to formulate classical mechanics in such a way that time, is treated on the same foot as the other physical variables, and not regarded as a special and independent variable.

Afterward, we analyze theories which propose quantization of time, such as the connection between discrete nature of time and causality formulated by Ronald Rieck, and the theory of a discrete time in QM presented by Stanley Bruce, where he proposes a discrete complex-time formulation of QM with a discrete form for Schrödinger's equation.

We continue with an analysis of some of the proposed time operators for non-relativistic and relativistic QM. At the end of that section, we discuss theories related with time in quantum gravity and then we briefly explore in the physics of time travel.

6.2 TIME IN THE CELLULAR AUTOMATON THEORY

According to Gerard't Hooft[3] to obtain *models* that adequately explain the way in which different events occur in our environment and their nature, it is essential to have an order of them. It is hard to imagine an adequate model of our universe without an ordering defined for the occurrence of these events. This allows us to explain the sequence in which they occur. An event A can be the cause of a subsequent event B, if the time variable of A is less than that of B. At least on the macroscopic scale in which the processes that mark the development of our life take place, it seems impossible to establish a model of our universe if it is not possible to describe some kind of order in which events occur. The notion of time would not be required if we consider only a collection of data, without any attempt to explain them and how they are correlated. For QM, this is also valid, since it also needs a coordinate associated with the temporal ordering of the events, as can be seen through the Schrödinger's equation and its predictions for the evolution of the state of a quantum system.

The interpretation of the theory of QM that follows from the theory of cellular automatom formulated by Hooft, consists in affirming that what today is known as QM can be the result of a vector space analysis of a classical system.

As G. Hooft points out, QM has so far been a successful theory, which explains a large part of the effects that occur on a microscopic scale, which cannot be interpreted by classical physics. However, in the tireless search for knowledge and deepening or generalization of the theories established so far, physics continues to advance, despite the fact that until now a sufficiently general theory that encompasses the known theories is not close. Although the predictions based on QM are validated by practically all the experiments that have been carried out so far, according to Hooft and from a point of view of consistency, this theory raises questions that still do not have an answer widely accepted by the scientific community. One of these questions is about the reality that QM describes, and several possible realities can exist whenever one tries to predict the outcome of one experiment. According to the Copenhague conclusion obtained in the early years of the 20th century, there is a wave function called a quantum state which is a vector in Hilbert space and in the case of non-relativistic quantum mechanics (NRQM), this wave function obeys the Schrödinger's equation.

The modules squared of the components of the vector can be used for obtaining probabilities if it is required to predict or interpret the results of an experiment. Nevertheless, the question about which is the reality one wishes to describe is still a matter of discussion.

With the suspicion that reality may be very different and simpler from what can be read in most text books, Hooft has shown that deterministic models reproducing QM behavior do exist. He shows examples that are essentially classical, but can be analyzed by the use of quantum techniques, arguing that even the standard model, incorporating gravitational interactions, can be considered a quantum mechanical approach to analyzing a system that could be classical at its core.[4] In several publications of this researcher, he has worked out some fundamental physical systems. Most of them, despite having characteristics that would require the use of a quantum treatment, can be acceptably described in classical form and some of his models are located in a domain of knowledge between classical mechanics and QM, a domain always thought to be empty.

One of the essential concepts in the interpretation of QM in the context of the theory of cellular automaton formulated by Hooft is that a classical system may be analyzed through the association of each state of the system to a vector, in such a way that the set of all states forms an orthonormal basis in a vector space called the Hilbert space. He postulates that what is called QM can be obtained from the analysis of vectorial space corresponding to the description of a classical system.

A cellular automaton can be regarded as a system of digital data located in a kind of network and updated at the beat of a clock. Information is spread with a velocity smaller than light velocity and all this information is classical.

In principle, the automaton can be in any of a huge number of states. If we assume one degree of freedom, that using Boolean algebra could be associated to a cube in which each side is a Planck length, roughly we would have a number of states equal to $20^{10^{99}}$ in every cm^3. This can be seen using the following reasoning:

If we imagine a cubic Planck length l_p (of the order of 10^{-33} cm) where there is some quantum system with just two degrees of freedom, the number of such cubes in one cm is $\mathcal{N} = \frac{1}{l_P^3} = 10^{99}$. Therefore, the number of combinations of states is $2^{\mathcal{N}} = 20^{10^{99}}$.

Only if we consider all these states, we could say that the system is deterministic. Since a system compression cannot be performed without loss of information, a better technique is required. According to Hooft, this technique is his cellular automaton interpretation (CAI) or "vector space analysis" as he calls it. In group theory, it is useful to give matrix representation of the elements, and therefore, in CAI a subset of the permutation group is considered. The elements of the set where permutations are made are represented as orthonormal vectors in the vector space.

Specifically, the evolution law is represented by a matrix expressed in terms of unit vectors that form an orthonormal basis in Hilbert space. Hooft refers to these states and the matrix that represents them as the ontological state and the ontological matrix, respectively. Applied to CAI, an ontological unitary matrix is defined as a matrix containing only one number equal to 1 with all others being zero in each row and column.

The CAI of QM is formulated using ontological states instead of wave functions. If the considered universe contains a certain state according to this interpretation, a 1 is assigned, or a 0 otherwise. There are no other wave functions, and neither superposition of functions representing states, as occurs in QM. This provides a solution to issues which still are a matter of discussion, such as the collapse of the corresponding wave function when an observable $\mathcal{O}$ is measured, appearing as if the act of measuring itself causes the wave function to acquire a new form and giving rise to what looks like a time-asymmetric process.

In this section, we discuss one example for illustrating the point of view of G. Hooft about time-reversal asymmetry for a deterministic system with loss of information. We consider a system that can be in just three states, designated as (1), (2), and (3). The law of evolution is that when a clock emits a pulse, state (1) evolves to state (2), (2) to (3), and (3) to (1), as shown in Figure 6.1. Let us consider that the clock beats with time intervals δt. Associating Dirac kets with these states, denoted as $\lceil 1\rangle$, $\lceil 2\rangle$, and $\lceil 3\rangle$, respectively, the evolution of these states is described by the following matrix:

$$U_{ev}(\delta t) = \begin{bmatrix} 0 & 0 & 1 \\ 1 & 0 & 0 \\ 0 & 1 & 0 \end{bmatrix} \tag{6.1}$$

In terms of the kets $\lceil 1\rangle$, $\lceil 2\rangle$, and $\lceil 3\rangle$, the eigen states corresponding to this matrix are:

$$\lceil 0\rangle_H = \frac{1}{\sqrt{3}}\left(e^{-i2\pi/3}\lceil 1\rangle + e^{i2\pi/3}\lceil 2\rangle + \lceil 3\rangle\right) \tag{6.2}$$

$$\lceil 1\rangle_H = \frac{1}{\sqrt{3}}\left(e^{i2\pi/3}\lceil 1\rangle + e^{-i2\pi/3}\lceil 2\rangle + \lceil 3\rangle\right) \tag{6.3}$$

$$\lceil 2\rangle_H = \frac{1}{\sqrt{3}}\left(\lceil 1\rangle + \lceil 2\rangle + \lceil 3\rangle\right) \tag{6.4}$$

Using these eigenvectors, we construct the matrix D that diagonalizes $U_{ev}\delta(t)$ as

$$D = \frac{1}{\sqrt{3}}\begin{bmatrix} e^{-i2\pi/3} & e^{i2\pi/3} & 1 \\ e^{i2\pi/3} & e^{-i2\pi/3} & 1 \\ 1 & 1 & 1 \end{bmatrix} \tag{6.5}$$

$$D^{-1}U_{ev}(\delta t)D = \begin{bmatrix} e^{i2\pi/3} & 0 & 0 \\ 0 & e^{-i2\pi/3} & 0 \\ 0 & 0 & 1 \end{bmatrix} = U_{ev}^{\,diag}(\delta t) \tag{6.6}$$

Under the action of $U_{ev}^{diag}\delta(t)$ after one beat of the clock, these states evolve as

$$U_{ev}^{diag}(\delta t)\begin{pmatrix}\lceil 0\rangle_H\\ \lceil 1\rangle_H\\ \lceil 2\rangle_H\end{pmatrix}=\begin{pmatrix}e^{i2\pi/3}\lceil 0\rangle_H\\ e^{-i2\pi/3}\lceil 1\rangle_H\\ \lceil 2\rangle_H\end{pmatrix}=\begin{pmatrix}e^{i2\pi/3}\lceil 0\rangle_H\\ e^{i4\pi/3}\lceil 1\rangle_H\\ \lceil 2\rangle_H\end{pmatrix}\tag{6.7}$$

From QM theory, we know that the evolution operator U can be written as a function of time and the Hamiltonian $\mathcal{H}$ as $U_{ev}(t)=e^{-i\frac{\mathcal{H}}{\hbar}t}$ so that for this case, we obtain that the Hamiltonian in the basis $\lceil 0\rangle_H$, and $\lceil 2\rangle_H$ can be written as

$$\mathcal{H}^{diag}=\frac{2\pi}{3\delta t}diag(1,2,3)\tag{6.8}$$

The eigenvalues of this Hamiltonian are $\frac{2\pi}{3\delta t}n$ with $n=1,2,3$. We can see that this deterministic model leads to quantized eigenvalues what has a similarity with the Zeeman splitting of the energy states of an atom when interacting with a homogeneous magnetic field. In fact, let us make a brief remind of this effect:

The magnetic moment vector of an atom arises from the combined spin and orbital angular momentum of the constituent electrons which can be described by quantum numbers j and m_j, respectively. In a weak magnetic field, the z component of the magnetic moment of an atom is

$$\mu_z=-g\frac{e\hbar}{2m_e}m_j\tag{6.9}$$

where m_e and e are the mass and charge of the electron, g is the Landé factor and m_j is a quantum number which can take $2j+1$ discrete values between $-j$ and $+j$ with j other quantum number corresponding to the orbital angular momentum.[5,6] The action of a magnetic field vector oriented in the z direction gives rise to an interaction between this field and the magnetic moment with a corresponding energy given by

$$E_{mag}=-\mu_z B=m_j\,g\frac{e\hbar}{2m_e}B\tag{6.10}$$

For $j=1$, this energy can take $2\cdot 1+1=3$ values with a difference $g\frac{e\hbar}{2m_e}B$ between adjacent levels. It can be concluded that the behavior of this atom corresponds to a deterministic system with three states, or a cogwheel, as it is called by G. Hooft.

It must be considered that all that was needed for the above-matrix manipulations in eqs 6.1–6.8 is that the deterministic law used as a starting point is time-reversible otherwise the matrix U would not be unitary. Thus, as observed by G. Hooft, it can be said that *"any deterministic time-reversible system can be described using a quantum mechanical Hilbert space, where states obey a Schrödinger equation"* (verbatim phrase extracted from Ref. [7]).

This assessment can be complemented with the following example:

Consider a discrete system that can be in any one of the states e_1, e_2, e_3, or e_4. Let there be an evolution law so that after every time step:

$$e_1 \rightarrow e_2, e_2 \rightarrow e_1, e_3 \rightarrow e_3, e_4 \rightarrow e_1 \tag{6.11}$$

This evolution is sketched in Figure 6.1 and is entirely deterministic.

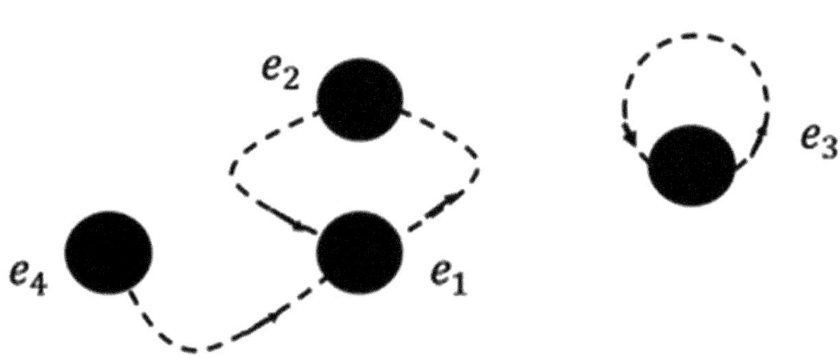

FIGURE 6.1 Transitions corresponding to eq 6.11.

Source: Adapted from Ref. [4]

It is useful to introduce the Hilbert space spanned by all four states, in order to statistically manage the evolution. Now, in this space, the one-time-step evolution operator would be

$$U_{ev} = \begin{bmatrix} 0 & 1 & 0 & 1 \\ 1 & 0 & 0 & 0 \\ 0 & 0 & 1 & 0 \\ 0 & 0 & 0 & 0 \end{bmatrix} \tag{6.12}$$

This can be proved considering that since we are dealing with a deterministic system, if its state at any given instant is e_j with $j = 1, 2, 3, 4$. This can be described by the column vector which has elements given by δ_{jk}. Therefore, if for example, the state is e_4 the action of the matrix U_{ev} would be

$$U_{ev}\begin{bmatrix}0\\0\\0\\1\end{bmatrix}=\begin{bmatrix}0&1&0&1\\1&0&0&0\\0&0&1&0\\0&0&0&0\end{bmatrix}\begin{bmatrix}0\\0\\0\\1\end{bmatrix}=\begin{bmatrix}1\\0\\0\\0\end{bmatrix} \tag{6.13}$$

which means that the state e_4 evolves to state e_1.

From Figure 6.1, it can be seen that there is not evolution for the state e_3.

This evolution operator is not unitary since the evolution rule (6.11) is not time reversible. After a short lapse of time, only states e_1, e_2, and e_3 can be reached. In this simple example, one could simply erase state e_4 and treat the upper left 3 × 3 part of the matrix U_{ev} given by eq 6.12 as the unitary evolution matrix. Thus, the quantum system corresponding to the new evolution law is three-dimensional, not four-dimensional, and it seems to be trivial to restore the property of time-reversibility and the unitary characteristic.

Nevertheless, in more complicated non-time-reversible evolving systems, the "genuine" quantum states and the false ones (the ones that cannot be reached from the distant past) are actually quite difficult to distinguish.

For this reason, G. Hooft has introduced the notion of *equivalence classes*. Two states are called equivalent if, after some finite time interval, they evolve into the same state. The system described above has three equivalence classes,

$$E_1=\{(2),(4)\},\ E_2=\{1\},E_3=\{3\} \tag{6.14}$$

and the reduced evolution operator in terms of the states E_1, E_2, and E_3 becomes

$$U^{red}=\begin{bmatrix}0&1&0\\1&0&0\\0&0&1\end{bmatrix} \tag{6.15}$$

For this equivalent system, one can build a Hamiltonian $\mathcal{H}$ such that $U(t)=e^{-i\frac{\mathcal{H}}{\hbar}t}$ as made before in eq 6.8 and this system would be in an eigenstate of this Hamiltonian. Since the phases of the states e_i are arbitrary, we can assume that this system can be in the state $|E_3\rangle$ or in the state $|\psi^+\rangle$ given by

$$|\psi^+\rangle=\frac{1}{\sqrt{2}}\left(|E_1\rangle+|E_2\rangle\right) \tag{6.16}$$

This is an example of the way by which the vector space analysis formulated by Hooft considers information loss as a source of time-reversal asymmetry.

One may simply ignore the "unreachable" states and notice that the universe ends up in a limit cycle that is indistinguishable from what a universe without information loss would do. It is the fact that the true universe is too large to end up in a limit cycle, and that we wish to understand small regions of this universe, nearly but no quite decoupled from the rest of the world, which imply that the introduction of information loss and equivalence classes is not trivial.

6.3 THE THERMAL TIME HYPOTHESIS

According to Carlo Rovelli, it is possible to formulate classical mechanics in a way in which the time variable is treated in a similar way to the other variables, and not singled out as the special independent variable.[8] He argues that this is the natural formalism for describing general relativistic systems and that it is possible to formulate QM in the same manner, which would be an effective advance in the formulation of a quantum theory that incorporates gravity.

Considering the context of the work of Rovelli, it is worth to make a revisit to what Galileo Galilei discovered in century XVI, where according to what has been known he watched the oscillation of a chandelier. Using his own pulse as a clock, Galileo discovered that there was the same number of pulses within each oscillation of the chandelier. This form of measuring oscillations has been used since many years ago and the question formulated by Rovelli is "how can we justify that the measurement of a clock corresponds to time, if we can only check it against another clock?"

According to Newton (*Principia*), the true time variable t is never measured. Rather, we always construct devices, the "clocks" indeed, that have observable quantities whose motion is proportional to the true time, within an approximation good enough for purposes of measuring time intervals. In non-relativistic mechanics what is measured is only relative evolution between variables, and it is implicitly assumed that there is a fundamental variable known as time, such that all observable quantities evolve with respect to it, and the equations look simple when written in terms of that variable.

The thermal time hypothesis

In a macroscopic system formed by a lot of molecules in a non-equilibrium thermal state, all thermodynamic variables are in a permanent change.

However, there is just one that does not vary: the total amount of energy that an isolated system has. As known, there exists a close link between energy and time. On the one hand, if we know the energy of a given system we can deduce its evolution. For example, if the evolution of our system is determined by the Hamiltonian $\mathcal{H}$, according to the equations of classical mechanics, the rate of variation of a quantity $\mathcal{A}$ is given by:

$\frac{d\mathcal{A}}{dt} = \{\mathcal{A}, \mathcal{H}\}$, where $\{\square\}$ correspond to the Poisson brackets.

On the other hand, if the total amount of energy of our system is constant, in his thermal agitation, the system passes only through all configurations which have the same energy.

The set of these configurations which macroscopically cannot be distinguished constitutes the macroscopic equilibrium state. In the usual interpretation of the relation between time and equilibrium, time is considered as an absolute and objective quantity and the energy is what rules the temporal evolution. This is sketched by Rovelli in his book The Order of Time, as the following sequence:

$$Time \Rightarrow Energy \Rightarrow Macroscopic\ State$$

But we can also conceive this relation regarding the macroscopic state as a corresponding configuration which conserves an energy, which in turns generates a time, as sketched in Figure 6.2.

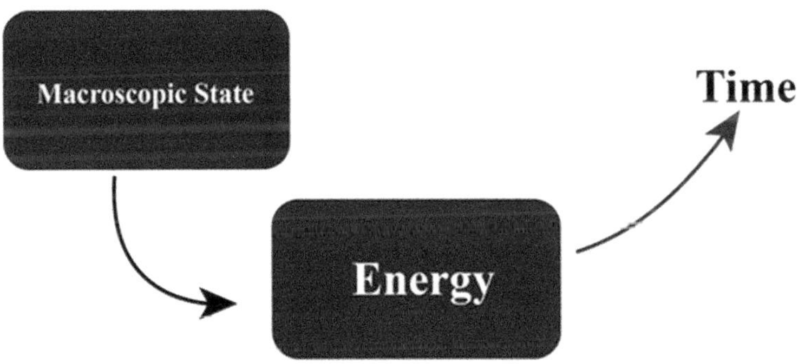

FIGURE 6.2 Time generated from a macroscopic state according to the thermal time hypothesis.

This last observation leads to Rovelli to formulate the hypothesis that in an elementary physical system, there is not any privileged variable which behaves as time and all the variables are at the same level of importance for the description of this system. In other words, for a macroscopic state it is possible to choose some variable having the characteristics of time. The

time so determined is called ***thermal time.*** From the microscopic point of view it has nothing in special, but macroscopically its nature is such that among several variables, it is that whose behavior is more similar to that corresponding to what we usually call time.

The concept of thermal time has been proposed by C. Rovelli and M. Smerlack, to obtain a thermodynamic theory that includes general relativity. The thermal time flow for an observable A is defined as follows[9]:

$$\frac{dA}{d\tau} = \frac{1}{k_B T}\frac{dA}{dt} \tag{6.17}$$

where T is the Kelvin temperature and k_B is the Boltzmann constant. In this equation, $\tau = \hbar/k_B T$ is defined as the *thermal time*. It is worth to note that in the standard path integral formalism, the quantity $\hbar/k_B T$ is considered as a formal time.[10]

Whatever the statistical state ρ is, there exists always a variable t_p, measured by the thermal clock (a device whose reading grows linearly with the thermal time flow), with respect to which the system is in equilibrium and physics is the same as in the conventional non-relativistic statistical case.

This observation leads to Rovelli to the following hypothesis:

In nature, there is no preferred variable to assign as time. Rather all variables are equivalent. The system can be found in an arbitrary state designated as ρ, with which we can single out a specific variable that we call time.

"In nature, there is no preferred variable to assign as time. There are no equilibrium states that can be chosen a priori. Rather all variables are equivalent. The system can be found in an arbitrary state designated as ρ, with which we can single out a specific variable, that we call time" (verbatim phrase extracted from Ref. [9]).

This is equivalent to say that the statistical state determines to which variable the physical time corresponds, the physical time, and not some hypothetical type of flow that brings the system to a selected statistical state. To say that a certain variable plays the role of time does not imply a statement related to the fundamental mechanical structure of reality. Rather, a statement is made about the statistical distribution used to describe the properties of the system described in macroscopic form. According to the thermal time hypothesis which is called time corresponds to the thermal time of the statistical state in which the world exists if it is described in terms of the chosen macroscopic parameters.

An example in which this hypothesis works fairly well corresponds to a radiation filled covariant cosmological model, with no preferred time variable and in which the cosmological background radiation is represented by a statistical state. In that case, the thermal time of this state turns out to be precisely the Friedmann time. Additionally, this hypothesis can be extended in a natural way to the quantum context, and even more to the quantum field theoretical context, leading also to a general abstract state-independent notion of time flow.

6.4 TIME AS A DISCRETE DYNAMICAL VARIABLE

Is time quantized?

According to John Baez (researcher of the Mathematics Faculty of the University of California Riverside), a brief answer is that up to now "nobody knows" and there are as yet no adequately established physical theories that incorporate a fundamental unit of time. There are no experimental results to verify the existence of a minimal unit of time. On the other hand, there is also no experimental evidence against it.

William G. Unruh (professor of the Department of Physics and Astronomy at the University of British Columbia) has a similar opinion about the issue of quantization of the time. He thinks that putting together the three constant G_N (Newton's constant of gravity), h (Planck's constant), and c (the velocity of light) in the form $t_{Plank} = \sqrt{G_N \hbar / c^5}$, one could get a minimum amount of time that makes sense and that is called the **Planck time,** with a value of about 10^{-44} s. For time intervals shorter than Plank's time, the concept of "time" would have no physical meaning.

For William G. Tifft (researcher of the Department of Astronomy at the University of Arizona), there is not at present any conclusive evidence that time is quantized. Nevertheless, some theoretical studies suggest that in order to unify general relativity (which adequately incorporates gravitational effects) with the theories of quantum physics that describe fundamental particles and its interactions, it could be necessary to quantize space and probably also the time.

In spite of the above opinions coming from prominent and contemporary physicists, in this section we discuss some of the relatively recent theories which postulate the quantization of time.

6.4.1 CAUSALITY AND DISCRETE NATURE OF TIME

Nearly 40 years ago, T.D. Lee (Nobel Prize on Physics, 1957) postulated that it is possible a discrete time formulation and also that time can be treated as a dynamical variable.[11] This is applicable to classical and non-relativistic QM, as also to relativistic quantum field theories.

These ideas have been extended by several other researchers, as for example, Ronal Riek, researcher of ETH, Zurich, which defines time as the metric of causality and demonstrates that time, must have a discrete nature.[12] In what follows we review the main results presented by R. Riek.

Starting with a description of classical Hamilton equations, the second law of Newton can be written as a difference equation including an infinitesimal time step $\Delta t \rightarrow 0$ but different from zero:

$$\frac{p_i(t+\Delta t)-p_i(t)}{\Delta t}=-\frac{\partial V}{\partial q_i} \tag{6.18}$$

This last equation considers the Hamilton function $H(\boldsymbol{q},\boldsymbol{p})=\frac{\boldsymbol{p}^2}{2m}+V(\boldsymbol{q})$ for a particle of mass m, where $\boldsymbol{q}=(q_1,q_2,q_3)$ and $\boldsymbol{p}=(p_1,p_2,p_3)$ is the canonical position and canonical momentum, respectively, in the generalized phase space. The eq 6.18 can be rewritten in a form which connects the cause described by the force $-\frac{\partial V}{\partial q_i}$ with the effect $p_i(t+\Delta t)$ as

$$p_i(t+\Delta t)=\left\{-\frac{\partial V}{\partial q_i}\Delta t\right\}+p_i(t) \tag{6.19}$$

The first term to the right of eq 6.19 and inside brackets is the cause, that is the force $-\frac{\partial V}{\partial q_i}$ acting during an infinitesimal time step $\Delta t_{\rightarrow 0}\neq 0$ (in kinematics is traditionally called an *impulse*). If Δt was zero there would not be any effect. This implies that a force can only produce an effect if time is discrete. This can be seen in terms of Poisson brackets as follows:

If two functions $f(\boldsymbol{q},\boldsymbol{p})$ and $g(\boldsymbol{q},\boldsymbol{p})$ commute so that $\{f,g\}=\sum_{j=1}^{n}\left(\frac{\partial f}{\partial q_j}\frac{\partial g}{\partial p_j}-\frac{\partial f}{\partial p_j}\frac{\partial g}{\partial q_j}\right)=0$ (with n being the number of degrees of freedom), then these functions are independent and there is not a causality relation between them. The Poisson bracket $\left\{p_i(t+\Delta t),-\frac{\partial V}{\partial q_i}\right\}$ gives

$$\left\{p_i(t+\Delta t),-\frac{\partial V}{\partial q_i}\Delta t\right\}=\left\{p_i(t)+-\frac{\partial V}{\partial q_i}\Delta t,-\frac{\partial V}{\partial q_i}\Delta t\right\}$$
$$=\left\{-\frac{\partial V}{\partial q_i}\Delta t,-\frac{\partial V}{\partial q_i}\Delta t\right\}+\left\{p_i(t),-\frac{\partial V}{\partial q_i}\Delta t\right\}$$

$$=\left\{p_i(t),-\frac{\partial V}{\partial q_i}\Delta t\right\}=\sum_{j=1}^{n}\left\{\frac{\partial p_i}{\partial q_j}\frac{\partial}{\partial p_j}\left(-\frac{\partial V}{\partial q_i}\Delta t\right)-\frac{\partial p_i}{\partial p_j}\frac{\partial}{\partial q_j}\left(-\frac{\partial V}{\partial q_i}\Delta t\right)\right\}$$

$$=\frac{\partial^2 V}{\partial q_i^2}\Delta t \tag{6.20}$$

In this last equation, it has been considered that momenta p_i are independent variables and therefore, they are independent of the coordinates q_j, and that $\frac{\partial p_i}{\partial p_j}=\delta_{ij}$.

If Δt was zero, this commutator would be zero for any potential, what would be a negation of the causality relation between $p_i(t+\Delta t)$ and the force $-\frac{\partial V}{\partial q_i}$ derived from Newton's second law. One could wander what happens with eq 6.20 in the case of a particle subject to a force corresponding to a gravitational potential of the form $-mgy$, since in this case the second derivative with respect to the canonical coordinates q_j would be zero, and therefore, the Poisson bracket of eq 6.24 would also be null, independent of the time step Δt, which would not require to be of a discrete nature. According to an explanation kindly given by Professor Riek, the second derivative of the gravitational potential can of course never be zero because it acts through space (only approximately on earth it is a constant g) and thus changes along the space coordinate.

Let us try to make an extension to non-relativistic QM of the interpretation about causality and discrete nature of time suggested by eq 6.20. For that purpose, we consider the Ehrenfest theorem:

$$\frac{d}{dt}\langle\vec{p}\rangle=-\langle\nabla V(\vec{r})\rangle \tag{6.21}$$

This last equation relates the time derivative of the expectation value of the momentum with the corresponding value of the force acting on the particle. From this equation, we can express the expectation value of the momentum at the instant $t+\Delta t$ as:

$$\langle\vec{p}\rangle(t+\Delta t)=-\langle\nabla V(\vec{r})\rangle\Delta t+\langle\vec{p}\rangle(t) \tag{6.22}$$

This last equation is very similar to eq 6.19 except that in this case, we are dealing with expectation values. Furthermore, if one uses the Heisenberg equations of motion for a particle subject to a potential we obtain[5]:

$$\frac{dp_i}{dt}=\frac{1}{i\hbar}\left[p_i,V(\vec{r})\right]=-\frac{\partial}{\partial x_i}V(\vec{r}) \tag{6.23}$$

This equation in a form similar to that of eq 6.21, except that in this case we are working with operators:

$$p_i(t+\Delta t)=\left\{-\frac{\partial V}{\partial x_i}\Delta t\right\}+p_i(t) \tag{6.24}$$

From eq 6.24, we obtain

$$\left[p_i(t+\Delta t),-\frac{\partial}{\partial x_i}V(\vec{r})\Delta t\right]=\left[p_i(t)-\frac{\partial}{\partial x_i}V(\vec{r})\Delta t,-\frac{\partial}{\partial x_i}V(\vec{r})\Delta t\right]$$

$$=\left[p_i(t),-\frac{\partial}{\partial x_i}V(\vec{r})\Delta t\right]$$

$$=-i\hbar\frac{\partial}{\partial x_i}\left(-\frac{\partial}{\partial x_i}V(\vec{r})\Delta t\right)=i\hbar\frac{\partial^2 V(\vec{r})}{\partial x_i^2}\Delta t \tag{6.25}$$

Similar arguments to those given with respect to eq 6.20 suggest that assuming valid the hypothesis of existence of causality in QM, the smallest interval of time possible cannot be zero and this is associated with the idea of a discrete nature. While this idea is an issue which has been postulated from several years ago, the subject of causality in QM seems losing relevance in the quantum realm as we have discussed in Chapter 3, for example, in the way the EPR paradox is described.

With respect to a quantized time, it is worth to mention that nearly 100 years ago there were various attempts to introduce a discrete space-time. In 1929, while physicists were frustrated by the problem of infinities predicted by quantum electrodynamics at the development stage at this moment, Werner Heisenberg suggested as one alternative of solution the idea that the space is subdivided into cells of finite dimensions and perhaps to consider small units of time. The essential concepts of this (unpublished) idea are contained in a letter to Niels Borh.[13] Several years later, T.D. Lee (Nobel Prize on Physics, 1957) postulated that it is possible a discrete time formulation and also that time can be treated as a dynamical variable.[11]

In relation to compliance by QM of the causality principle, this is a controversial issue. In particular, non-relativistic QM seems to violate causality since apparently superluminal propagation of signals is possible in its theoretical frame, as occurs, for example, in the EPR paradox. Several solutions to this paradox have been given in several books and publications (see e.g., Ref. [14]). Nevertheless, current research in the field of quantum causality leads to think that the notion of defined causal structures will not be achieved in a theory in which gravity, and therefore, the metric field and the distances in space-time between events, are ruled by the laws of QM. Recent research reveals that causal relationships can be placed in quantum superposition states where A influences B and B influences A.[15]

6.4.2 DISCRETE TIME VALUES IN QUANTUM MECHANICS

Throughout the development of quantum mechanics, time is always considered as a continuous parameter. Nevertheless, the same does not occur for the position. For example, in Feynman's path integration, the corresponding path can be regarded as that over the whole range of eigenvalues of the position operator $\hat{r}$. This shows a clear difference between $\hat{r}$ as an operator and t as a parameter.

In relativistic quantum field theory, space and time must be treated symmetrically due to Lorentz invariance. The usual approach is to regard the position and time as parameters, while the operators are field variables.

In a publication of 2001, Stanley Bruce, researcher of the University of Andes, Chile,[16] has proposed a discrete complex-time formulation of QM with a discrete form for Schrödinger equation containing a discrete derivative with respect to a complex variable defined as:

$$s = -\frac{i}{\hbar}\left(t + i\upsilon\right) \tag{6.26}$$

In this equation, υ is a real parameter and t is the usual variable corresponding to the parameter time in standard QM so that the complex variable s has units $[Energy]^{-1}$ and the time evolution is given by the complex-time propagator $U(s) = exp(s\mathcal{H})$ which for s real gives the known evolution operator $U(t) = exp\left(-\frac{i}{\hbar}\mathcal{H}t\right)$, if one assumes that the Hamiltonian $\mathcal{H}$ is hermitian and time-independent.

Considering a particular physical system with a set of commuting observables $\hat{\alpha}$, one can take their simultaneous eigenkets as basic kets: $|\psi_\alpha(s)\rangle$, with $\alpha = (\alpha_1, \alpha_2, \ldots\ldots)$ where α_i is the eigenvalue of the observable $\hat{\alpha}_i$.

Starting from the Schrödinger equation, we obtain that in terms of the complex variable s, the evolution for this system is given by

$$i\hbar\frac{\partial|\psi_\alpha(s)\rangle}{\partial t}=\mathcal{H}|\psi_\alpha(s)\rangle\Rightarrow i\hbar\frac{\partial|\psi_\alpha(s)\rangle}{\partial s}\frac{\partial s}{dt}$$

$$=\mathcal{H}|\psi_\alpha(s)\rangle\Rightarrow\frac{\partial|\psi_\alpha(s)\rangle}{\partial s}=\mathcal{H}|\psi_\alpha(s)\rangle \tag{6.27}$$

In this last expression, we have considered that according to eq 6.26 $\frac{\partial s}{\partial t}=-\frac{i}{\hbar}$. Considering a discrete distribution $0, s_1, s_2 \ldots \ldots s_N$ of the variable s in the complex plane and a fundamental interval s, Schrödinger's equation takes the form:

$$\mathcal{H}_D|\psi_\alpha(s)\rangle=\frac{\delta_\lambda}{\delta s}|\psi_\alpha(s)\rangle \tag{6.28}$$

In this last equation, $\mathcal{H}_D$ represents the equivalent discrete Hamiltonian corresponding to a continuous Hamiltonian $\mathcal{H}$ and s corresponds to given s_j. The subindex α designates the eigenvalue α_i of the observable $\hat{\alpha}_i$ and λ is a complex-valued function of s and δs in the complex plane λ.

The state discrete difference $\delta_\lambda|\psi_\alpha(s)\rangle$ is defined as $\delta_\lambda|\psi_\alpha(s)\rangle = |\psi_\alpha(s+\delta_s)\rangle - |\psi_\alpha(s)\rangle$ which can be expanded in a Laurent series as

$$\delta_\lambda|\psi_\alpha(s)\rangle=\frac{d}{ds}|\psi_\alpha(s)\rangle\delta s+\sum_{n=2}^{\infty}\frac{1}{n!}\frac{d^n}{ds^n}|\psi_\alpha(s)\rangle\delta s^n \tag{6.29}$$

Combining the last three equations and considering the expansion $e^x=1+x+\frac{x^2}{2!}+\ldots$, this last equation can be expressed as

$$\delta_\lambda|\psi_\alpha(s)\rangle=e^{\delta s\mathcal{H}}-1=2e^{\frac{\delta s\mathcal{H}}{2}}\sinh\left(\frac{\delta s\mathcal{H}}{2}\right) \tag{6.30}$$

$$\mathcal{H}_D=\frac{2}{\delta s}e^{\frac{\delta s\mathcal{H}}{2}}\sinh\left(\frac{\delta s\mathcal{H}}{2}\right) \tag{6.31}$$

This last equation is in agreement with eq 6.13 of Ref. [16] if one considers that the function $\lambda_s=\lambda(s,\delta s)$ there considered is taken as 1.

It is instructive to discuss two examples of application of the results obtained in the quoted paper to illustrate how the standard time–energy commutation relation is modified (Example 6.1) and the shift of the inertial mass of a relativistic spin −0 free particle (Example 6.2).

EXAMPLE 6.1

From eq 6.35 with $\mathcal{H} = i\hbar\frac{\partial}{\partial t}$, we obtain that the discrete Hamiltonian $\mathcal{H}_D$ can be written as:

$$\mathcal{H}_D = \frac{2}{\delta s} e^{\frac{\delta s \mathcal{H}}{2}} sinh\left(\frac{\delta s \mathcal{H}}{2}\right) = \frac{1}{\delta s}\left(e^{\delta s \mathcal{H}} - 1\right) = \frac{1}{\delta s}\left(e^{i\hbar\delta s \frac{\partial}{\partial t}} - 1\right) = \frac{1}{\delta s}\left(e^{\delta s \frac{\partial}{\partial s}} - 1\right)$$

$$= \frac{1}{\delta s}\sum_{n=1}^{\infty} \frac{(\delta s)^n}{n!}\frac{d^n}{ds^n} = \sum_{n=1}^{\infty} \frac{(\delta s)^{n-1}}{n!}\frac{d^n}{ds^n} \tag{6.32}$$

Using this last expression, we calculate the commutator $[\mathcal{H}_D,s]$ operating over a ket $|\psi_\alpha(s)\rangle$ as follows:

$$[\mathcal{H}_D, s]|\psi_\alpha(s)\rangle = (\mathcal{H}_D s - s\mathcal{H}_D)|\psi_\alpha(s)\rangle$$

$$= \sum_{n=1}^{\infty} \frac{(\delta s)^{n-1}}{n!}\frac{d^n}{ds^n}\left(s|\psi_\alpha(s)\rangle\right) - s\sum_{n=1}^{\infty} \frac{(\delta s)^{n-1}}{n!}\frac{d^n}{ds^n}\left(|\psi_\alpha(s)\rangle\right) \tag{6.33}$$

Since $\frac{d^n|s\psi_\alpha(s)\rangle}{ds^n} = n\frac{d^{n-1}|\psi_\alpha(s)\rangle}{ds^{n-1}} + s\frac{d^n|\psi_\alpha(s)\rangle}{ds^n}$ after some algebra, we obtain

$$[\mathcal{H}_D,s] = I + \delta s \mathcal{H}_D \tag{6.34}$$

where I is the identity operator.

With $\delta s = \frac{1}{i\hbar}\tau$ where τ is a finite time element, eq. (6.38) takes the form

$$[\mathcal{H}_D, t] = i\hbar I + \tau \mathcal{H}_D \tag{6.35}$$

which implies a modification to the time–energy commutation relation $[\mathcal{H}, t] = i\hbar I$ proposed by Heisenberg in his famous article on the Uncertainty Principle.[17] This uncertainty relation between energy and time has been very controversial and several publications discuss its validity as that of the relation $\Delta E \cdot \Delta t \geq \hbar$. As discussed by J. Hilgevoord,[18] there are many instances in physics where an uncertainty principle of some sort for energy and time does hold, the principal being the relation between the energy width and the lifetime of a quantum state. In addition, it must be considered that the time–energy commutation relation $[\mathcal{H}, t] = i\hbar I$ can be deduced directly even the quantity t is considered not an operator but only a parameter. This is shown as follows if this commutator is made to act over any ket $|\psi\rangle$

$$[\mathcal{H},t]|\psi\rangle = (\mathcal{H}t - t\mathcal{H})|\psi\rangle = i\hbar\frac{\partial}{\partial t}(t|\psi\rangle) - ti\hbar\frac{\partial}{\partial t}(|\psi\rangle)$$
$$= i\hbar\left\{|\psi\rangle + t\frac{\partial}{\partial t}(|\psi\rangle) - t\frac{\partial}{\partial t}(|\psi\rangle)\right\}$$
$$= i\hbar|\psi\rangle \tag{6.36}$$

Therefore, even if t is considered only as a number instead of an operator, the relation $[\mathcal{H}, t] = i\hbar I$ is fulfilled though it is not possible to derive from it the relation $\Delta E \cdot \Delta t \geq \hbar$. This does not occur with the uncertainty relation $\Delta x_i \cdot \Delta p_i \geq \hbar/2$ between one of the components of the position vector and the corresponding component of the momentum whose origin is the general expression relating the uncertainty in the expectation values of any two observables A and B that do not commute, that is,

$$\Delta A \, \Delta B \geq \frac{1}{2}\left|\left\langle\left[\hat{A}, \hat{B}\right]\right\rangle\right| \tag{6.37}$$

Not having a time operator has caused a concern among quantum theory researchers. We will come back to this subject in Section 6.5.

EXAMPLE 6.2

We consider a relativistic spin-0 free particle of mass m and momentum $\vec{p}$. We will verify one important result obtained with this quantum discrete model, concerned with the possibility of superluminal propagation. From eq 6.31, we can write:

$$\mathcal{H}_D = \frac{2}{\delta s} e^{\frac{\delta s \mathcal{H}}{2}} \sinh\left(\frac{\delta s \mathcal{H}}{2}\right) = \frac{1}{\delta s}\left(e^{\delta s \mathcal{H}} - 1\right) \tag{6.38}$$

where for this case, the continuous Hamiltonian operator $\mathcal{H}$ is given by

$$\mathcal{H} = \sqrt{c^2 \vec{p}^2 + m^2 c^4} \tag{6.39}$$

with $\vec{p} = -i\hbar\nabla$, the continuous momentum operator. In the discrete model, the components of the momentum operator are defined as

$$\hat{p}_j \equiv \left(\frac{\mathcal{H}_D}{c^2}\right)\frac{\delta \hat{q}_j}{\delta s} \tag{6.40}$$

In this last equation, the term $\frac{\mathcal{H}_D}{c^2}$ is associated to a mass M which in this context is postulated as a relativistic invariant quantity, while $\frac{\delta \hat{q}_j}{\delta s}$ which can be considered as equivalent to the velocity of the particle in this model with $\hat{q}_j$ the components of the position operator. In analogy with the Heisenberg equation of motion, we have for this case:

$$\frac{\delta \hat{q}_j}{\delta s} = \frac{i}{\hbar}\left[\mathcal{H}_D, \hat{q}_j\right] = \frac{\partial \mathcal{H}_D}{\partial p_j} = \mathcal{H}^{-1} e^{\delta s \mathcal{H}} c^2 p_j \tag{6.41}$$

For obtaining eq 6.41, we have used eq 6.32 and the chain rule for derivation, so that $\frac{\partial \mathcal{H}_D}{\partial p_j} = \frac{\partial \mathcal{H}_D}{\partial \mathcal{H}} \frac{\partial \mathcal{H}}{\partial p_j}$, where the Hamiltonian $\mathcal{H}$ can be expressed in terms of the momenta operators p_j as $\mathcal{H} = \sqrt{\sum_j c^2 p_j^2 + m^2 c^4}$.

Combining eqs 6.40 and 6.41, we have

$$\hat{p}_j = \mathcal{H}_D \mathcal{H}^{-1} e^{\delta s \mathcal{H}} p_j \tag{6.42}$$

In the limit $E \to c\,|\vec{p}|$ as could be the case of a massless or an ultrarelativistic particle, from eq 6.41 it can be seen that this model predicts the possibility of superluminal propagation. In fact in this limit and taking $\delta s = \tau/\hbar$ with τ a finite time element, we would have:

$$\upsilon_j = \frac{\delta \hat{q}_j}{\delta s} = \frac{1}{E} e^{\delta s E} c^2 p_j = c \frac{e^{\left(\frac{\tau}{\hbar} c|\vec{p}|\right)}}{|\vec{p}|} p_j \tag{6.43}$$

From this last equation, it can be observed that the factor $\frac{e^{\left(\frac{\tau}{\hbar} c|\vec{p}|\right)}}{|\vec{p}|} p_j$ even for very small time step τ could be greater than 1, for instance as is the case of one dimensional motion with $p_j = |\vec{p}|$.

Other result of this model for the case of a relativistic spin-0 free particle we are considering, is a shift in value of the inertial mass. In fact, if the existence of a relativistic invariant mass M is postulated, we can write the following condition

$$E_D^{\,2} - c^2 \hat{p}^2 = M^2 c^4 \tag{6.44}$$

where E_D is the energy eigenvalue of the operator $\mathcal{H}_D$ so that from eq 6.38:

$$E_D = \frac{1}{\delta s}\left(e^{\delta s E} - 1\right) \tag{6.45}$$

Expressing eq 6.42 in terms of energies, we have

$$c^2 \hat{p}^2 = \frac{E_D^{\,2}}{E^2} e^{2\delta s E} c^2 \vec{p}^2 \tag{6.46}$$

From the relativistic expression for the energy of a particle of mass m and momentum $\vec{p}$, we have $c^2\vec{p}^2 = E^2 - m^2c^4$ which combined with eqs. (6.45) and (6.46) gives

$$E_D{}^2 - c^2\hat{p}^2 = E_D{}^2\left\{1 - exp\left[2\delta sE\right]\left(1 - \left(\frac{mc^2}{E}\right)^2\right)\right\} = M^2c^4 \quad (6.47)$$

For $\vec{p} = 0$ and from eq 6.47, we obtain

$$M = \frac{E_D}{c^2} = \frac{1}{c^2\delta s}\left(e^{\delta sE} - 1\right) \quad (6.48)$$

If we consider again $\delta s = \tau/\hbar$ and making a series expansion of the exponential factor $e^{\delta sE}$, we get:

$$M = \frac{\hbar}{c^2\tau}\left(\frac{\tau}{\hbar}mc^2 + \frac{\tau^2}{2\hbar^2}m^2c^4 + \mathcal{O}\left(\frac{\tau^2}{\hbar^2}\right)\right) \approx m + \frac{\tau}{2\hbar}m^2c^2 \quad (6.49)$$

where we have neglected terms with powers of $\frac{\tau}{\hbar}$ greater than 2. The second term of eq 6.49 represents a shift in value of the inertial mass m. The validity of the approximation used to obtain eq 6.49 can be justified if we consider the mass m of the order of the mass of the electron, and τ of the order of the Planck's time, giving $\tau mc^2/\hbar \sim 10^{-23}$.

6.4.3 QUANTIZATION OF SPACE–TIME

Lucas C. Céleri and Vasileios I. Kioses have proposed a way of encoding acceleration directly into the fields of quantum theory, setting up the accelerated quantum fields, as they are usually called, and which have some very interesting properties.[19] The most important property is related to the fact that they provide a mathematically consistent form that allows the quantization of the space–time in the same way as energy and momentum are subjected to a quantization process in established quantum field theories.

The description of a classical system made from a non-inertial reference frame exhibits fictitious forces. These forces are not originated by any physical interaction, being solely a consequence of the non-inertial feature of the reference frame itself. The notion of fictitious forces also emerges in general theory of relativity. Einstein used accelerated systems and their accompanying fictitious forces, as a guide in order to comprehend the nature of gravity in classical physics. Einstein's motivation was the observation that mass never appears in the equations of motion under fictitious forces, a feature which is also encountered with gravity, already known from Galileo's famous experiments.

Several difficulties arise from the application of the usual prescription of quantum field theory to the quantization of the gravitational field as described by general relativity. This fact has led to the theorists to consider more radical approaches to the problem of quantum gravity, among which it is worth noting the theory of loop quantum gravity (LQG) which provides a novel theory of geometry described from a quantum point of view. This theory formulates the Hamiltonian (or canonical) quantization of general relativity using variables different from the standard metric variables. According to LQG, the structure of space is composed of finite loops woven into an extremely fine network. The networks composed of such loops are called spin networks. The main prediction of loop quantum gravity (LGG) is that geometrical operators such as area and volume have a discrete spectrum of eigenvalues.

Einstein concluded that space–time and the gravitational field represent the same physical quantity after recognizing the equivalence between fictitious forces and gravitational forces. Due to the importance of fictitious forces in the geometrization of gravity, one could believe that fictitious forces could play a key role also in quantization of gravity. A natural question that comes up is the role of these forces in quantum field theory and whether Einstein's idea can also be applied in quantum theory.

As discussed in Chapter 4, accelerated systems can be analyzed using special relativity. Since quantum field theories are subject to the rules of special relativity and QM, we can ask if there is any quantum field theory capable of embracing acceleration as a fundamental characteristic of the fields. This answer is affirmative by defining the accelerated quantum fields. The structure of accelerated quantum fields permits us to interpret them as fictitious in the sense that they disappear in an inertial reference frame.

The definition of accelerated fields given in reference[19] rests on the kinematical description of the particles and the assumption that their kinematics is described by a different wave function. The procedure used is close to the standard quantization of a free scalar field in flat space–time.

One of the main results obtained in Ref. [19] is the argument that it is possible to define accelerated fields by canonically quantizing the momentum space. Therefore, it is necessary to proceed beyond the standard canonical field quantization in Minkowski space–time, with the establishment of the field equations and the commutation relations and, then, with the construction of the associated framework of observables and quantum states. In the theory of accelerated fields, the wave-particle duality appears in the association of the wave vector and angular frequency with space and time, respectively. Here, space and time are quantized, in an analogous way as occurs for quantities like energy and momentum in usual quantum field theories.

As a result of the quantization of the space-time using the concept of accelerated fields, the same authors have shown an alternative way to show the Unruh effect. It must be considered that from the point of view of an accelerating observer or detector, empty space contains a gas of particles at a temperature which is proportional to the acceleration. From quantum field theory applied to a quantum space-time, Céleri and Kioses have shown that Unruh effect can be obtained without changing the reference frame.[20] Therefore, in the framework of accelerated fields, the observational confirmation of Unruh effect could be associated to the existence of quantum properties of space-time.

In a paper of 1974, Hawking predicted that a black hole should radiate with a temperature $T = \hbar g/2\pi k_B c$ where g is the gravitational acceleration at the surface of the black hole, k_B is the Bolttzmann's constant and c is the speed of light in vacuum.[21] Shortly, thereafter it was shown separately by Davies and Unruh[22,23] that a uniformly accelerated detector in vacuum responds just as though it were in a thermal field of temperature $\hbar a/2\pi k_B c$ where a is the acceleration in the instantaneous rest frame of the detector.

6.5 QUANTUM OPERATOR FOR THE TIME IN NON-RELATIVISTIC QUANTUM MECHANICS

6.5.1 INTRODUCTION

In quantum theory, time would be expected to have an associated operator. However, from the first years of QM, it is known that in general, time cannot be represented by a self-adjoint operator.[24,25]

Time in QM appears as a classical parameter in the Schrödinger's equation. Physically, it represents the time shown by a "classical" clock in the laboratory. Even though this is acceptable for all practical purposes, a fully quantum description of time is still an open question.

In non-relativistic QM and in particular in the Schrödinger's equation, time is considered as a parameter, and not as an operator, as is the case of the momentum or the position. The objection formulated by Pauli about the validity of a time–energy uncertainty relation analogous to the position-momentum one, seemed to exclude the existence of such an operator.

One of the consequences of this is, for example, that one cannot deduce the Heisenberg uncertainty relation for time and energy from a kinematical point of view because time does not belong to the algebra of observables. Despite this, the relation $\Delta E \cdot \Delta t \geq \hbar$ is commonly used to study characteristic times for deformation of wave packets or decay of excited states, as in the following example taken from traditional books used in introductory courses to QM.

EXAMPLE 6.3

Let us consider a pulse that has the shape outlined in Figure 6.3. The spectral composition of this pulse can be obtained by means of the Fourier integrals and it is shown in Figure 6.4. From these figures, it can be shown that the following relation holds for the relation between the time of duration of this pulse and its spectral width $\Delta\omega$: $\Delta\omega \cdot T \sim 1$. For the case of photons $\Delta\omega = \Delta E/\hbar$ so that theses relations lead to the uncertainty time–energy inequality $\Delta E \cdot \Delta t \geq \hbar$ with $\Delta t = T$.

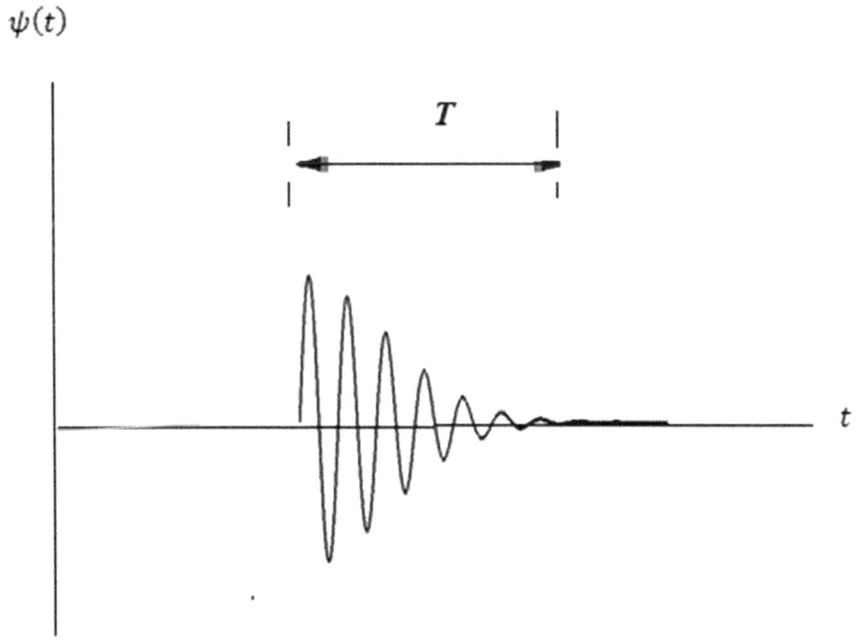

FIGURE 6.3 Light pulse emitted during the decay of a set of excited atoms.

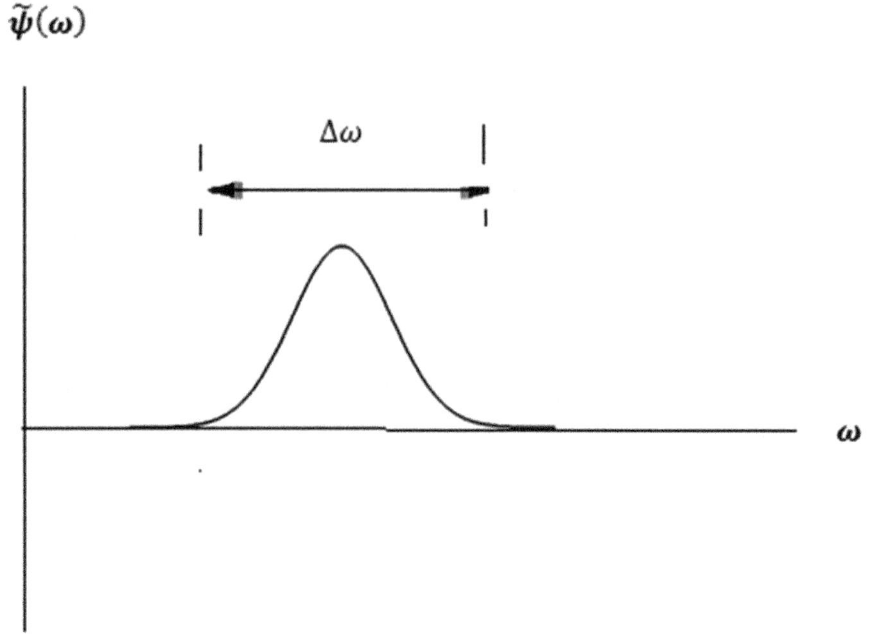

FIGURE 6.4 Fourier transform of the pulse shown in Figure 6.3.

Consider now that the pulse shown in Figure 6.3 corresponds to the radiation emitted by a set of excited atoms all in an unstable state of energy E_1 which after an average time τ decay to a ground state with energy E_o, with a spectrum as given in Figure 6.5 in terms of wavelength of the measured radiation. Since $\Delta\omega = \Delta E/\hbar$ the usual interpretation of the relation $\Delta\omega \cdot T \sim 1$ is that $\Delta E \cdot T \sim \hbar$, what it is commonly called as the time–energy uncertainty.

Determine the average time of duration τ of the atoms in the excited state.

SOLUTION

The spread $\Delta\lambda$ of the measured wavelengths corresponds to a spread $\Delta\omega = \Delta E/\hbar$ in the frequencies of the photons emitted by the atoms during their radioactive decay. The relation between frequency and wavelength is $\omega = 2\pi c/\lambda$ implies that

$$\Delta\omega = \frac{2\pi c}{\lambda_o^2}\Delta\lambda$$

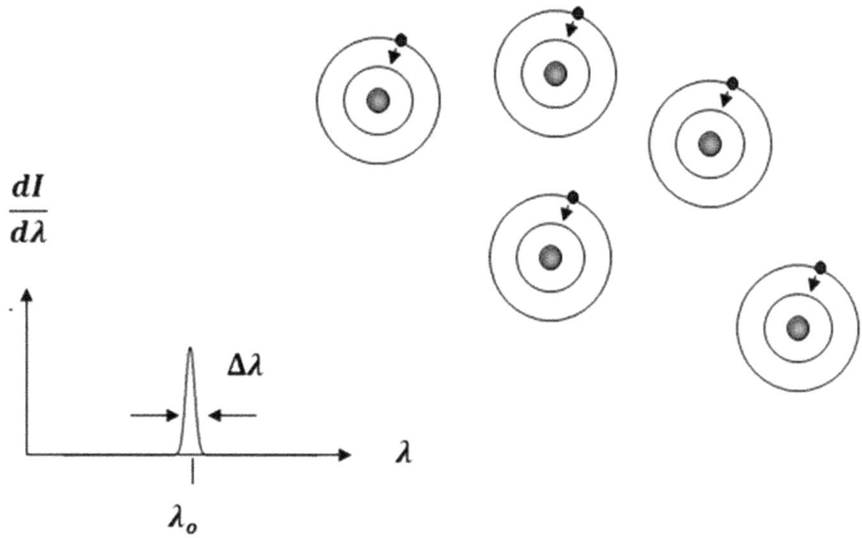

FIGURE 6.5 Spectrum of measured wavelengths for the light emitted during the decay of a set of excited atoms.

That means an uncertainty in a measurement of energy given by

$$\Delta E = \hbar\Delta\omega = \frac{hc}{\lambda_o^2}\Delta\lambda$$

Therefore, using the so called uncertainty time–energy, the time interval τ can be obtained from

$$\Delta \mathrm{E}\cdot\Delta t \sim \hbar \Rightarrow \tau = \frac{\lambda_o^2}{2\pi c\Delta\lambda}$$

The interpretation of this result is that the lower is the time duration of an excited state the higher is the uncertainty in the measurement of energy what is expressed as the width of the spectral line.

Despite the interpretation of the time–energy relationship given in the above example, this cannot be derived in a form similar to the uncertainty relation position-momentum, since up to now there is not an accepted time operator. However, starting from Aharonov and Bohm in 1961, several time operators have been proposed. In this section, in particular we discuss two approaches for time operators applied to non-relativistic QM and in Section 6.6, we discuss some of the time operators proposed for relativistic QM. One of these approaches is proposed by Recami, Olkhowsky, and Maysanyuk, where the time operator is expressed in a form analogous to a Hamiltonian that can be used for the study of collisions or movements in the case of a continuous energy spectrum.[26] Other approaches come from the work of Prvanović who proposes a time operator acting as the generator that allows the energy to change.[27,28]

6.5.2 TIME OPERATOR AS AN ANALOGOUS "HAMILTONIAN"

In non-relativistic quantum theory, the energy operator can be expressed in two forms: (i) as $i\hbar\frac{\partial}{\partial t}$ in the time representation, and (ii) $\mathcal{H}(\hat{x},\hat{p}_x,\ldots\ldots)$ in the best known Hamiltonian formalism. These two forms are dual as can be deduced from the Schrödinger's equation itself, $\mathcal{H}\Psi = i\hbar\frac{\partial\Psi}{\partial t}$. A similar duality can be introduced for the case of time. One form is expressed the time operator as $-i\hbar\frac{\partial}{\partial E}$ in the energy-based representation that is applicable to any physical system in the range of continuous energy spectral values. The other form is presented in Ref. [26] where the time operator is expressed in a "Hamiltonian form": replacing the operators energy $\hat{E}$ and time $\hat{t}$ by $\mathcal{H}(\hat{x},\hat{p}_x,\ldots\ldots)$ and $T(\hat{x},\hat{p}_x,\ldots\ldots)$, respectively, and using a commutation relation formally equal to that of eq 6.36, that is

$$[\hat{E},\hat{t}\,] = i\hbar \leftrightarrow [\hat{\mathcal{H}},\hat{T}\,] = i\hbar \tag{6.50}$$

for a given specific Hamiltonian, one can obtain the corresponding explicit expression for $T(\hat{x},\hat{p}_x,\ldots\ldots)$. As a simple illustrative case, let us consider a free particle of mass m moving in the x direction whose Hamiltonian in the momentum representation is

$$\mathcal{H} = \frac{p_x^{\;2}}{2m} \tag{6.51}$$

We can verify that in the momentum representation, the following Hamilton-type time operator satisfies eq 6.50:

$$\hat{T} = -\frac{m}{2}\left(p_x^{\;-1}\hat{x} + \hat{x}p_x^{\;-1} + i\hbar / p_x^{\;2}\right) \tag{6.52}$$

Applying this operator to a plane-wave of the form $\varphi(x) = exp\left(\frac{i}{\hbar}p_x x\right)$ and with the operator $\hat{x} = i\hbar\frac{\partial}{\eth p_x}$, we obtain

$$\hat{T}\varphi(x) = -\frac{m}{2}\left\{\frac{i\hbar}{p_x}\frac{\partial}{\partial p_x}\varphi(x) + i\hbar\frac{\partial}{\partial p_x}\frac{\varphi(x)}{p_x} + \frac{i\hbar}{p_x^{\,2}}\varphi(x)\right\} = \frac{mx}{p_x}\varphi(x) \quad (6.53)$$

To simplify the analysis, we assume that the concept of velocity in QM can be considered in an analogous way as in classical physics so that we can write $p_x = m\upsilon_x$ and the above equation becomes

$$\hat{T}\varphi(x) = \frac{x}{\upsilon_x}exp\left(\frac{i}{\hbar}p_x x\right) \quad (6.54)$$

where x/υ_x can be classically interpreted as the arrival time at $x = 0$ of a freely moving particle having initial position x and momentum p_x. However, caution is required for this interpretation considering that in QM it is not clear what velocity means and it is better to consider the expectation value of the velocity which is the same as the time derivative of the expectation value of the position, that is $\langle\upsilon_x\rangle = d\langle x\rangle/dt$.

6.5.3 TIME AND ENERGY OPERATORS IN AN ENLARGED HILBERT SPACE

As we have seen in Chapter 3, the Hamiltonian $\mathcal{H}$ of the Schrödinger equation determines the time evolution of the state of a quantum system. In fact, if a quantum system is initially in a state $|\psi(0)\rangle$ at time t it will be in the state $|\psi(t)\rangle = exp\left(-i\frac{\mathcal{H}}{\hbar}t\right)$. This expression is valid for a time-independent Hamiltonian.

However, in general, it is not easy to determine from the Schrödinger equation how the state of the quantum system changes if it gains or loses energy due to the interactions with the environment. In order to solve this problem, an equation is required that allows to differentiate with respect to the energy, rather than with respect to time as expressed in the Schrödinger equation: $i\hbar\frac{\partial|\psi(t)\rangle}{\partial t} = \mathcal{H}|\psi(t)\rangle$.

A proposal to satisfy this requirement is presented in the work of Slovodan Prvanović, researcher of the University of Belgrado.[27,28] In this section, we analyze that work, where the proposed operator for time plays the role of an energy change generator, while the operator of energy (conjugate of the time operator), generates the time evolution.

In order to fulfill the requirement coming from the general relativity theory (GRT), that space and time be treated equally, in that work a Hilbert space $\mathcal{H}_t$ is introduced where the operator of time $\hat{t}$ acts together with the operator of energy denoted here as $\hat{s}$ which is conjugate to time, so that $[\hat{t}, \hat{s}] = i\hbar$. Considering for simplicity one degree of freedom, in the quoted references, it is considered that the operators $\hat{q}$ and the conjugate momentum $\hat{p}$ act in a Hilbert subspace $\mathcal{H}_q$, while operators $\hat{t}$ and $\hat{s}$ act in $\mathcal{H}_t$, so that the following commutation relations hold in the enlarged Hilbert space $\mathcal{H}_q \otimes \mathcal{H}_t$:

$$\frac{1}{i\hbar}\left[\hat{q}\otimes\hat{I}, \hat{p}\otimes\hat{I}\right] = \hat{I}\otimes\hat{I} \tag{6.55}$$

$$\frac{1}{i\hbar}\left[\hat{I}\otimes\hat{t}, \hat{I}\otimes\hat{I}\right] = \hat{I}\otimes\hat{I} \tag{6.56}$$

Where, I is the identity operator.

The Hamiltonian $\mathcal{H}(\hat{q}, \hat{p})$ and the operator of energy $\hat{s}$ act in different Hilbert domains, but there is a subspace of the complete Hilbert space, in which

$$\hat{s}|\psi\rangle = \mathcal{H}(\hat{q}, \hat{p})|\psi\rangle \tag{6.57}$$

In an analogous way, one should expect that

$$\hat{t}|\psi\rangle = G(\hat{q}, \hat{p})|\psi\rangle \tag{6.58}$$

where $G(\hat{q}, \hat{p})$ represents the dynamical counterpart of the time operator which for the example of the harmonic oscillator is given by S. Prvanović in Ref. [27], can be chosen as a function proportional to the Hamiltonian $\mathcal{H}(\hat{q}, \hat{p})$ and with units of time, so that:

$$\mathcal{H}(\hat{q}, \hat{p}) \Leftrightarrow G(\hat{q}, \hat{p}) = \alpha\mathcal{H}(\hat{q}, \hat{p}) \tag{6.59}$$

where α is a factor with units [time/energy] chosen in this reference as $\alpha = \frac{\hbar}{\left(mc^2\right)^2}$. In this way, the eigenkets corresponding to the solutions of eq 6.58 are $|\psi_n\rangle \otimes |t_n\rangle$ with n a positive integer and $|\psi_n\rangle$ corresponds to the spatial part of the solutions of the harmonic oscillator, so that the eigenvalues of the time operator $\hat{t}$ are given by

$$t_n = \frac{\hbar^2\omega}{\left(mc^2\right)^2}\left(n + \frac{1}{2}\right) \tag{6.60}$$

This last relation can be written in a form easier to interpret where time appears as quantized if it is expressed in terms of the discrete values E_n of the energy of the harmonic oscillator:

$$t_n = \left(\frac{\lambda_C / 2\pi}{c}\right)\left(\frac{E_n}{mc^2}\right) \quad (6.61)$$

where $\lambda_C = h/mc$ is the Compton wavelength corresponding to the mass m undergoing the harmonic oscillation. The first factor of this expression has units of time representing the time that a photon requires to travel a distance equivalent to the Compton reduced wavelength, while the second factor is dimensionless and corresponds to the quantized energy $E_n = \hbar\omega\left(n+\frac{1}{2}\right)$ expressed in terms of the rest energy mc^2.

FURTHER DISCUSSION

While this section was being developed, correspondence has been maintained with Dr. Prvanović, author of papers quoted here as,[27,28] from whom valuable help has been received to get a better understanding about his work. One of the issues discussed with Dr. Prvanović is concerned with the commutator $[\hat{s}, \hat{t}] = i\hbar I$ and its compatibility with eqs 6.61 and 6.62. He has explained that these operators act, by definition, in one rigged Hilbert space denoted as $\mathcal{H}_t$ and that they are equivalent to the operators of coordinate and momentum ($\hat{q}$ and $\hat{p}$, respectively), which act in the separate rigged Hilbert space $\mathcal{H}_q$. In particular, the commutator $[\hat{s}, \hat{t}]$ in $\mathcal{H}_t$, just like $[\hat{q}, \hat{p}]$ in $\mathcal{H}_q$, does not vanish.

This researcher refers in his work to eqs 6.57 and 6.58 as constraint equations indicating that they have as their second part belonging to $\mathcal{H}_t$, eigenvectors associated to energy and time, respectively. This holds independently of the concrete expressions for $\mathcal{H}(\hat{q},\hat{p})$ and $G(\hat{q},\hat{p})$, so that even if these functions are proportional between them (meaning that they commute and that the first parts of the solutions to these equations are equal), the seconds parts are different and, since they belong to the Hilbert subspace $\mathcal{H}_t$, that is the place where $[\hat{s}, \hat{t}] \neq 0$.

6.6 QUANTUM OPERATOR FOR TIME IN RELATIVISTIC QUANTUM MECHANICS

According to Pauli, the finite lower bound for the energy spectrum makes mathematically impossible to find a Hermitian operator that is canonically conjugate to the Hamiltonian.

An alternative to solve this problem is to enlarge the Hilbert space, either continuing the energy spectrum to negative energies, or equivalently introducing a spin-like quantity such that two states can be associated with each positive value of energy. In Ref. [29], a quantum description of phase and angle is given and Heisenberg inequalities are derived, breaking the usual rule of quantum theory which assigns a Hermitian operator to every physical observable, as is the case of the momentum or the position of a particle.

As shown by Mariano Bauer, a researcher of the Autonomous National University of Mexico,[30] Dirac's quantum formulation for the relativistic electron allows us to introduce a dynamical operator for time that is self-adjoint (i.e., based on a dynamic observable, namely the position, and dependent on the parameter t that appears in the dynamic equations of quantum mechanics).

The time operator proposed by M. Bauer generates energy shifts within the positive (or negative) energy degenerate spectrum through displacements in momentum space where there is no bound, while remaining either above (or below) the $2m_oc^2$ energy gap. In this way, Pauli's objection is at least circumvented.

This section presents a summary analysis of the work of M. Bauer with respect to his proposed time operator for a free relativistic quantum particle. In a form analogous to the Dirac Hamiltonian for a free particle a dynamical self-adjoint operator for time has been proposed in the preceding quoted reference which is given by:

$$\hat{T} = \vec{\alpha} \cdot \frac{\hat{r}}{c} + \beta \tau_o \tag{6.62}$$

where α_i (i=1,2,3) and β are the 4×4 Dirac matrices satisfying the following relations:

$$\alpha_i \alpha_j + \alpha_j \alpha_i = 2\delta_{ij},\ \alpha_i \beta + \beta \alpha_i = 0,\ \beta^2 = 1 \tag{6.63}$$

In this formulation, τ_o is an invariant quantity in the (r,τ) space so that

$$\tau_o^2 = \tau^2 - (\boldsymbol{r}/c)^2 \tag{6.64}$$

Equation 6.64 is analogous with the dynamical relation between mass, energy, and momentum in special relativity theory:

$$\left(m_o c^2\right)^2 = E^2 - \left(c\boldsymbol{p}\right)^2 \tag{6.65}$$

Demonstration of eq 6.64

Writing eq 6.62 as $\hat{T}-\vec{\alpha}\cdot\frac{\hat{r}}{c}-\beta\tau_o=0$ and multiplying by the left by $\hat{T}+\vec{\alpha}\cdot\frac{\hat{r}}{c}+\beta\tau_o$, we obtain

$$\left(\hat{T}+\vec{\alpha}\cdot\frac{\hat{r}}{c}+\beta\tau_o\right)\left(\hat{T}-\vec{\alpha}\cdot\frac{\hat{r}}{c}-\beta\tau_o\right)=(\hat{T})^2+\frac{\alpha_j\hat{x}_j}{c}\hat{T}+\tau_o\beta\hat{T}$$

$$-\hat{T}\frac{\alpha_j\hat{x}_j}{c}-\frac{(\alpha_j\hat{x}_j)^2}{c^2}-\tau_o\beta\frac{\alpha_j\hat{x}_j}{c}$$

$$-\tau_o\hat{T}\beta-\tau_o\frac{\alpha_j\hat{x}_j}{c}\beta-\beta^2\tau_o{}^2 \quad (6.66)$$

Since the time operator $\hat{T}$ does not act over matrices α_j and β and neither over the components $\hat{x}_j$ of the position operator $\hat{r}$, we can simplify terms as indicated in eq 6.66 where the Einstein rule of summation has been used for terms involving $\vec{\alpha}\cdot\frac{\hat{r}}{c}$.

Therefore, eq 6.66 can be rewritten as

$$\left(\hat{T}+\vec{\alpha}\cdot\frac{\hat{r}}{c}+\beta\tau_o\right)\left(\hat{T}-\vec{\alpha}\cdot\frac{\hat{r}}{c}-\beta\tau_o\right)=\left(\hat{T}\right)^2-\frac{\left(\alpha_j\hat{x}_j\right)^2}{c^2}$$

$$-\frac{\tau_o}{c}\left(\beta\alpha_j+\alpha_j\beta\right)\hat{x}_j-\beta^2\tau_o{}^2=0 \quad (6.67)$$

The second term to the right of this last equation gives

$$\frac{\left(\alpha_j\hat{x}_j\right)^2}{c^2}=\frac{1}{c^2}\left\{\sum\nolimits_j\alpha_j{}^2\hat{x}_j{}^2+\sum\nolimits_{jk}\left[\alpha_j\alpha_k+\alpha_k\alpha_j\right]\hat{x}_j\hat{x}_k\right\} \quad (6.68)$$

The second term at the right of eq 6.68 is valid for $j > k$. From eqs 6.63 and 6.68 and considering that $\alpha_j^2 = I$, we obtain:

$$\left(\hat{T}+\vec{\alpha}\cdot\frac{\hat{r}}{c}+\beta\tau_o\right)\left(\hat{T}-\vec{\alpha}\cdot\frac{\hat{r}}{c}-\beta\tau_o\right)=\left(\hat{T}\right)^2-\frac{\hat{r}^2}{c^2}-\tau_o{}^2=0 \quad (6.69)$$

Therefore, we obtain that the time operator $\hat{T}$ satisfies the following relation:

$$\hat{T}^2-\frac{\hat{r}^2}{c^2}=\tau_o^2 \quad (6.70)$$

This is a relation between the operators' time and position but its form is similar to that expressing the invariance of the quantity $t^2 - \frac{\vec{r}^2}{c^2}$ in the space-time.

Following a similar algebra to that developed for the Dirac Hamiltonian, the following eigenvalue equation is obtained

$$\hat{T}|\tau\rangle = \tau|\tau\rangle \tag{6.71}$$

The corresponding eigenvalues are

$$\tau = \pm\left[(r/c)^2 + \tau_o^2\right]^{1/2} \tag{6.72}$$

which is in agreement with eq 6.70.

It must be noted from eq 6.72 that the eigenvalue spectrum has two continuous branches, a positive branch and another negative which are separated by a $2\tau_o$ gap, as shown in Figure 6.6.

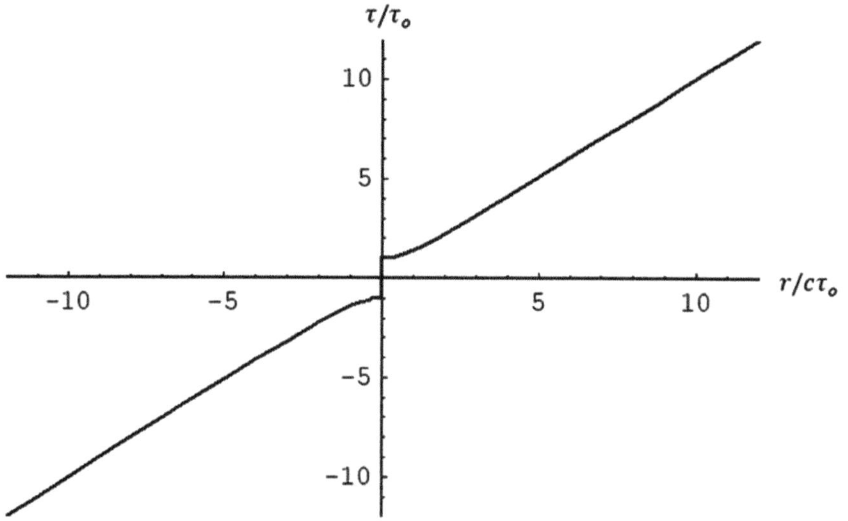

FIGURE 6.6 Eigenvalues spectrum of the relativistic time operator.

What about the time interval τ_o?

Bauer has shown that the invariant interval of time τ_o is equal to the de Broglie period h/mc^2, equivalent to the time required for a photon to travel a distance equal to the Compton wavelength corresponding to the mass m.[31]

This is obtained considering that the time operator proposed by Bauer produces the Lorentz boost that originates the de Broglie wave, with the corresponding de Broglie period h/mc^2. As expressed by William Baylis, researcher of the Windsor University: **"What in the rest frame is a synchronized oscillation in time is seen in the laboratory to be a wave in space"** (phrase taken verbatim from Ref. [32]). In the last cited Ref. [32], he shows that any in-phase oscillations of an extended system in the rest frame is equivalent to de Broglie-like waves in other inertial frames. These waves result from the relativistic property that time is not universal.

Relation between proper time τ_o and the de Broglie period

The dynamical time operator $\hat{T}$ can be considered as the generator of a unitary transformation given by

$$U(\in) = \hat{U}_T(\in) = e^{-i\in\hat{T}/\hbar} = e^{-i\delta\left\{\alpha\cdot\frac{\hat{r}}{c}+\tau_o\beta\right\}/\hbar} \tag{6.73}$$

where $\in$ is a real quantity with dimensions of energy.

For an infinitesimal transformation $\delta\in \ll 1$, Bauer has shown in his work that $\hat{U}_T(\in)$ generates displacements in momentum and a change in phase given by

$$\delta\phi = \beta\tau_o\delta\in/\hbar \tag{6.74}$$

To maintain the fundamental indeterminacy $n2\pi$ (with n an integer) in the phase of the complex eigenfunctions for a finite change of energy $\Delta\in = mc^2$, it is necessary to set for $n = 1$:

$$\tau_o = \frac{2\pi\hbar}{(\beta\Delta\in)} = \frac{h}{mc^2} \tag{6.75}$$

which is the de Broglie period (for a better understanding of this result, remember that the de Broglie wavelength is $\lambda = h/mc$ and the relation $\lambda = c\cdot$ *period*. For an electron, the de Broglie wavelength is 1.2 nm and the corresponding period is $4 \cdot 10^{-18}$ sec).

6.7 TIME IN QUANTUM GRAVITY

In this section, we provide a brief overview of the aspects related to the conception of time in the formulation of a quantum theory of gravity.

6.7.1 INTRODUCTION

Both QM and classical mechanics consider time as a parameter independent of the dynamics of the system. On the other hand, the GRT, radically changes this picture by understanding time as a relevant component of the space-time, which it interacts with—and thus gets acted upon by—the physical systems it "contains." This fact cannot be ignored to obtain a theory of quantum gravity which includes both relativistic and quantum effects. Therefore, a fundamentally new conception of time must be introduced, and as discussed in Ref. [33], the Philosophy of Science can make a great contribution for that purpose.

For making an analysis about the nature of time in quantum gravity, it must be considered that there are phenomena in which both quantum and gravitational effects play a crucial role. This is the case of the conditions which existed at the very early universe, and also is the case of black holes where there are high energy densities combined with strong gravitational fields. To carry out this analysis, three possible ways to approach them can be considered:

- A first way to obtain a quantum theory that properly incorporates gravity starts out from the up to now the best classical theory of gravity, the GRT, and applies a quantization procedure, which mathematically consists of a rigorous transposition of the classical theory into a quantum theory.
- The second approach is based on the quantum theory of matter that should be extended to take into account strong gravitational fields and their interactions with the fields associated to the matter.
- The third approach consists of introducing into the theoretical frame new principles radically different from those considered so far, in order to formulate a fundamental theory of quantum gravity.

Nevertheless, all these approaches are only at an initial stage of his development and so far they have not been consolidated as a new theory in this regard.

6.7.2 THE CONCEPT OF TIME IN GENERAL RELATIVITY THEORY

According to John Earman,[34] there are two traditions in the philosophy of time which share two assumptions: first, that time presupposes change; and, second, that genuine change requires Becoming. Both traditions have ancient roots. One, inspired by the reasoning of the ancient Greek philosopher Parmenides, denies the reality of change and time by rejecting Becoming; the other, can be traced to Aristotle, and upholds the reality of change and time by claiming to find Becoming at work in our real world.

All fully established theories of physics, from classical to quantum physics and including relativity theory, require a time parameter, with respect to which the dynamics manifest.

Completely independent of whether time is regarded as something that exists by itself or as a parameter associated with the changes of a physical system, in general, there is agreement that time allows ordering the events that occur in our universe and in our own lives, indicating, for example, which of two events occur before the other, whether or not they are causally connected.[3,4]

Time in fully established theories of physics has a metric that allows not only to determine the sequence with which events occur but also the duration and interval between successive events.

In GRT, space-time can have causal loops associated to closed time like curves, in the context of the model of rotating universe formulated by Gödel (see Section 6.8). This means that a pair of events can be placed in it without there being a temporal relationship between them, or also with a temporal relationship in both directions so that, without being simultaneous, each one of them precedes the other.

A central foundational aspect of General Relativity (GRT) is the so-called "background independence," a concept which can be explained as follows. The GRT establishes the equivalence of inertial and gravitational mass of all physical bodies. GRT generalizes this principle by establishing that the effects due to gravitational fields and those produced by inertia are caused by one and the same field, called the inertio-gravitational field usually described by the metric field g. This inertio-gravitational field has different gravitational and inertial components in different frames, just as is the case with the electromagnetic field in electrodynamics (see Section 4.12.2 of Chapter 4). This is in sharp contrasts to Newtonian Physics, where a variable gravitational field that is placed on a fixed space-temporal background is what gives rise to inertial effects. This background is fixed in the sense that it affects the physical phenomena that occur in it without being affected by them.

The generalized equivalence principle does not allow a fixed inertial background and thus leads to a theory "background independent." This kind of independence is formally implemented in GRT by demanding that the theory be "generally covariant." Following the arguments given by Einstein,[35] this is equivalent to postulate that the laws of nature must be expressed by equations valid for any coordinate system.

6.7.3 SPACE–TIME QUANTIZATION CONSIDERING CLASSICAL GRAVITY

One of the approaches to quantize space-time considering classical gravity comes from the work of S. Doplicher, K. Fredenhagen, and J. E. Roberts. These researchers propose a quantization of space-time formulating uncertainty relations based on Heisenberg's uncertainty principle and on Einstein's theory of classical gravity.[36] They consider that the quantum deviation of space-time from its classical structure can be understood using the basic principles underlying QM and general relativity. Moreover, those deviations should be manifested only at the Plank scale, while the large scale structure of quantum space–time should be the same as for the usual Minkowski space.

Considering idealized states describing few interacting elementary particles in the absence of strong external gravitational fields, ordinary space–time can be regarded as translation parameters, which make possible to formulate time evolution and wave equations.

6.7.4 TIME IN A QUANTUM THEORY OF GENERAL RELATIVITY

Canonical quantization is one of the most used ways for obtaining a quantum theory from a classical theory. For this, the Hamiltonian formulation must be used, which considers the physical system described as spatially extended and evolving in time. This procedure presupposes the concept of time represented by a quantity that is somehow imposed externally.

As expected, the dynamical equations of the Hamiltonian formulation do not alone show equivalence with the Einstein's field equations of the general theory of relativity. To achieve this equivalence, additional "constraint equations" must be considered in the Hamiltonian formulation. One of the constraint equations includes the Hamiltonian function $\mathcal{H}$, which in the standard QM theory represents the total energy of the closed system and generates the evolution of the system through the Hamilton equations. In the Hamiltonian $\mathcal{H}$ that must be used to combine with GRT, it turns out that $\mathcal{H}$ itself represents a constraint equation, whose action over any ket gives a result equal to zero. The absence of an external time with respect to which the spatial system could evolve thus leads to the vanishing of the generator of the evolution, dynamically "freezing" the system. Therefore, the *"problem of time"* appears already at the purely classical level.

Consequently, the Hamiltonian constraint $\hat{H}$ acting over any quantum state gives

$$\hat{H}|\psi\rangle = 0 \tag{6.76}$$

Only those quantum states which satisfy this constraint equation are considered physical and constitute the physical Hilbert space of all possible states of the quantum system. Every operator representing a physical magnitude ("observable") has to commute with the Hamiltonian $\hat{H}$. Therefore, these operators are constants of motion, that is, quantities which are conserved through motions and thus over "time." The above implies that all genuine physical magnitudes do not change, but are dynamically "frozen."

The dynamical eq 6.76 also called the "Wheeler-DeWitt equation," has a form similar to the Schrödinger equation of non-relativistic QM,

$$\hat{H}|\psi\rangle = i\hbar\frac{\partial}{\partial t}|\psi\rangle \tag{6.77}$$

with the essential difference that the right-hand side of eq 6.76 contains a "zero" instead of the temporal derivative as in eq 6.77. One interpretation of the above is that both the time parameter t is absent in the "dynamical" Wheeler-DeWitt eq 6.76, and that there is no time evolution.

6.7.5 DISCUSSION

If one accepts the neo-Parmenidean conception of time previously mentioned in this section, this implies to accept the absence of time (or at least of change) in the fundamental physics. In that case, it is necessary to seek other ways for doing physics in a world without the notions of change and its association with time. This becomes particularly complex in the case of QM, where there are numerous interpretational issues concerned with systems—such as our universe—for which the dynamical Schrödinger eq 6.77 degenerates into the Wheeler-DeWitt eq 6.76.

The Italian Physicist Carlo Rovelli proposes evolving constants of motion as intent to show an alternative way to do physics in the absence of time and change.[8] He suggests that a good strategy for understanding quantum gravity is to use a picture of the physical world where the notion of time does not play any role, formulating the following conclusions:

1. It is possible to formulate classical mechanics in a way where the time variable is considered in a similar form to that of the other physical variables, and not identified as the special independent variable. This is the natural formalism for describing general relativistic systems.
2. The above conclusion can be applied to reformulate quantum mechanics in the same manner and this may be the effective formalism for quantum gravity.
3. The special properties of the time variable have thermodynamic nature, and can be suitably described by the thermal time hypothesis, which establishes that what we understand by "time" is referred to the thermal time of the statistical state in which the world is placed, when it is described in terms of the macroscopic parameters we have chosen.
4. In order to formulate a quantum theory of gravity, the better strategy is to forget at all the notion of time, and to define a quantum theory with the capacity to predict the possible correlations between partial observables.

Maudlin,[37] proposes two arguments concerned with the nonexistence of change in Hamiltonian GRT. The first one is related with the need to "sop up" indeterminism (to use Earman's expression), while the second results of an analysis of "observables." With respect to the latter, as Earman emphasizes, the problem of time strengthens the importance of giving answer to the question about the interpretation of the genuine physical magnitudes given by GRT.

6.7.6 SPACE–TIME IN STRING THEORY

As known, quantum particle physics has provided a theory which has successfully described electromagnetic and nuclear interactions in terms of the exchange of force carrying particles, such as photons or bosons. On the other hand, gravity is not a feature of dynamical space–time geometry, but a force in a fixed space–time. Thus it would be expected to describe it in terms of the exchange of force-bearing "gravitons." Unfortunately, such a theory is apparently "non-renormalizable," meaning in broad terms that it is incompatible with the interactions at a point assumed by particle physics.

String theory provides a solution to the non-renormalizability problem, by considering a space-time with more than four dimensions. This theory, which was formulated more than 50 years ago, but to date, has no experimental evidence, states that the fundamental particles are excitations of one dimensional string, instead of point-like point particles. The strings can be either open or closed, depending on the characteristics of the particle to be replaced by the string, whose length is of the order of 10^{-33} cm. For example, an electron is considered not as a like-point particle but rather as a string, vibrating and rotating on a length scale too small to be proved even with the most advanced particle accelerators available to date.[38]

For the quantization of a string in flat space-time more than four dimensions are required. In string theory, at least six additional dimensions remain undetected since they are wrapped up in a compact manifold.[39] The extra dimension allows interactions to be extended in space-time, avoiding the problem of non-renormalizability.

According to Ref. [38], from the way strings move and interact, properties of space and time can be deduced, by means of an approach called world sheet. In this approach, space–time represents the list of labels in the description of the motion of the string (it is analogous to the GPS record of where a moving object, such as a car, at a given instant is in a place on a road). These labels have quantum fluctuations that can be analyzed only if the space-time obeys the equations of general relativity.

6.8 EXPLORING IN THE PHYSICS OF TIME TRAVEL

Is it possible to travel to the past? And move to the future? Why do we only remember the past?

These are questions frequently raised in numerous books and reflections on the philosophy and physics of time, for which there seems to be no convincing answer. In this section, we will try to get as close as possible an answer to these questions from the point of view of contemporary physical theories.

According to the special theory of relativity, traveling into the future is in principle possible, it suffices to travel fast enough, or to spend little time sufficiently near to a black hole horizon.

But can we travel to the past? The general theory of relativity suggests that this is theoretically possible, since it allows solutions with closed time-like curves, like those formulated by Kurt Gödel. This mathematician found a solution to Einstein's equations of general relativity that describes closed time like curves (CTC). He also discovered that based on the gravitational field equations it is possible to describe a rotating model of the universe.[40]

Gödel's rotating universe, allows the existence of space-time trajectories that return to their starting point. These trajectories are namely CTCs and remain confined locally to their future light cones. An object traveling along a CTC never moves faster than the local speed of light, so that CTCs do not violate the laws of special relativity, while representing possible paths for material objects. This is a surprising result since it implies that a traveler could leave for the future but would arrive in the past. It also causes incompatibilities with the principle of causality, which establishes a fixed temporal sequence between cause and effect, and with the fact that according to our perception of reality, it is not possible to modify the past.

In the rotating universe of Gödel, there is no expansion, which clearly contradicts the observations of the redshifts of distant galaxies and the cosmic background radiation. According to Gödel, from the moment when an observer starting from a given point in space-time, travels along a trajectory allowed by the principle of relativity—that is, confined inside light cones—to his chronological past, time exists relative only to the observer and stops having an independent reality. From his point of view, he considered his model of rotating universe, whether realistic or not, as a *possible* universe to the extent that it was an exact mathematical solution of Einstein's field equations. And for him, if time was an illusion in a possible universe, this had also to be true in the real universe.

Another manifestation of the CTC possibility has been suggested by Morris and Thorne, who have suggested a kind of short cut in space-time, known as a wormhole. They have shown that how the two ends of a wormhole could be moved so as to form a CTC.[41,42]

Other famous solution containing CTCs is the Kerr's space-time which describes a rotating black hole.[43] This space-times includes regions devoid of CTCs, and one or more regions that contain CTCs; their common outline is usually called the chronological horizon.[42] The Kerr solution includes two event horizons denoted as $r+$ and $r-$, respectively. The regions $r > r+$ and $r+ > r > r-$ are chronal since it does not include CTCs, while the region $r < r-$ is achronal since it includes CTCs; the inner event horizon r—is equivalent to a chronological boundary. A time like curve from a given point of the chronal region outside the black hole can successively cross the two event horizons $r+$ and $r-$, pass through the achronal region $r < r-$, and exit from the black hole to arrive in the chronological past. In this way, a CTC of the type referred to as a wormhole is achieved.

But the idea of traveling to the past has always raised puzzlement. It is commonly said that traveling to the past would generate logical paradoxes. For example, if that would be possible a crazy person could travel to the past to kill his parents before he was born. Carlo Rovelli argues that this and similar paradoxes disappear if one examines with care the full thermodynamic and statistical physics along these curves.[44] An event on a closed time like curve could be both in the future and in the past of itself, what would not allow a formulation of dynamics according to an "initial condition" problem.

According to the quoted paper of Rovelli, the phenomenology that we commonly associate to the forward passage of time (memory, decisions, cumulative counting of the periodic pulses emitted by a luminous clock, and so on) depends entirely on the second principle of thermodynamics. This implies that the future direction is determined by the derivative of a state function, which can be associated to the entropy.

As known, it is not possible for a function to increase uniformly around a circle. In an analogous way, we can never "travel to the past" in the sense of arriving to the past knowing the future and acting in a different way than what we have done before. What's done is done. These are all thermodynamic phenomena (that require irreversibility), and the laws of thermodynamics in this sense do not allow travel into the past.

For David Deutsh and Michael Lockwood, if time travels is impossible, then the reasons has yet not been discovered.[45] Based on this last reference, two paradoxes concerned with time travel to the past are briefly discussed here with some changes that do not alter their essence.

PARADOX 1: INCONSISTENCE PARADOX

Suppose that Juan is a famous medical doctor 70 years old who at June 2022 lives in a country of South America. By means of a time machine Juan travels back to meet his earlier self. At this meeting, Juan tells to his younger self 15 years old that when he reaches the age of 18 he will have to choose university career and only two options will be available for him: medicine or musician. Juan says to his earlier self that at the moment of his decision the young Juan will have to choose the career of medicine marking this option in a computer program that will process his application to the university. This young records what his old self tells him, and at the corresponding time, having already reached the age of 18, he deliberately tries to choose something different. Can Juan travel to the past and change the decision he made when he was young, so that in 2022 instead of being a prestigious doctor he is a renowned musician?

According to classical physics and above all, to our perception of reality that makes us clearly distinguish (at least for most of us) the possible from the impossible, the answer is a categorical No. There must be something that the prevent Juan from deviating from what has already happened and made by him in the past.

Our perception of reality tells us that everything that happens around our lives is consistent with the fact that it is not possible to act on the past to modify the present. From the point of view of the philosophy of science, this consistency is related to the principle of autonomy that establishes that it is only possible to create in our environment configurations of matter allowed by the laws of physics.

PARADOX 2: KNOWLEDGE PARADOX

A contemporary writer author of several books about the history of physics visits a physician of the 19th century who currently is considered as one of scientists who contributed to give the grounds for radio communication systems. Studying the results of his experiments, the writer shows to this scientist books and papers which make reference to his work. What would happen if this scientist—who is thrilled with looking the result which his work will have several years later—decides to share this knowledge with the scientific community of this time so that radio communication system begins to operate before the enormous contribution provided by Guglielmo Marconi at the beginning of the 20th century?

This last story is of course a violation of common sense and according with Deutsh and Lockwood is called a knowledge paradox. This kind of paradox violates the principle that knowledge can begin its existence only as a result of a process of solving problems as occurs in the history of technological developments. There is not any known classical analysis that can wholly eliminate this paradox.

Independent of the type of paradox resulting of the idea of time travel to the past, up to now we are a very long way from proving the existence of any of these CTCs and there is not any experimental evidence of their existence. Actually nobody knows if one day it will be possible locate or create navigable CTCs.

6.9 CONCLUSIONS AND EPILOGUE

For the physicists, it is hard to imagine an adequate model of our universe without an ordering defined for the occurrence of these events. Physics and the Science in general, would not need any notion of time if we consider only a collection of data, without any attempt to explain them and how they are correlated.

We have discussed some of the most recent publications related with physical theories about time, starting with the analysis of the cellular automaton interpretation of QM formulated by Gerard Hooft, who states that Physics requires the concept of time and concludes that information loss can be considered as a source of time-reversal asymmetry. A totally different point of view is the one raised by Carlo Rovelli, who asserts that it is possible to formulate classical mechanics in such a way that the time variable is treated on the same footing as the other physical variables, and not singled out as the special independent variable.

Theories which propose quantization of time and time operators have been analyzed in this chapter. Furthermore, we have included a conceptual discussion about theories related with time when quantum and gravitational effects must be considered, and also a brief exploration in the physics of time travel.

We can see that up to now there is not a theory about time widely accepted and which could be considered as adequate to give account of several questions concerned with time in theories that still are under research, such as quantum gravity.

As an Epilogue

Of everything exposed in this chapter and in the previous ones of this book, it is worth pointing out some reflections as indicated below.

Why we are used to think that time flows only forward?

There is no doubt that according to the way our lives go by, in all our perception of time, it goes from the past to the future and not the other way around. We remember the past but we do not conceive of the possibility of "remembering the future." This in itself denotes an asymmetry in the representation we give ourselves for our concept of time.

However, as we have discussed in the previous chapters, the basic laws of physics at the microscopic level show symmetry with respect to time reversal, except when this symmetry is broken, for example, by applying an external disturbance such as magnetic field to quantum system.

For us time begins with our birth, but it is the same for the universe as we understand it? Will the time for the universe have an end? Is it possible that other cycle will come from a new big bang?

It seems to exist a close relation between light and time, as expressed, for example, with the representation of the causality between events using the light cone in a space-time diagram. As we have discussed, the theory of relativity links time with space in such a way that these concepts cannot be uncoupled, although their properties or characteristics are represented to us differently. For example, we can travel between two geographical points, say two cities or countries, in both directions. But the same does not happen with two instants of our life, each of which only live once. It is not possible for us to live again moments as happy as when we were children. We can recreate them but it's not the same anymore. Is this a biological limitation of ours as human beings? Furthermore, even if we could relive a moment already lived several years ago, there is a powerful principle called causality, which would prevent us from performing any action that changes our present in a hypothetical trip to the past. This impossibility of traveling into the past to modify the present is compatible with the so-called Chronological Protection Conjecture formulated by Stephen Hawking in a paper published in Physical Review D in 1992. According to Hawking, closed trajectories cannot exist in space-time and one of their arguments is that otherwise we would be invaded by tourists from the future.

However, the equations of the theory of relativity provide solutions that in principle allow the existence of closed loops in space-time, which in theory would allow time travel, but with many limitations that make it impossible for now, at least for humans, as we have discussed for wormholes.

On the other hand, the colossal dimensions and distances between our earth and the planets of our galaxy and even more, the distances that separate us from the so-called exoplanets (planets of other galaxies), make manned trips impossible to see what is beyond our galaxy. This would require ships that travel at a speed considerably greater than the speed of light, so that the duration of the trip is compatible with the maximum number of years that a human being can live. Will there be other civilizations in distant galaxies that have already overcome all these limitations and that perhaps can approach our earth? This sounds like a science fiction question, but it is worth asking ourselves if the works of science fiction writers such as Jules Verne, Isaac Asimov, Arthur Clark, among others, have influenced or stimulated in some way the development of contemporary physics theories. For example, let's mention that the idea of teleportation in science fiction dates back to the end of the 19th century with the book "the man without a body" by the writer David Page Mitchell, in which he recounts the transmission of the atoms of a cat through the wires of a Telegraph. This idea, which seemed only from the area of science fiction, was put into practice in 2005, when it was possible to teleport a pair of quantum-entangled photons over a distance of 600 m. In 2007, quantum communication was achieved at 144 km. away. Perhaps in not many more years, it will be possible the teleportation of the atoms that form the body.

Regarding time travel, this idea exists in science fiction literature since 1881 with the book "the clock that was going backwards" by the writer Edward Page Mitchell and later reinforced in the book by H. G. Wells "The Time Machine." As we have said, at least for now this last idea is not feasible with current technology. The best we can get for now is a view of the universe through a powerful telescope such as the James Web which has recently shown us images of the first galaxies that formed after the big bang and which are at a distance of about 13 billion of light years. This is the deepest view, we have of our universe so far and the light captured by the telescope is still traveling even though these galaxies probably disappeared millions of years ago.

This book has attempted to provide a panoramic view of the theories related to time, which, due to the impossibility of experimental verification, will continue to be theories. However, this has not stopped efforts to further develop them, presenting a huge and exciting challenge for research in theoretical physics.

KEYWORDS

- **spacetime**
- **arrow of time**
- **causality**
- **Hamiltonian**
- **Hilbert spacc**
- **Planck scale**
- **hermiticity**

REFERENCES

1. Vaccaro, J. A. Quantum Asymmetry Between Time and Space. *Proc.R., Soc. A* 2016, *472*, 20150670.

2. Rovelli, C, Quantum Mechanics Without Time: A Model. *Phys. Rev. D* **1990,** *42*, 2638–2647.
3. Hooft, G. Time, The Arrow of Time and Quantum Mechanics. Front. Phys **2018,** *6*. Article 81.
4. Hooft, G. t. The Cellular Automaton Interpretation of Quantum Mechanics: A View on the Quantum Nature of our Universe, Compulsory or Impossible? 2015. https://www.researchgate.net/publication/262145006
5. Sakurai, J. J. *Modern Quantum Mechanics*, Rev. ed.,Tuan, S. F., Editor; Wesley Publishing Company: Addison, 1994.
6. Griffiths, D. J. *Introduction to Quantum Mechanics*. Prentice Hall: Upper Saddel River, 1994.
7. Hooft, G. Determinism and Dissipation in Quantum Gravity, 2000. SPIN-2000/07, hep-th/0003005.
8. Rovelli, C. Forget Time. Essay Written for the FQXi Contest on the Nature of Time, August 2008.
9. Rovelli, C.; Smerlak, M. Thermal Time and Tolman-Ehrenfest Effect: Temperature as the Speed of Time. *Classical Quant. Gravity* May **2010**.
10. Feynman, R. P.; Hibbs, A. R. *Quantum Mechanics and Path Integrals*; McGraw Hill: New York,1965.
11. Lee, T. Can Time Be a Discrete Dynamical Variable? *Phys. Lett.* **1983,** *122*, 217–220.
12. Riek, R. Entropy Derived from Causality. *Entropy* **2020,** *22*, 647.
13. Carazza, B.; Kragh, H. Heisenberg's Lattice World: The 1930 Theory Sketch. *Am. J. Phys*. **1995,** *63*, 595.
14. Lee, J. EPR Paradox Solved by Special Theory of Relativity. *Acta Physica Polonica A* **2014,** *125*, 1107.
15. Brukner, Č. Causality in a Quantum World. *Physics Today*, March 2018.
16. Bruce, S. Discrete Time in Quantum Mechanics. *Phys. Rev. A* **2001,** *64*, 014103-1.
17. Heisenberg, W. The Physical Content of Quantum Kinematics and Mechanics. *Z. Phys.* **1927,** *43*, 172–198.
18. Hilgevoord, J. The Uncertainty Principle for Energy and Time. *Am. J. Phys.* **1996,** *64*, 1451.
19. Céleri, L. C.; Kioses, V. I. Canonical Field Quantization in Momentum Space and Quantum Space-Time, 2017. https://arxiv.org /abs /1712.05206
20. Céleri, L. C.; Kioses, V. I. Unruh Effect as a Result of Quantization of Spacetime. *Phys. Lett. B* **2018,** *781*, 611–615.
21. Hawking, S. W. Black Hole Explosions? *Nature* **1974,** *248*, 30.
22. Davies, P. C. W. Scalar Production in Schwarzschild and Rindler Metrics. *J. Phys. A: Math. Gen.* **1975,** *8*, 609.
23. Unruh, W. G. Notes on Black-Hole Evaporation. *Phys. Rev. D* **1976,** *14*, 870.
24. Pauli, W. *Handbuch der Physik*, vol. 5/1 Ed. S. Fluegge); Berlin, 1926; p 60.
25. Pauli, W. *General Principles of Quantum Theory*, Springer: Berlin, 1980.
26. Recami, E.; Olkhovsky, V. S.; Maysanyuk, S. P. On a Time-Space Operator (and Other Non-Selfadjoint Operators) for Observables in QM and QFT. Int. J. Modern Phys. A **2010,** *25*, 1785–1818.
27. Prvanović, S. Operator of Time and Generalized Schrödinger Equation. *Adv. Math. Phys.* **2017,** *208*.
28. Prvanović, S. Operator of Time and Properties of Solutions of Schrödinger Equation for Time Dependent Hamiltonian, 2017. arXiv:1701.07076 (quant-ph).

29. Lévy-Leblond, J-M. Who Is Afraid of Nonhermitian Operators? A Quantum Description of Angle and Phase. *Ann. Phys.* **1976,** *101*, 319.
30. Bauer, M. A Dynamical Time Operator in Dirac's Relativistic Quantum Mechanics. *Int. J. Modern Phys. A* **2014,** 29, 1450036.
31. Bauer, M.; Clock, de B. Electron Channeling and Time in Quantum Mechanics. *Can. J. Phys.* **2018**.
32. Baylis, W. E. De Broglie Waves as an Effect of Clock Desynchronization. *Can. J. Phys.* **2007,** *85*, 1317.
33. Huggett, N.; Vistarini, T.; Wüthrich, C. Time in Quantum Gravity, arXiv:1207.1365, Journal Reference: In *A Companion to the Philosophy of Time*; Bardon, A., Dyke, H., Eds.; Wiley-Blackwell: Chichester, 2013; pp 242–261.
34. Earman, J. Thoroughly Modern McTaggart: Or What McTaggart Would Have Said If He Had Read the General Theory of Relativity. *Philosophers' Imprint* **2002,** *2* (3).
35. Einstein, A. Die Grundlage der allgemeinen Relativitätstheorie. *Annalen der Physik* **1916,** *49*, 769–822.
36. Doplicher, S.; Fredenhagen, K.; Roberts, J. E. Spacetime Quantization Induced by Classical Gravity. *Phys. Lett. B* **1994,** *331*, 39–44.
37. Maudlin, T. Thoroughly Muddled McTaggart: Or How to Abuse Gauge Freedom to Generate Metaphysical Monstrosities. *Philosophers' Imprint* **2002,** 2 (4).
38. Gubser, S. S. *The Little Book of String Theory*; Princeton University Press: Princeton and Oxford, 2010.
39. Horowitz, G. T. Spacetime in String Theory. *New J. Phys.* **2005,** *7*, 201.
40. Gödel, K. An Example of a New Type of Cosmological Solutions of Einstein's Field Equations of Gravitation. *Rev. Mod. Phys.* **1949,** *21*, 447.
41. Morris, M. S.; Thorne, K. S. Wormholes in Spacetime and Their Use for Interstellar Travel: A Tool for Teaching General Relativity. *Am. J. Phys.* **1988,** *56*, 395–412.
42. Thorne, K. Do the Laws of Physics permit Closed Timelike Curves? *Ann, NY Acad. Sci.* **1991,** *631*, 182–193.
43. Carter, B. Black Hole Equilibrium States, Part I: Analytical and Geometric Properties of the Kerr Solutions. *Gen. Relativ. Grav.* **2009,** *41*, 873.
44. Rovelli, C. Can We Travel to the Past? Irreversible Physics Along Closed Time Like Curves, 2019. https://arxiv.org/abs/1912.04702
45. Deutsch, D.; Lockwood, M. The Quantum Physics of Time Travel. *Scientific American*, March 1994.

Index

R

S

T

U

V

W